Metal Ions and Complexes in Solution

Coordination Chemistry Fundamentals Series

Series editors:
Shigehisa Akine, *Kanazawa University, Japan*
Yasuhiro Funahashi, *Osaka University, Japan*
Osamu Ishitani, *Tokyo Institute of Technology, Japan*
Hiroshi Nakazawa, *Osaka Metropolitan University, Japan*
Hidetaka Nakai, *Kindai University, Japan*
Yasuhiro Ohki, *Kyoto University, Japan*
Toshio Yamaguchi, *Fukuoka University, Japan/Qinghai Institute of Salt Lakes, CAS, China*

Titles in the series:
1: Organometallic Chemistry
2: Metal Ions and Complexes in Solution

How to obtain future titles on publication:
A standing order plan is available for this series. A standing order will bring delivery of each new volume immediately on publication.

For further information please contact:
Book Sales Department, Royal Society of Chemistry, Thomas Graham House, Science Park, Milton Road, Cambridge, CB4 0WF, UK
Telephone: +44 (0)1223 420066, Fax: +44 (0)1223 420247
Email: booksales@rsc.org
Visit our website at books.rsc.org

Metal Ions and Complexes in Solution

Edited by

Toshio Yamaguchi
Fukuoka University, Japan
Email: yamaguch@fukuoka-u.ac.jp

and

Ingmar Persson
Swedish Agriculture University, Sweden
Email: ingmar.persson@slu.se

ROYAL SOCIETY
OF **CHEMISTRY**

Coordination Chemistry Fundamentals Series No. 2

Print ISBN: 978-1-83916-809-3
PDF ISBN: 978-1-83916-960-1
EPUB ISBN: 978-1-83916-961-8
Print ISSN: 2635-1498
Electronic ISSN: 2635-1501

A catalogue record for this book is available from the British Library

Translation from the Japanese language edition:
Youeki no Sakutai Kagaku (*Solution Complex Science*), published by Sankyo Shuppan,
edited by Haruhiko Yokoyama and Masaaki Tabata
Copyright © 2012 Japan Society of Coordination Chemistry
All Rights Reserved

The Royal Society of Chemistry is a charity, registered in England and Wales,
Number 207890, and a company incorporated in England by Royal Charter
(Registered No. RC000524), registered office: Burlington House, Piccadilly,
London W1J 0BA, UK, Telephone: +44 (0)20 7437 8656.

For further information see our website at www.rsc.org

Preface

This book is in the series Coordination Chemistry organized by the Japan Society of Coordination Chemistry (JSCC) with the aim to improve the knowledge of coordination chemistry in general and especially to middle to final year chemistry students in undergraduate courses in inorganic and physical chemistry. The Japanese language version of this book named 錯体の溶液化学 (*Sakutai no Youeki Kagaku = Solution Chemistry of Complexes*) has been used as a textbook of advanced inorganic and physical chemistry with a special emphasis on metal ions and complexes in solution in Japan since 2012, and has good reputation. It has now been decided that the Royal Society of Chemistry will publish several books from this series in English, but updated to present scientific knowledge. The content of some chapters in the original Japanese version has been merged into chapters with a broader content, and some new chapters on new and more applied areas have been added.

This book consists of four sections: (1) fundamental physical and theoretical chemistry methods for the study of solutions, (2) properties and structure of solvents and solvation of metal ions and complexes in solution, (3) formation of metal complexes in solution, and (4) applications of solutions containing metal complexes. The first section describes in more detail the mostly used physical and theoretical chemistry methods applied in studies of solutions containing metal ions and complexes. The aim of this chapter is to give the reader a solid overview and background of the physical and theoretical chemistry methods used. In this respect, it is important to emphasise that several different methods need to be applied to provide a detailed understanding of the chemical system under study. The second section describes the properties and structures of different kinds of solvents and some proposed scales to evaluate solvent properties. This chapter aims to provide the reader with a background to be able to make the best possible

Coordination Chemistry Fundamentals Series No. 2
Metal Ions and Complexes in Solution
Edited by Toshio Yamaguchi and Ingmar Persson
© Japan Society of Coordination Chemistry 2024
Published by the Royal Society of Chemistry, www.rsc.org

choice of solvents for specific applications. The third section describes how metal ions interact with other chemical species, other than the solvent, to form complexes in solution. This includes complex formation, substitution reactions and reduction–oxidation (redox) reactions in solution as well as in living organisms. Furthermore, the properties of metal complexes under extreme conditions such as high temperature and pressure and metal complexes being surface active are described and discussed. Recent developed liquids consisting of only ions, ionic liquids, as well as their possible applications are described. The fourth section covers some industrially applied systems where metal complexes play a major role, *e.g.* in separating elements from each other in spent nuclear fuel, and of highly concentrated metals containing solutions with unique properties.

This book was edited by T. Yamaguchi and I. Persson with contributions from R. Kanzaki (Chapter 1), T. Yamaguchi (Chapters 2, 3, 7 and 8), I. Persson (Chapters 3, 9 and 10), T. Umecky (Chapters 4 and 13), H. Torii (Chapter 5), H. Sato (Chapter 6), S. Iuchi (Chapter 6), Y. Umebayashi (Chapter 10), K. Ishihara (Chapters 11 and 14), M. Inamo (Chapter 11), H. Takagi (Chapter 12), T. Takamuku (Chapter 15), M. Iida (Chapters 15 and 16), A. Odani (Chapter 17), C. Ekberg (Chapter 18), P. Sipos (Chapter 19), B. Kutus (Chapter 19), Q. Zou (Chapter 20), Z. Liu (Chapter 20), and M. Wang (Chapter 20). The authors would like to thank the many people who have contributed to the preparation and the content of this book.

Author Biography

Christian Ekeberg is a professor in both Nuclear Chemistry as well as Industrial Materials Recycling at Chalmers University of Technology, Sweden. He is an elected member of the Royal Swedish Academy for Engineering Sciences as well as The Royal Society for Arts and Sciences. The scientific focus has varied over the years from modelling of aqueous chemistry through determination of thermodynamic constants to applications of hydrometallurgy using mainly solvent extraction. In most cases, a special focus has been on heavy elements such as actinides. In these areas, Ekberg has published about 250 journal papers and graduated 25 PhD students.

Masayasu Iida is a professor emeritus at Nara Women's University (Japan). He received his Doctor of Science degree in inorganic chemistry in 1982 from Nagoya University (Japan). He was a postdoctoral researcher in Simon Fraser University (Canada) with Alan S. Tracey, where he worked on metal complex ions in lyotropic liquid crystals using NMR. He was an associate professor at Nara Women's University from 1988 to 1995. He was then a full professor from 1995 to 2016. His research interests focus on the aggregations and functionalities of amphiphilic metal complexes in solution.

Coordination Chemistry Fundamentals Series No. 2
Metal Ions and Complexes in Solution
Edited by Toshio Yamaguchi and Ingmar Persson
© Japan Society of Coordination Chemistry 2024
Published by the Royal Society of Chemistry, www.rsc.org

Masahiko Inamo was born in Kagawa Prefecture, Japan in 1958. He obtained his PhD in 1987 under the supervision of Professor Motoharu Tanaka from Nagoya University. In 1985, he started independent research projects at the Department of Chemistry, Aichi University of Education. In 1989, he became an associate professor and has been a full professor at the same institute since 2004. He stayed at the laboratory of Professor A. E. Merbach, the Institute of Inorganic and Analytical Chemistry, University of Lausanne from 1988 to 1989 as a visiting researcher.

Koji Ishihara completed his PhD in 1983 at Nagoya University (Japan) under the direction of Professor Motoharu Tanaka. Then, he worked for four years as a Research Associate in Professor M. Tanaka's laboratory. During this period, he worked as a postdoctoral fellow in Professor Thomas W. Swaddle's laboratory at the University of Calgary (Canada) for one year. In 1988, he started his career at Waseda University (Tokyo) as an Assistant Professor and was promoted to an Associate Professor in 1990 and a Professor in 1995. His research interest is focused on inorganic reaction chemistry in solution.

Satoru Iuchi is an assistant professor at the Graduate School of Informatics, Nagoya University. He received his doctor's degree from Kyoto University in 2005. After working at the University of Utah as a postdoc and working at Kyoto University, he joined Nagoya University as an assistant professor in 2010. His current research interests include excited state properties of transition metal complexes in solution.

Ryo Kanzaki is an associate professor at Kagoshima University. He received his PhD in 2002 from Kyushu University under the supervision of Professor S. Ishiguro. He worked at AIST, Tsukuba, and became an Assistant Professor at Kyushu University in 2003. He moved to Kagoshima University in 2009. His research interest lies in the thermodynamics of chemical reactions in electrolyte solutions, especially in ionic liquids. Nanoparticle colloids have been included in his research interests since his visiting research at Université Pierre et Marie Curie (now Sorbonne Université) from 2014 to 2015.

Bence Kutus has joined the Material and Solution Structure Research Group at the University of Szeged in 2010, where he studied complexation equilibria in strongly alkaline media. He received his PhD degree in Chemistry in 2018 with his thesis 'Complex formation and lactonization reactions in aqueous solutions containing Ca^{2+} or Nd^{3+} ions and sugar-type ligands'. Later, he joined the Liquid Dynamics Group at the Max Planck Institute of Polymer Research as a Humboldt postdoctoral fellow, where he focused on the dynamics of water as well as ion pairing. In 2022, he returned to his alma mater in Szeged, where he is currently involved in numerous projects related to solution chemistry.

Zhong Liu received his BS and MS degrees from the China University of Geosciences (Wuhan) and his PhD from the Institute of Coal Chemistry (CAS) in 2012. He is a professor at the Qinghai Institute of Salt Lakes (CAS), and his research focuses on optimizing the surface and interface properties of adsorbents used to separate abundant ions from salt lakes by the traditional and electrically assisted separation technologies.

Akira Odani received his Doctor's degree in pharmaceutical sciences in 1981 from Osaka University. He joined Professor Yamauchi's group at Kanazawa University as an Assistant Professor in 1981. He moved to the Department of Chemistry, Nagoya University in 1987 and was promoted to an Associate Professor in 1995, an Associate Professor of Research Center for Materials Science in 1999, and an Associate Professor of Graduate School of Sciences in 2006. In 2008, he moved to Kanazawa University as a Professor in the Faculty of Pharmaceutical Sciences and retired in 2019. His research interests include noncovalent interactions in metal complexes, solution chemistry of albumin–drug interactions, and platinum chemistry in medicine.

Ingmar Persson received his PhD degree in inorganic chemistry at Lund University in 1980. After a postdoc visit to Stanford University in the group of Professor Henry Taube, he held a position as an Assistant Professor at Lund University. In 1989, he held a position as a professor of inorganic and physical chemistry at the Swedish University of Agricultural Sciences in Uppsala. One of his research interests throughout his career has been structure determination of hydrated and solvated metal ions and anions in solution in a range of solvents, mainly applying the EXAFS and large angle X-ray scattering (LAXS) techniques.

Hirofumi Sato was born in Funabashi, Japan, in 1968. He received his Master's degree (1993) and PhD (1996) in chemistry under the supervision of Professor Shigeki Kato from Kyoto University (Japan). Before joining Kyoto University in 2002, Sato was an assistant professor at the Institute for Molecular Science (IMS) with Professor Fumio Hirata. Since 2010, he has been a Professor at the Department of Molecular Engineering, Kyoto University. He is also the director of the Fukui Institute for Fundamental Chemistry, Kyoto University. His research interests include chemical reactions and processes in the solution phase.

Pàl M. Sipos is a professor of Chemistry at the Department of Inorganic and Analytical Chemistry, Faculty of Science and Informatics, University of Szeged. In 2008, together with Professor István Pálinkó, he founded, and since then has been leading, the Materials and Solution Structure Research group. He holds the Doctor of the Hungarian Academy degree. His research interest has been in inorganic solution chemistry, in particular in extremely concentrated aqueous electrolytes. Under his (co)supervision, 22 PhD dissertations have been completed and 7 are in progress. He has (co)authored 300+ refereed research papers and has *ca.* 3000 independent citations with an H index of 28.

Hideo D. Takagi was born in Japan in 1954. He obtained his BSc degree in 1978 and PhD in 1983 from Tokyo Institute of Technology. He began his career as an exchange student at the University of East Anglia, UK in 1981 with Professor Roderick D. Cannon. He worked with Professor Thomas W. Swaddle at the University of Calgary, Canada from 1985 to 1992 as a PDF/Research Associate. He became an Associate Professor at Nagoya University, Japan in 1992. He is the author of two earlier books: *Inorganic Chromotropism* (Fukuda ed. 2007, Springer) and *Inorganic Chemistry Based on Quantum Mechanics* (Nagoya University Press, 2018).

Toshiyuki Takamuku is a Professor of Faculty of Science and Engineering at Saga University, Japan. In 1987, he obtained his Master of Science degree from Tokyo Institute of Technology. He worked as a researcher in the Research and Development Division of TOTO Ltd. during 1988–1996 and received his Doctor of Science degree from Fukuoka University in 1993. He worked as a research assistant in the Faculty of Science, Fukuoka University in 1996. He moved to the Faculty of Science and Engineering, Saga University as an associate professor in 1997 and became a professor in 2012. His research interests are in the structures of liquids and solutions on both microscopic and mesoscopic scales.

Hajime Torii is a professor at the Department of Applied Chemistry and Biochemical Engineering, Faculty of Engineering, and the Department of Optoelectronics and Nanostructure Science, Graduate School of Science and Technology, Shizuoka University. He received his PhD from the University of Tokyo in 1992. After working as a research associate at the University of Tokyo, he moved to Shizuoka University in 2001 as an associate professor (PI) and was promoted to a professor in 2008. His present research interests focus on theories and computations on intermolecular interactions, structures, dynamics, and spectroscopic properties of liquids and biomolecules.

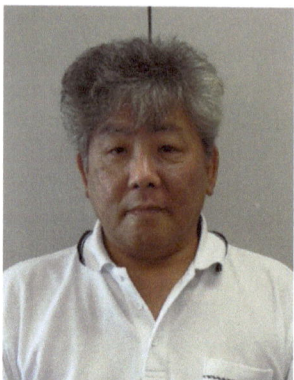

Yasuhiro Umebayashi obtained his PhD at Kyushu University in 1997. He worked as a research associate at Kyushu University from 1998 under a JSPS postdoctoral fellowship. He was promoted to an associate professor in 2002. He moved to Niigata University as a professor in 2012. His scientific interest is in the structure and dynamics of ions in aqueous and non-aqueous solutions and ionic liquids, particularly ion conduction mechanisms.

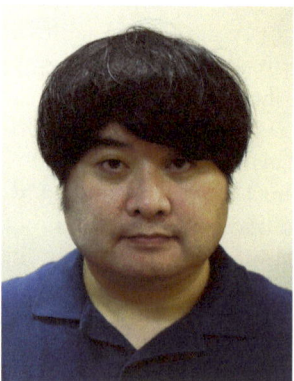

Tatsuya Umecky received his PhD degree in 2003 from Tohoku University. After completing postdoctoral research fellowships at the National Institute of Advanced Industrial Science and Technology and at Chuo University, he became an assistant professor at Saga University in 2011, followed by a promotion to an associate professor in 2016 at the same institution. He is familiar with solvation structures and dynamics using nuclear magnetic resonance spectroscopy. His scientific interests are focused on the chemical absorption reactions of carbon dioxide in green solvents as well as the functionalization of ionic liquids and deep eutectic solvents, including their microscopic mechanisms.

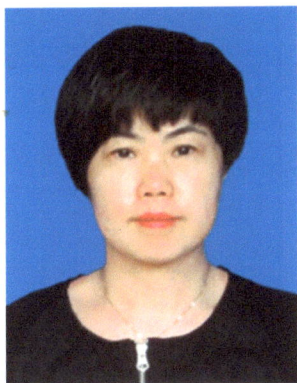

Min Wang is a Professor of Qinghai Institute of Salt Lakes, CAS. Her research interests include the comprehensive development, utilization and industrialization of salt lake potassium, lithium, boron, and magnesium resources. She proposed the innovative "gradient coupled membrane separation technology" and developed a new technology for lithium separation and extraction from salt lake brine with a high magnesium–lithium ratio. She has more than 40 authorized invitation patents. She has published 50 papers and proposed 5 national standards and 1 group standard.

Toshio Yamaguchi is a Professor at Qinghai Institute of Salt Lakes, Chinese Academy of Sciences. He received his BSc degree from Nagoya Institute of Technology (1973) and PhD from Tokyo Institute of Technology (1078). After working as a Research Associate at Tokyo Institute of Technology (1978–1979), a Postdoc at Gothenburg University (1979–1982), and an Assistant Professor at Tokyo Institute of Technology (1982–1986), he was an Associate Professor at Fukuoka University (1986), a Full Professor (1994), the Dean of Faculty of Science (2013–2017), and a Professor Emeritus (2020). His research focuses on the structure and dynamics of liquids and solutions by X-ray and neutron scattering and X-ray absorption spectroscopy.

Yongquan Zhou is a professor at Qinghai Institute of Salt Lakes, Chinese Academy of Sciences. His research interests include the structure and dynamics of disordered materials (liquid or amorphous) by X-ray/neutron scattering and computer simulations, and the fundamentals in rare element identification and separation from salt lakes. He has chaired over 15 research projects and has published more than 60 SCI-indexed papers. He acts as an international steering committee member of the International Conference on Solution Chemistry of IUPAC, and as an editorial board member of *Computer and Applied Chemistry* and *Journal of Salt Lake Research*.

Contents

Coordination Chemistry Fundamentals Series No. 2
Metal Ions and Complexes in Solution
Edited by Toshio Yamaguchi and Ingmar Persson
© Japan Society of Coordination Chemistry 2024
Published by the Royal Society of Chemistry, www.rsc.org

Chapter 3   X-ray Absorption Spectroscopy   40
Ingmar Persson

Chapter 4   Nuclear Magnetic Resonance   49
Tatsuya Umecky

Chapter 5   Vibrational Spectroscopy   62
Hajime Torii

CHAPTER 1

Thermodynamic Measurements: Potentiometry, UV/Vis Spectrometry and Calorimetry

RYO KANZAKI

Kagoshima University, Korimoto, Kagoshima, 890-0065, Japan
Email: kanzaki@sci.kagoshima-u.ac.jp

1.1 Introduction

In this chapter, the physico-chemical relationships, structure, and behavior of metal complexes in solution will be described and discussed. If only a single type of metal complex is present in solution, this investigation may be realized without too much difficulty because the performance of the solution can be attributed to the complex. Once the complex is isolated as a single crystal, its structure may help image the aspect in solution. However, a part of metal complexes in solution can load and release ligand species *via* ligand exchange reactions to convert into other chemical forms with different coordination numbers and ligand species, which behave as independent components with distinct individual properties. This occurs dynamically depending on the solution composition, and the performance of the solution appears as a superposition. Therefore, the primary aim of "thermodynamic measurements" is to predict the formation distribution of complex species. This can be achieved based on their stability constants. This information enables us to assign the performance of the solution to the individual constituent complex species. It also helps to find the special compositions to maximize the availability of the solution owing to the specific metal complex. This chapter starts by describing stability constants and their

Coordination Chemistry Fundamentals Series No. 2
Metal Ions and Complexes in Solution
Edited by Toshio Yamaguchi and Ingmar Persson
© Japan Society of Coordination Chemistry 2024
Published by the Royal Society of Chemistry, www.rsc.org

use to derive the formation distribution of metal complex species. In addition, the stability constants and the simultaneously obtained thermodynamic parameters provide essential insights into the metal ions and complexes in solution. The stability constants are directly linked to the reaction Gibbs energies of complex formation, allowing quantification of the reactivity or preferential formation of the complex species. The UV/Vis absorption spectra of individual complex species from spectrometry (spectrophotometry) and the complex formation enthalpies and entropies from calorimetry are closely related to the solvation state of the metal complexes in solution. In particular, energy-based parameters provide quantitative information about the driving force of complexation, which can hardly be obtained by structural techniques. This is discussed at the end of this chapter.

1.2 Formation Distribution of Metal Complexes

This section starts with an introduction of complex formation equilibria and the definition of the stability constants. The concentrations of all metal complex species formed in solution of known composition can be calculated from respective stability constants, and consequently the formation fractions of the components are obtained. Readers who have already learned this field may skip this part. After that, an expansion to multinuclear and ternary complexes is made. Lastly, data analysis is discussed. However, the data fitting procedure will be described in detail with examples in the following section on potentiometric titrations.

1.2.1 Complexation Equilibria

Here, let us consider the formation of a mononuclear complex $[ML_n]$ as an example of complexation equilibrium (charges are omitted hereafter), where L and M present the ligand and metal ion, respectively. The value of the co-ordination number (n) depends on the metal, ligand, and solvent, however, now we assume that all complexes of coordination numbers ranging from 1 to N are formed. Then the equilibrium for the formation of each complex and the corresponding equilibrium constant are expressed as follows:

$$M + L \rightleftarrows [ML] \quad K_1 = \frac{[ML]}{[M][L]} \tag{1.1.1}$$

$$[ML] + L \rightleftarrows [ML_2] \quad K_2 = \frac{[ML_2]}{[ML][L]} \tag{1.1.2}$$

$$\vdots$$

$$[ML_{n-1}] + L \rightleftarrows [ML_n] \quad K_n = \frac{[ML_n]}{[ML_{n-1}][L]} \tag{1.1.3}$$

The square brackets represent the complex ions and, at the same time, the concentration of each component in the equation of formation constants. The equilibrium constants, expressed as K, are called stepwise (or successive) stability (or formation) constants, whereas the overall (or cumulative) stability constants, denoted by β, may be appropriate for calculating formation distribution.

$$\beta_1 = \frac{[ML]}{[M][L]} = K_1 \quad M + L \rightleftarrows [ML] \tag{1.2.1}$$

$$\beta_2 = \frac{[ML_2]}{[M][L]^2} = K_1 K_2 \quad M + 2L \rightleftarrows [ML_2] \tag{1.2.2}$$

$$\vdots$$

$$\beta_n = \frac{[ML_n]}{[M][L]^n} = K_1 K_2 \ldots K_n \quad M + nL \rightleftarrows [ML_n] \tag{1.2.3}$$

The respective stability constants correspond to the equilibrium constants of the reactions shown in the right-hand column. However, it should be noted that the complexation reactions that actually occur in the solution are always addition reactions of one ligand at a time to the metals or complexes. Using the overall stability constants, the concentration of each complex ion could be expressed only in terms of the concentrations of free (unbound) metal ions and ligands.

$$[ML_n] = \beta_n [M][L]^n \tag{1.3}$$

The total concentration of the metal ion, C_M (sometimes referred to as analytical, formal, prepared, or conditioning concentration, depending on the situation), which contains all complexation forms of the metal ion, can be expressed as

$$\begin{aligned} C_M &= [M] + [ML] + [ML_2] + \ldots + [ML_N] \\ &= [M] + \beta_1 [M][L] + \beta_2 [M][L]^2 + \ldots + \beta_N [M][L]^N \end{aligned} \tag{1.4}$$

Thus, the formation fraction of a certain complex ion, $\alpha_n = \dfrac{[ML_n]}{C_M}$, can be expressed as

$$\alpha_n = \frac{[ML_n]}{C_M} = \frac{\beta_n [L]^n}{1 + \beta_1 [L] + \beta_2 [L]^2 + \ldots + \beta_N [L]^N} \tag{1.5}$$

As shown, for mononuclear complexation, the formation distribution depends only on the free ligand concentration [L] and is independent of C_M. Note that using $\beta_0 = 1$ enables [M] to be treated consistently as the complex with $n = 0$ in these equations. This is valid for the equations hereinafter.

To generalize, the multinuclear and partially protonated complexations are described as $[M_mH_pL_n]$ and their stability constants are expressed as follows:

$$\beta_{mpn} = \frac{[M_mH_pL_n]}{[M]^m[H]^p[L]^n} \tag{1.6}$$

The total concentrations are

$$C_M = [M] + \sum m\beta_{mpn}[M]^m[H]^p[L]^n \tag{1.7}$$

$$C_H = [H] + \sum p\beta_{mpn}[M]^m[H]^p[L]^n \tag{1.8}$$

$$C_L = [L] + \sum n\beta_{mpn}[M]^m[H]^p[L]^n \tag{1.9}$$

In this representation, reactions with $m = 0$ are relevant to the acid–base equilibria of the ligands, and the species with $p < 0$ correspond to hydroxides. If the stability constants of all possible complex species in solution are known, the concentration of each complex species can be calculated from the total concentrations by solving multi-order equations with three unknowns, *i.e.*, [M], [H] and [L]. It is sufficient to obtain a numerical solution with enough accuracy using computer programs, whereas an analytical solution is not necessarily required. Because the stability constants, once determined, are applicable to any system, a database of stability constants for various metal–ligand combinations has been compiled.[1,2] Note that values are generally listed as logarithms ($\log \beta$), sometimes without annotations.

1.2.2　Determination of Complexes Formed

If any of the concentrations of the free or complex species can be measured, unknown stability constants can be estimated. A sufficient dataset is needed for practical accuracy. The procedure will be described in detail in the following section on potentiometry with examples. Before that, it should be assured whether the system contains only mononuclear complexes or multinuclear complexes coexist, and the maximum coordination number of ligands in the complexes should be determined. To do this, experiments of a sufficiently wide concentration range of both the metal ion and ligand are required. After collecting a sufficient dataset, the stability constants that reproduce the experimental values are searched for, taking into account the pattern of possible complexation that may occur in solution. To omit a certain complex species, experiments should be conducted with extreme care that can observe all complexation equilibria (*i.e.*, increasing and decreasing formation fractions of all complex species appear). For example, in *N*-methylformamide, up to four-fold chloro-complexes of cobalt(II) are formed, while $[CoCl_2]$ formation is skipped.[3] It should be noted that, even in this case, assuming the formation of $[CoCl_2]$ apparently results in a better fit. This is merely due to the extra tunable parameter to explain the experimental data. Such misinterpretations must be carefully eliminated. We should decide whether the pattern of complexation is appropriate based on statistic parameters such as simplified Hamilton's R-factor,

$$R = \sqrt{\frac{\sum (y_{calc} - y_{obs})^2}{\sum y_{obs}^2}} \qquad (1.10)$$

(y_{obs} and y_{calc} are the measured and calculated values, respectively, of a certain observed quantity), the standard deviation of the stability constants, and the formation distribution according to the finally obtained stability constants. If necessary, additional experiments should be performed under conditions that maximize the formation fraction of relatively minor complex species. In particular, this is an extensive work for systems containing multinuclear complexes. Separate experiments and *a priori* perspectives from other solvents, ligands, and metal ions are combined to draw the most reasonable conclusions.

1.3   Potentiometric Titration

In the following sections, some titrimetric methods to determine stability constants are presented, including briefly the measurement principle and procedure, points to be noted during experiments, and what kinds of physical quantities can be concomitantly obtained. These may help readers understand and interpret published information, not only when trying to carry out these measurements. In the beginning, potentiometric titration is introduced. The procedure to determine the stability constants from the experimental data is also described in this section. As potentiometry is responsive to concentrations, data fitting directly provides stability constants. However, a similar procedure is applicable to other techniques.

1.3.1   Electrodes and Cells

Potentiometric titration uses electrodes that respond to specific free ions or other ionic species. Once the concentration of free (unbound) metal ions is obtained using an ion-selective electrode, the remaining metal ions correspond to those in the complex species. Thus, the formation fraction of the complex and, consequently, its stability constant are obtained. It is important to choose the suitable electrode that responds exclusively to a single metal ion or ligand in the concentration range examined. Solid metal electrodes, such as copper, silver, palladium, and platinum, work sufficiently. Electrodes composed of the combination of a metal and its sparingly soluble salt are also utilized. For example, the Ag/AgCl electrode is valid for both Ag^+ and Cl^-. Ion-selective electrodes are equipped with an ion-exchange membrane for metal ions or anionic ligands. Although they expand the types of ions that can be measured, it is necessary to confirm the reliability in the solvent and the concentration range used. In addition, a glass electrode (pH electrode) can also be used to indirectly determine the stability constants. This can be useful in terms of accuracy and convenience when the ligands have acid–base equilibria. In non-aqueous solvents, there are limitations in the use of ion-selective electrodes. Glass electrodes can work in mixtures of water and organic solvents of moderate permittivity, but not in completely dehydrated non-aqueous solvents. Under such a condition, electrodes using an ion-selective field effect transistor (IS-FET) have been found to respond well to pH. In addition to the indicator electrode, it is critical to choose a reference electrode, whose role is to provide constant potential throughout the titration. In general, a small half-cell connected to the sample solution through a porous material, such as frit, to avoid contamination of the internal solution and the test solution mutually is used. Notably, it is not necessary to know the numerical value of the reference electrode potential, as long as it remains constant during the potentiometric titration. Calomel and amalgam electrodes were occasionally used in the past. Although they are not currently preferable considering mercuric pollution, there is the case that they are the best choice. In addition, supporting electrolytes are also required. Although their original purpose is to reduce the impedance when using solvents of low electric conductivity, the addition of a certain concentration of such electrolytes is expected to keep the activity coefficients of the complex ions constant and to maintain the electrode potential of the reference electrode when it is immersed directly into the solution. Typically, KCl and NaCl are used in aqueous solutions. In non-aqueous solvents with low permittivity, halides of tetra-alkyl-ammonium or perchlorates and triflates (trifluoromethanesulfonates) of alkali Li^+ and Na^+ are used depending on the system and purpose.

1.3.2 Determination of Stability Constants

Here, the procedure to derive the stability constants using an Ag/AgCl electrode is presented as an example. The electromotive force (emf) of this electrode, in combination with a given reference electrode, depends on the concentration of free chloride ions as expressed by the following Nernst equation:

$$\mathrm{emf} = E_0 + \frac{RT}{nF} \ln \left(\frac{[Cl^-]}{mol\,dm^{-3}} \right) \tag{1.11}$$

where R is the gas constant, F is the Faraday constant, and T is the absolute temperature. E_0 is the formal potential, involving the potential of the reference electrode, standard redox potential of Ag/AgCl, activity coefficient of chloride, liquid junction potential (LJP), and other potential differences.

As n is the number of electrons in the cell reaction and $n = 1$ is employed for oxidation of chloride ions, the following equation is obtained:

$$\text{emf} = E_0 - \frac{RT}{F} \ln\left(\frac{[Cl^-]}{\text{mol dm}^{-3}}\right) \tag{1.12}$$

Once E_0 is obtained using a chloride solution of a known concentration, the unknown $[Cl^-]$ in the sample solution can be given as long as the E_0 value remains unchanged. Because the total concentrations of metal and chloride ions are known from the initial concentration conditions, the average coordination number of chloride ions (ligand, L) per metal ion M^+, denoted as n_L, is given by the following equation:

$$n_L = \frac{[ML] + 2[ML_2] + \ldots + N[ML_N]}{C_M} \quad \text{(definition)} \tag{1.13}$$

$$n_{L,\text{obs}} = \frac{C_L - [L]}{C_M} \quad \text{(observed)} \tag{1.14}$$

On the other hand, the formation distribution of mononuclear complexes depends only on $[L]$, as described above, and n_L of $[L]$ is calculated using the stability constants.

$$n_{L,\text{calc}} = \frac{\beta_1[L] + 2\beta_2[L]^2 + \ldots + N\beta_N[L]^N}{1 + \beta_1[L] + \beta_2[L]^2 + \ldots + \beta_N[L]^N} \tag{1.15}$$

Using arbitral values of stability constants, $n_{L,\text{calc}}$ can be calculated. Then, they are compared with $n_{L,\text{obs}}$, and the stability constants are adjusted to minimize the difference. The set of finally converged values of the stability constants has the highest likelihood. In the least-squares method, the sum of the squares of the deviations at all titration points is used as the evaluation function, U, to be minimized.

$$U = \sum \left(n_{L,\text{obs}} - n_{L,\text{calc}}\right)^2 \tag{1.16}$$

Sufficient data of $[Cl^-]$ as a function of the set of total concentrations (C_L, C_M) need to be collected to obtain satisfactorily accurate stability constants. The experimental conditions, including the titration plan and the concentrations of metal ions and ligands in the initial sample solution and titrant, should be adequately set up such that changes in the formation fractions of all complex species can be observed.

1.3.3   Complexation and Acid Dissociation

In this section, the use of a glass electrode to determine the stability constant is described as another example. Figure 1.1 shows an example of

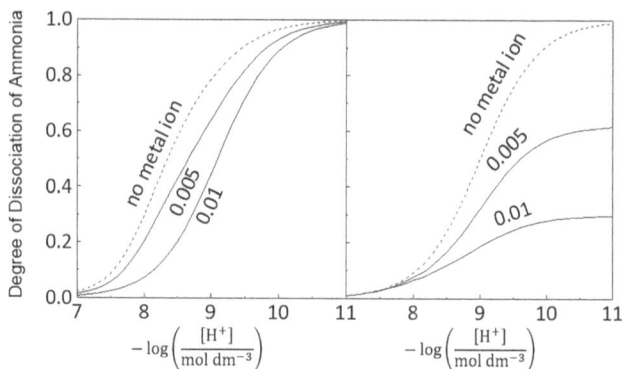

Figure 1.1 Theoretical potentiometric titration curves in solution containing Zn^{2+} and 0.05 mol dm^{-3} NH$_3$. The numbers denote the concentration of Zn^{2+} in mol dm^{-3} (the dashed line is for no Zn^{2+}). Left: The apparent degree of ionization of ammonia, obtained as follows:

$$\alpha_{app} = \frac{[L]}{[HL]+[L]} = 1 - \frac{C_H - [H^+] + \dfrac{K_W}{[H^+]}}{C_L}.$$

Right: The intrinsic degree of ionization of ammonia, obtained as follows:

$$\alpha_{int} = \frac{[L]}{C_L} = \frac{\left(C_H - [H^+] + \dfrac{K_{AP}}{[H^+]}\right) \times \dfrac{K_a}{[H^+]}}{C_L}.$$

the pH response in zinc-ammine complex formation. Ligands are Lewis bases and are likely Brønsted bases, such that the complexation reaction is competitive with (the reverse reaction of) the acid dissociation reaction. In such cases, a linear relationship between the Gibbs energies of complexation and acid dissociation, or log β and pK_a, is often found for ligands with acid–base equilibria, the so-called linear free energy relationship (LFER).

Glass electrodes generally have a double-tube structure. The outer space is filled with an inner solution, in which a reference electrode (typically Ag/AgCl) is immersed, and separated from the sample solution by a small frit. Some of them have an additional cell inside the outer space to allow the inner solution to be replaced. Otherwise, typically a concentrated (or approximately 3 mol dm^{-3}) KCl aqueous solution is enclosed. It should be noted that this type of glass electrode is severely damaged by the precipitation of potassium salts of perchlorate and triflate, which are often used as supporting electrolytes in non-aqueous solvents. The emf of a glass electrode is expressed as follows:

$$\text{emf} = E_0 + g\log\left(\frac{[H^+]}{\text{mol dm}^{-3}}\right) \tag{1.17}$$

where g corresponds to $\dfrac{RT \ln 10}{F}$ and equals 59.16 mV at 25 °C. After appropriate calibration to determine E_0, $[H^+]$ can be obtained precisely at moderate pH (almost 3–10, even at pH = 2–11 with some correction). For example, if ligand L is the conjugate base of monoacid HL, [L] can be calculated independent of the coexisting metal ions as follows:

$$[L] = \left(C_H - [H^+] + \frac{K_{AP}}{[H^+]} \right) \times \frac{K_a}{[H^+]} \tag{1.18}$$

where K_{AP} is the autoprotolysis constant. As previously mentioned, the free concentration of the ligand [L] leads to an estimation of the stability constants. Notably, pH is not exactly the same as $-\log\left(\dfrac{[H^+]}{\text{mol dm}^{-3}}\right)$. Therefore, the pH standards are not suitable for calibrating glass electrodes in the abovementioned analyses, and solutions of strong acids of known concentration should be used instead. In particular, this inconsistency is problematic in non-aqueous solvents and when comparing the stability constants in different solvents. Having said that, some information is available regarding the comparison of molality and pH in aqueous solutions[4–6] and pH standards in non-aqueous solutions.[7,8] In addition, glass electrodes need correction of the liquid junction potential in high H^+ concentration solutions and the so-called alkali errors in low H^+ concentration (basic) solutions. However, little information is available on these corrections in non-aqueous solvents. Due to them, performing potentiometric measurements in non-aqueous solutions is constrained.

1.4 UV/Vis Spectrometry

Most transition metal ions in solution have absorption bands in or around the visible-ultraviolet wavelength region. Because this originates from the ligand field induced by the primary coordination shell, the absorption (electronic) spectrum of a metal complex is sensitive to the ligand and coordination number, including the coordination atom geometry. Hence, the absorption spectra can distinguish the complex species formed in the solution. Spectrometry (spectrophotometry) is a method to determine simultaneously the stability constants and the individual absorption spectra of the complex species from changes in the absorption spectrum of the sample solution with varying solution compositions.

1.4.1 Measurements

In the batch method, a series of solutions with different compositions are prepared in advance to record the absorption spectra one by one. This procedure allows for the preparation of solutions over a wide concentration range, as well as highly viscous, highly hygroscopic, and air-sensitive solutions that are difficult to handle. In contrast, titrimetric methods can vary the composition of the solution successively without touching the measurement cell and can obtain a large dataset by automation, thus improving the quality of the experimental data. Using a syringe pump or burette of 1–2 cm^3, the titrant is loaded directly into the measurement cell inside the cell room of the spectrometer (spectrophotometer), which can reduce the total volume of the solvent required throughout the experiment. Otherwise, titration is carried out in a beaker outside the spectrometer, and the solution is circulated using a peristaltic pump to be introduced into the flow-type cell. In this case, other analytical techniques, such as potentiometry, can also be combined. These advantages and disadvantages should be carefully compared to select the most suitable method for the system of interest. The concentration of metal ions should be adjusted such that the absorbance is approximately 0.1–2, depending on the spectrometer, because the quantitative response is not ensured at higher absorbance. An absorbance of 3 indicates that the light intensity is attenuated to 1/1000. Some organic solvents, even though they appear transparent, have an intense absorption in the UV region and hinder the absorption of the target metal complexes. Particular attention should be paid while using such organic solvents because this fact may be blinded by baseline measurements. It should also be noted that the transmittance in the UV region greatly depends on the cell material.

1.4.2   Determination of Stability Constants and Individual Spectra

In the spectrometer, a monochromatic (single wavelength) beam extracted by a monochromator is guided into the sample cell, and the light intensity after passing through the sample solution is detected by a photomultiplier. The transmittance T_λ and absorbance A_λ at each wavelength λ recorded are

$$T_\lambda = I_\lambda / I_{\lambda,0} \tag{1.19}$$

$$A_\lambda = -\log T_\lambda = \log(I_{\lambda,0}/I_\lambda) \tag{1.20}$$

where $I_{\lambda,0}$ and I_λ are the light intensities of the incident and transmitted beams, respectively. In practice, $I_{0,\lambda}$ is recorded as the transmitted light intensity of the reference cell that is filled with the same solvent. Absorbance is recorded by scanning λ to obtain the absorption spectrum of the sample solution. In a double-beam spectrometer, the reference cell is loaded in pairs with the sample cell. The monochromatic beam is split into two, and each illuminates the sample and reference cells to detect I_λ and $I_{\lambda,0}$ simultaneously. This can eliminate the influence of fluctuations in light source intensity.

   According to Beer's law, the absorbance can be expressed in terms of the concentrations of the complex species $[ML_n]$ (including the solvated metal ion as $n = 0$) and their molar absorption coefficients $(\varepsilon_{n,\lambda})$ at each wavelength.

$$A_\lambda = \varepsilon_{0,\lambda}[\mathrm{M}] + \varepsilon_{1,\lambda}[\mathrm{ML}] + \varepsilon_{2,\lambda}[\mathrm{ML}_2] + \ldots + \varepsilon_{N,\lambda}[\mathrm{ML}_N] \qquad (1.21)$$

Now, let us assume that ML alone is formed in the solution to help visualize the procedure. Using the formation fraction $\left(n_\mathrm{L} = \dfrac{[\mathrm{ML}]}{C_\mathrm{M}}\right)$, the following linear relationship is given:

$$\frac{A_\lambda}{C_\mathrm{M}} = (1 - n_\mathrm{L})\varepsilon_{\mathrm{M},\lambda} + n_\mathrm{L}\varepsilon_{\mathrm{ML},\lambda} \qquad (1.22)$$

If β_1 was already known, n_L could be calculated. The $\dfrac{A_\lambda}{C_\mathrm{M}}$ vs. n_L plot would result in a straight line and intercepts at $n_\mathrm{L} = 0$ and $n_\mathrm{L} = 1$ would give $\varepsilon_{\mathrm{M},\lambda}$ and $\varepsilon_{\mathrm{ML},\lambda}$, respectively. Conversely, if β_1 is unknown, its value is adjusted such that the $\dfrac{A_\lambda}{C_\mathrm{M}}$ vs. n_L plot becomes a straight line. Figure 1.2 shows this concept graphically. Hence, to determine β_1 accurately, $\varepsilon_{\mathrm{M},\lambda}$ and $\varepsilon_{\mathrm{ML},\lambda}$ need to be sufficiently different from each other. In theory, the $\varepsilon_{0,\lambda}$ and $\varepsilon_{1,\lambda}$ values can be determined even when unknown; however, it is better to fix them to the values obtained in advance: $\varepsilon_{\mathrm{M},\lambda}$ from the solution of the metal ion

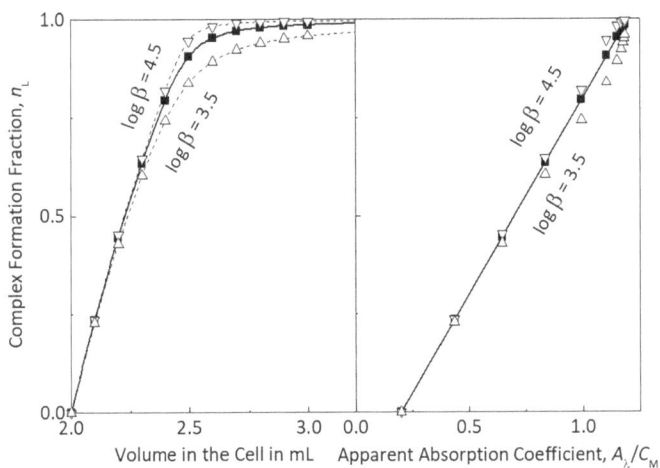

Figure 1.2 Theoretical spectroscopic titration curve calculated assuming that only ML is formed, where $\beta = 4.0$, $\varepsilon_{\mathrm{M},\lambda} = 20$ and $\varepsilon_{\mathrm{ML},\lambda} = 120$. Left: The formation fraction (n_L) increases with progress of titration, depending on β; a smaller β value than 4.0 underestimates and a larger β overestimates n_L. Right: The apparent absorption coefficient at λ, A_λ/C_M, is marked on the x-axis, where the altitude is n_L displayed on the left. The marks are aligned on a straight line when $\beta = 4.0$.

without ligands and $\varepsilon_{ML,\lambda}$ with sufficient ligands, giving practically the $n_L = 1$ condition. The same operation is performed for each wavelength with using common stability constants. This ensures the accuracy of the stability constant values. If there is a wavelength at which $\varepsilon_{M,\lambda} = \varepsilon_{ML,\lambda}$, an isosbestic point is observed. Although the absorbance data at this wavelength are useless in determining stability constants, the presence of an isosbestic point is a decisive fingerprint of stepwise complexation equilibrium in the corresponding concentration range. If $\varepsilon_{M,\lambda}$ and $\varepsilon_{ML,\lambda}$ are close at all wavelengths recorded, *i.e.*, [ML] has an absorption spectrum similar to that of M, the stability constants of [ML] cannot be determined with sufficient accuracy.

The same procedure is applicable when higher-order or multinuclear complexes are formed and/or ligands undergo light absorption. If the stability constants of all complex species formed are known, the absorbance is a linear function of the molar absorption coefficients.

$$A_\lambda = \sum_{m,p,n} \varepsilon_{mpn,\lambda}[M_m H_p L_n] = \sum_{m,p,n} \varepsilon_{mpn,\lambda} \beta_{mpn}[M]^m [H^+]^p [L]^n \quad (\varepsilon_{mpn,\lambda} > 0) \quad (1.23)$$

Therefore, ε_{mpn} can be determined analytically using the least-squares method. Conversely, to obtain the stability constants, all unknown parameters (β_{mpn} and $\varepsilon_{mpn,\lambda}$ of all λ) are optimized iteratively using a non-linear least-squares process, where the evaluation function includes the absorption at all titration points and wavelengths. By this way, the individual absorption spectrum of each complex species can be extracted even if the formation fraction does not reach 100%. Owing to the large number of data points, the standard deviations of the stability constants tend to be small; however, as shown in the previous example, the reliability may be less than that of potentiometric titrations. In such cases, a significant error appears in the molar absorption spectra of the complex species and should be confirmed.

1.4.3 Coordination Structure

The individual absorption spectra provide significant information about the coordination structure, coordination number, and ligand field of the metal complex species. For example, nickel chloro complexes can form up to $[NiCl_4]^{2-}$ in dimethylformamide (DMF). The absorption spectra of the Ni^{2+}, $[NiCl]^+$, and $[NiCl_2]$ are very different from those of $[NiCl_3]^-$ and $[NiCl_4]^{2-}$, indicating a change in the coordination structure from a six-fold coordinating octahedral structure of $[Ni(dmf)_6]^{2+}$, in which the d–d transition is forbidden, to a four-fold coordinating tetrahedral structure of $[NiCl_3(dmf)]^-$, in which the d–d transition is permissive.[9] In the case of the copper chloro complexes in an ionic liquid, 1-butyl-3-methylimidazolium trifluoromethanesulfonate, $[CuCl_3]^-$ and $[CuCl_4]^{2-}$ show similar spectra, indicating both are in a four-fold structure.[10]

1.5 Calorimetric Titration

Calorimetric titration uses the heat of reaction to determine the stability constants. Since most complexation reactions involve the heat of reaction, calorimetric titration has the advantage of being applicable to metals, ligands, and complexes that have no suitable electrodes or no light absorbance. Naturally, the complexation enthalpy $(\Delta H_{mpn}°)$ and consequently the complexation entropy $(\Delta S_{mpn}°)$ can be obtained simultaneously. Hence, calorimetric titration can independently obtain three thermodynamic parameters of complexation at once.

$$\Delta S_{mpn}° = \frac{\Delta H_{mpn}°}{T} + R\ln\beta_{mpn} \tag{1.24}$$

Another way to determine the reaction enthalpies is to use the van't Hoff relationship. However, this is not applicable when temperature-induced conformational changes (such as thermochromic complexes) or the temperature dependence of enthalpy (corresponding to a change in heat capacity during complexation) occurs. In addition, it is difficult to prepare a satisfactorily wide temperature range. Thus, calorimetric titration appears to be an ideal technique for exploring complexation mechanisms. Unfortunately, commercial instruments are not always tuned for observing metal ion complexation. In particular, very limited calibration methods in non-aqueous solvents make it difficult to ensure the reliability of the observed enthalpy. It is therefore essential for observers to be acquainted with the principles of the instruments and to design the experiments carefully. In this section, a brief overview of the instrumentation and data analysis procedures is presented at the beginning.

1.5.1 Calorimeter

An isoperibol (isothermal environment) is a type of calorimeter. A reaction vessel is placed in contact with a heat bath with a sufficient heat capacity to maintain a constant temperature during the measurement. When a titrant solution is added, heat of reaction is generated and the temperature of the solution varies. The heat then drains into the heat bath and the solution temperature returns to the equilibrium temperature again. This process is sometimes referred to as a thermal anomaly. The integral of the temperature variation from the baseline (the area of the peak) is proportional to the heat quantity. The proportionality factor is determined by referring to the Joule heat. A tiny resistor is immersed in the solution and a constant current is applied. Electric devices whose electric properties depend on temperature, such as metal wires (*e.g.*, platinum resistors) and thermoelectric elements (*e.g.*, Peltier elements), are used to detect temperature variations. In general, these instruments are designed as twin-type to eliminate the temperature fluctuations. That is, two or more vessels of the same equipment are placed in contact with the same thermostated bath, and the "difference in difference" in the temperatures between each vessel and the heat bath is detected. After generating the heat of reaction, the temperature returns to the baseline, obeying Newton's law of cooling. It is simply an exponential decay and curve fitting in this region provides the thermal conductivity. Once this value is obtained, the hypothetical adiabatic thermogram (enthalpogram) can be estimated and the heat of reaction can be obtained if the practical heat capacity of the vessel is given, *e.g.*, in terms of Joule heat. For diffusion-controlled reactions, the value thus obtained is more precise than the value based on the peak area. Because the subsequent thermogram is no longer needed, the temperature of the solution can be heated or cooled back to the baseline, which can save time. In contrast, most commercially available titration calorimeters are designed as an isothermal type. This type of calorimeter has a pair of sample and reference vessels. When a titrant solution is added into the sample vessel, heat of reaction is generated and the temperature varies. Then the reference cell is heated in order to compensate the generated heat to keep both cells isothermal. The heat of reaction can be obtained directly as the heat added during this process. However, since the heater is typically placed outside the vessel, it is impossible to transfer the entire heat to the solution. Therefore, some type of device factor is necessary, and to ensure reliability may be problematic when using uncommon solvents owing to the unusual heat capacity and viscosity, *etc.*, of the solution. Calibration is a crucial issue for any type of calorimeter. It is most reliable to refer to the Joule heat generated by a resistor. Also, some reference materials are suggested;[11–13] however, information on non-aqueous solvents is very limited. Thus, the difficulty in calorimetric titration is that there is no absolute reference, so the observer must consider the certainty of the results. In addition, there are many other considerations, such as the heat of dilution and slower reactions than the measuring time scale.

1.5.2 Determination of Reaction Enthalpies

Assuming that the data are obtained properly, we proceed to the analysis phase. In a given titration point, the addition of a titrant causes one or more complexations and decompositions (complex species are consumed). The observed heat of reaction involves all the heat generated during these processes. This can be simplified as follows:

$$q = \sum \Delta_\beta H_{mpn}^\circ \left\{ v \left[M_m H_p L_n\right] - (v - v_0) \left[M_m H_p L_n\right]_T - v_0 \left[M_m H_p L_n\right]_0 \right\} \quad (1.25)$$

where q is the observed heat of reaction, the subscript 0 denotes the volume and concentration in the solution before loading the titrant, and $[M_m H_p L_n]_T$ is the concentration of the complex species $[M_m H_p L_n]$ in the titrant. $\Delta_\beta H_{mpn}^\circ$ is the standard enthalpy of overall complexation, corresponding to the following reaction:

$$m M + p H + n L \rightarrow \left[M_m H_p L_n\right] \quad (1.26)$$

The subscript β is used to indicate the value corresponding to the overall complexation. Stepwise complexation enthalpies are more useful for investigating the complexation mechanism. In such cases, they can be derived from the overall enthalpies. For instance, the stepwise enthalpy to form the n-fold complex $[ML_n]$ corresponding to the reaction:

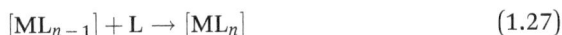

$$[ML_{n-1}] + L \rightarrow [ML_n] \tag{1.27}$$

can be derived as

$$\Delta H_n^\circ = \Delta_\beta H_{10n}^\circ - \Delta_\beta H_{10(n-1)}^\circ \tag{1.28}$$

When the titrant is a ligand solution, experimental data are generally represented by the heats of reactions divided by the added amount of ligand as a function of the ligand/metal molar ratio. Assigning a negative sign to this value corresponds to the apparent complexation enthalpy, as shown in Figure 1.3 In this case, unlike potentiometric and spectroscopic titrations, the experimental curve depends slightly on the experimental conditions. It may also be expressed as the accumulated heat of reaction divided by the amount of metal ions.

The procedure to determine the stability constants is similar to that for spectrometry. If all β_{mpn} values have been already known, the heat of reaction

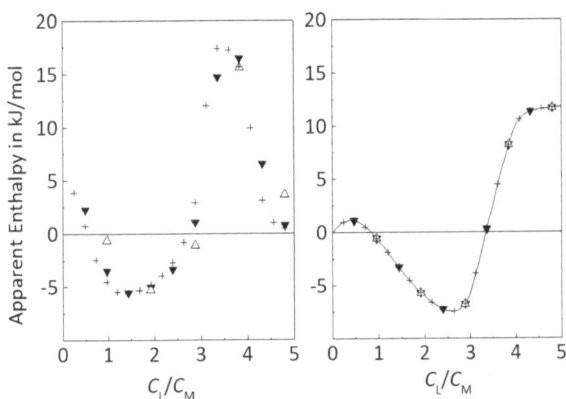

Figure 1.3 Theoretical simulation of calorimetric titration of copper chloride in 1-butyl-3-methylimidazolium trifluoromethanesulfonate. The apparent enthalpies are plotted against the molar ratio of the ligand (Cl^-) to the metal ion (Cu^{2+}). Left (differential type): The heat of reaction at each titration point is divided by the added amount of ligand; despite being theoretically calculated, dependence on the titration series (different titrant volumes per titration point) appears. Right (integral type): The accumulated heat of reaction is divided by the amount of metal ions. Although a single theoretical line can describe the experimental data, experimental errors tend to be accumulated in the later stages of the titration.

is a linear function of the $\Delta_\beta H_{mpn}°$ values, which can be analytically determined. To obtain the stability constants, conversely, the set of β_{mpn} and $\Delta_\beta H_{mpn}°$ to give the best fit is determined by means of the non-linear least-squares procedure. However, if available, it is advisable to determine the stability constants in advance using other techniques, such as potentiometric and spectroscopic titrations.

1.5.3 Thermodynamic Quantities of Complexation

Complexation converts the chemical form of a complex species and alters its chemical properties such as charge, hydrodynamic radius, and solvent affinity. These are the direct driving forces of complexation, along with the metal–ligand interactions. The complexation enthalpy exhibits these contributions in energy units because the reaction enthalpy in solution involves almost exclusively the change in the internal energy. However, the stability constants are by no means solely dependent on the internal energy, but are always affected by the entropy. This is caused, for instance, by the gain of degrees of freedom of solvent molecules released from the primary solvation shell and by changes in the thickness of the secondary solvation shell, which is more structured than the bulk. In the case of copper(II)–chloro complexes, the complexation enthalpies in organic solvents like dimethylsulfoxide (DMSO) and N,N-dimethylformamide (DMF) are positive (*i.e.*, endothermic, or unfavorable in terms of internal energy), whereas the large and positive reaction entropies are the driving forces of the complexation.[14] This is attributed to the release of solvent molecules constrained in the primary solvation shell. In other words, what promotes the complexation is not the electrostatic interaction between copper and chloride but the desolvation, which does not appear to be involved. On the other hand, copper–chloro complexation is unfavorable in aqueous solutions. This is because the released water molecules do not gain sufficient entropy in the bulk, where a higher-order network structure is formed. Thus, solvent effects play a key role in the complexation, which is mainly through the entropy. The superposition of two independent effects, represented by enthalpy and entropy, in the complexation is the complexation Gibbs energy that determines the stability constant.

$$\Delta_\beta G_{mpn}° = -\log \beta_{mpn} = \Delta_\beta H_{mpn}° - T\Delta_\beta S_{mpn}° \qquad (1.29)$$

In this context, the complexation Gibbs energy is no more than the quantification of the apparent behavior of the complex, such as stability or favorability. When the interaction energy between the metal and ligands, as well as the solvents, increases (the reaction enthalpy decreases), they tend to be constrained more intensely, resulting in a decrease in entropy and a reduced increase in the stability constant. This is called the enthalpy–entropy compensation effect, whereby the Gibbs energy no longer reflects the metal–ligand interactions. In such cases, the Gibbs energy needs to be decomposed into enthalpic and entropic contributions to be evaluated separately. Thus, precise information on enthalpy is quite useful.

Problem 1

Choose the most appropriate technique to determine the stability constants of the following combinations, along with the reason, from potentiometry, UV-Vis spectrometry, and calorimetry. Consider the other techniques.

(1) K^+ & 18-crown-6 (18C6)
(2) Cu^{2+} & chloride
(3) Zn^{2+} & 1,10-phenanthroline

Problem 2

(1) Derive eqn (1.18).
(2) What is the best transmittance (and absorbance) for the least error in the concentration determination? Consider error propagation from the transmittance to the concentration.
(3) Derive the van't Hoff relationship from eqn (1.24).

Answer to Problem 1

(1) Potentiometry is available by using a K^+-selective electrode. Calorimetry is also applicable. However, UV-Vis spectrometry is not applicable because Na^+, 18C6, and their complex are colorless.
(2) While all techniques are applicable, potentiometry simultaneously using copper and Ag/AgCl electrodes may bring higher reliability. Depending on the situation, the absorption spectra of the respective complexes obtained by UV-Vis spectrometry or the enthalpy and entropy by calorimetry are also useful.
(3) In the aqueous phase, potentiometry using a glass electrode is the first choice. Otherwise, calorimetry is also applicable. Although Zn^{2+}, the ligand, and their complexes are colorless, ultraviolet absorption may be available for the determination of the stability constants.

Answer to Problem 2

(1) For monoacids, $C_H = [H^+] + [HL] - [OH^-]$ and $K_a = [H^+][L]/[HL]$ hold (the charges of ligands are omitted). Thus, $C_H = [H^+] + [H^+][L]/K_a - K_{AP}/[H^+]$,

derived to eqn (1.16). Note that the following equation holds for diacids ($K_{a1} = [H^+][HL]/[H_2L]$, $K_{a2} = [H^+][L]/[HL]$):

$$[L] = \frac{\left(C_H - [H^+] + \dfrac{K_{AP}}{[H^+]}\right)}{\dfrac{2[H^+]^2}{K_{a1}K_{a2}} + \dfrac{[H^+]}{K_{a2}}}$$

(2) From Beer's law, $c = \dfrac{A}{\varepsilon} = -\log\dfrac{T}{\varepsilon}$ (c: the concentration of a certain chemical species). The error propagation from T (ΔT) to c (Δc) is written using the (partial) derivative: $\Delta c^2 = \left(\dfrac{\partial c}{\partial T}\right)^2 \Delta T^2$, where $\dfrac{\partial c}{\partial T} = \dfrac{c}{T\ln T}$ is given. To find the minimum $\left(\dfrac{\partial c}{\partial T}\right)^2$, solve $\dfrac{d}{dT}(T\ln T) = 0$, obtaining $T = e^{-1} = 0.368$, or $A = 0.434$. The value may not be so significant because it depends on the noise source. However, it is important to know that there is a theoretically optimal absorbance between 0 and 1.

(3) $\ln \beta_{mpn} = -\Delta H^\circ/T + \Delta S^\circ/R$. Note that this is a linear function between $\ln \beta_{mpn}$ and $1/T$. This equation can be used to convert ΔH° and the temperature dependence of the stability constant to each other.

References

1. R. M. Smith and A. E. Martell, *Critical Stability Constants*, Springer, New York, Second Supplement, 1989.
2. Chemistry International – Newsmagazine for IUPAC, 2006, 28, 26. The IUPAC Stability Constants Database, https://old.iupac.org/publications/scdb/index.html (accessed August 2022).
3. K. Fujii, Y. Umebayashi, R. Kanzaki, D. Kobayashi, R. Matsuura and S. Ishiguro, *J. Solution Chem.*, 2005, **34**, 739.
4. M. Filomena Camões, M. J. Guiomar Lito, M. I. A. Ferra and A. K. Covington, *Pure Appl. Chem.*, 1997, **69**, 1325.
5. C.-Y. Chan, Y.-W. Eng and K.-S. Eu, *J. Chem. Eng. Data*, 1995, **40**, 685.
6. J. I. Partanen and A. K. Covington, *J. Chem. Eng. Data*, 2006, **51**, 777.
7. S. Rondinini, P. R. Mussini and T. Mussini, *Pure Appl. Chem.*, 1987, **59**, 1549.
8. T. Mussini, A. K. Covington, P. Longhi and S. Rondinini, *Pure Appl. Chem.*, 1985, **57**, 865.
9. S. Ishiguro, K. Ozutsumi and H. Ohtaki, *Bull. Chem. Soc. Jpn.*, 1987, **60**, 531.
10. R. Kanzaki, S. Uchida, H. Kodamatani and T. Tomiyasu, *J. Phys. Chem. B*, 2017, **121**, 9659.
11. I. Wadsö and R. N. Goldberg, *Pure Appl. Chem.*, 2001, **73**, 1625.

12. R. Sabbah, A. Xu-wu, J. S. Chickos, M. L. Planas Leitão, M. V. Roux and L. A. Torres, *Thermochim. Acta*, 1999, **331**, 93.
13. K. Ozutsumi, K. Kohyama, K. Ohtsu and T. Kawashima, *J. Chem. Soc., Dalton Trans.*, 1995, 3081.
14. S. Ishiguro, H. Suzuki, B. G. Jeliazkova and H. Ohtaki, *Bull. Chem. Soc. Jpn.*, 1986, **59**, 2407.

CHAPTER 2

X-ray and Neutron Scattering

TOSHIO YAMAGUCHI[a,b]

[a] Key Laboratory of Salt Lake Resources Chemistry of Qinghai Province, Qinghai Institute of Salt Lakes, Chinese Academy of Sciences, Xining, Qinghai 810008, P. R. China; [b] Professor Emeritus, Department of Chemistry Faculty of Science, Fukuoka University, 8-19-1 Nanakuma, Jonan, Fukuoka 814-0180, Japan
Email: yamaguch@fukuoka-u.ac.jp

2.1   Introduction

When X-rays and neutrons are radiated to solutions containing metal complexes, X-rays are scattered by the electrons around the nuclei, whereas the nuclei scatter neutrons. The number of electrons is proportional to the atomic number; thus, the X-ray scattering intensity is proportional to the atomic number. In neutron scattering, on the other hand, the nuclei of atoms have characteristic scattering lengths as given in Table 2.1, *e.g.* the scattering length of H is comparable to that of Ni. Thus, neutron scattering has an advantage in locating the position of light atoms, such as H and Li. Another merit of neutron scattering is that the isotope has different coherent scattering lengths b_c. As shown in Table 2.1,[1] the coherent scattering lengths for some elements or isotopes demonstrate the large difference for H and D, ^{6}Li and ^{7}Li, ^{35}Cl and ^{37}Cl, ^{60}Ni and ^{62}Ni, *etc.* This property of neutron scattering enables us to use the isotopic substitution method to extract the structure of metal complexes in solution, described in the subsequent section.

It is ideal to use both X-ray and neutron scattering techniques complementarily. Both scattering techniques give us the one-dimensional structural

Coordination Chemistry Fundamentals Series No. 2
Metal Ions and Complexes in Solution
Edited by Toshio Yamaguchi and Ingmar Persson
© Japan Society of Coordination Chemistry 2024
Published by the Royal Society of Chemistry, www.rsc.org

Table 2.1 Thermal neutron coherent scattering lengths and incoherent cross sections for some elements.[1]

Element or isotope	Scattering length b_c (10^{-12} cm)	Incoherent cross-section σ_{inc} (barns)
H	-0.37390	80.26
D	0.6671	2.05
^{N}Li	-0.190	0.92
^{6}Li	0.174	0.46
^{7}Li	-0.222	0.78
^{N}O	0.5803	0.0008
^{N}Na	0.363	1.62
^{N}Cl	0.95770	5.3
^{35}Cl	1.165	4.7
^{37}Cl	0.308	0.001
^{N}Ti	-0.3438	2.87
^{N}V	-0.03824	5.08
^{N}Ni	1.03	5.2
^{60}Ni	0.28	0
^{62}Ni	-0.87	0
^{N}Zr	0.716	0.02

information related to all atom pair interactions in a solution. Thus, in a multicomponent solution, such as metal complexes in solution, it is very complicated to determine the structure of ion solvation and metal complexes in solution. To overcome this difficulty and extract the site–site partial structure factor, the pair distribution functions, the coordination number distribution, the angle distribution, and the three-dimensional structure of a solution, an empirical potential structure refinement (EPSR) method has been developed by applying a Monte Carlo method combined with the pair-wise potentials and the experimental structure factors obtained by X-ray and neutron scattering of solutions.[2,3]

2.2 Theory of X-ray and Neutron Scattering of Solutions

The fundamental theory for X-ray and neutron scattering from solutions has been described elsewhere,[4] and only the essentials will be reported here. The structure of solutions is described by the pair distribution function (pdf) $g_{pq}(r)$, which is the probability of finding atom p at a distance r from atom q. Since the structure of a solution is disordered and space-averaged, the number of atom q, dn_q, within a sphere in radius r and thickness dr around a central atom p is expressed by

$$dn_q = n_q V^{-1} 4\pi r^2 g_{pq}(r) dr \qquad (2.1)$$

where n_q denotes the number of atom q in volume V. The radial distribution function $g(r)$ is obtained by Fourier transform of the partial structure factor (psf) $S_{pq}(Q)$ as

$$S_{pq}(Q) = 1 + 4\pi n_q V^{-1} Q^{-1} \int_0^\infty [g_{pq}(r) - 1] r \sin(rQ) dr \qquad (2.2)$$

$$g_{pq}(r) = 1 + V(2\pi^2 n_q r)^{-1} \int_0^\infty [S_{pq}(Q) - 1] Q \sin(rQ) dQ \qquad (2.3)$$

Here, Q is the amplitude of the wave vector given by $Q = \left(\dfrac{4\pi}{\lambda}\right) \sin\theta$ (λ is the wavelength of X-rays and neutrons and 2θ is the scattering angle). When a solution contains i kinds of atomic species, the total structure factor $S(Q)$ consists of $i(i+1)/2$ psfs. X-ray and neutron scattering techniques measure the differential scattering cross-section $d\sigma/d\Omega$, which is described by

$$\frac{d\sigma}{d\Omega} = \sum x_i a_i^2 + \sum\sum n_p n_q a_p a_q [S_{pq}(Q) - 1] \qquad (2.4)$$

where x_i is the atomic fraction of atom i, a_i is the scattering factor of atom i (for X-rays, $a_i = fi(Q)$ is the atomic scattering factor of atom i; for neutrons, $a_i = b_i$ is the coherent scattering length of the ith nucleus). The first term corresponds to the self-scattering of atom i and the second term is the structure sensitive function consisting of the intramolecular and inter-molecular structures of ions or molecules. The structure function $i(Q)$ is defined as

$$i(Q) = \left[\frac{d\sigma}{d\Omega} - \sum x_i a_i^2\right] = \sum\sum n_p n_q a_p a_q [S_{pq}(Q) - 1] \qquad (2.5)$$

The total radial distribution function $D(r)$ is given by Fourier transform of $i(Q)$ as

$$D(r) = 4\pi r^2 \rho_0 + 2r\pi^{-1} \int_{Q_{min}}^{Q_{max}} Qi(Q)M(Q)\sin(rQ)dQ \qquad (2.6)$$

Here, ρ_0 is the atomic number density of a solution, Q_{min} and Q_{max} are the minimum and the maximum Q value available in experiments. $M(Q) = 1/(\sum x_i a_i)^2$ is the correction term of the phase shift of the atomic scattering factor for X-rays and is thus not necessary for Q-independent scattering length for neutrons.

Since X-ray and neutron scattering from solutions is limited to one-dimensional information of a space- and time-averaged structure, an empirical

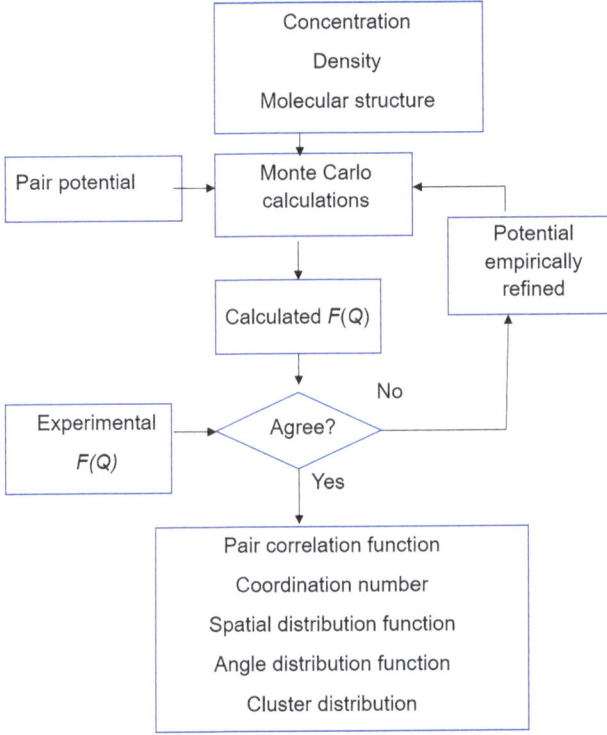

Figure 2.1 A flow chart of empirical potential structure refinement modeling.

potential structure refinement (EPSR) modeling method has been de-veloped.[2,3] The scheme of EPSR is shown in Figure 2.1.

The interference function $F(Q)$ used in the EPSR calculations and the corresponding radial distribution function $G(r)$ are obtained by eqn (2.8) and (2.9).

$$S(Q) = \frac{\left[\frac{d\sigma}{d\Omega} - \left\{\sum b_i^2 - \left(\sum b_i\right)^2\right\}\right]}{\left(\sum b_i\right)^2} \tag{2.7}$$

$$F(Q) = \left[\left(\sum x_i b_i\right)^2 (S(Q) - 1)\right] \bigg/ \left(\sum x_i b_i^2\right) \tag{2.8}$$

$$G(r) = 1 + \frac{1}{2\pi^2 r \rho_0} \int_{Q_{min}}^{Q_{max}} QF(Q)\sin(Qr)dQ \tag{2.9}$$

EPSR calculations are performed in a unit cell to reproduce the composition of a sample solution and the density. Firstly, Monte Carlo calculations are performed using the pair potentials of the Lennard-Jones parameters, the atomic mass, and the charge for the constituent atoms to equilibrate the system. The potential energy between atoms of types i and j separated by distance r is expressed by eqn (2.10).

$$U_{ij}(r) = U_{intra}(r) + U_{inter}(r) + U_{EP}(r) \tag{2.10}$$

$$U_{inter}(r) = \sum_{ij} \left(4\varepsilon_{ij}\left[\left(\frac{\sigma_{ij}}{r_{ij}}\right)^{12} - \left(\frac{\sigma_{ij}}{r_{ij}}\right)^{6}\right] + \frac{q_i q_j}{4\pi\varepsilon_0 r_{ij}} \right) \tag{2.11}$$

$$\varepsilon_{ij} = (\varepsilon_i \varepsilon_j)^{\frac{1}{2}}, \quad \sigma_{ij} = \frac{1}{2}(\sigma_i + \sigma_j). \tag{2.12}$$

$U_{intra}(r)$ is the harmonic potential for the intramolecular structure, ε_{ij} and σ_{ij} are Lennard-Jones parameters for the potential well depth and effective atom size, respectively. ε_0 is the vacuum permittivity, r_{ij} is the distance of the interaction, and q_i and q_j are the charges of atoms i and j. $U_{EP}(r)$ is the empirical potential derived during EPSR calculations.

Secondly, the pair potentials are empirically improved from the old ones by comparing simulated and experimental interference functions. The details of the procedure are described elsewhere.[2,3] The above calculations are repeated until the simulated interference functions reproduce the experimental ones well. Once good agreements are obtained, the pair correlation functions, the coordination number distributions, the triangular distributions, and the spatial density functions (SDF), *i.e.*, three-dimensional structures, are calculated.

The coordination number CN of atom j around atom i is calculated from the corresponding pair correlation function $g_{ij}(r)$ using eqn (2.13):

$$CN_{ij} = 4\pi\rho_j \int_{r_{min}}^{r_{max}} g_{ij}(r) r^2 dr \tag{2.13}$$

where ρ_j is the number density of atom j, and r_{min} and r_{max} are the lower and upper limits of integration. In the present study, r_{max} is the r-value at the first minimum of the pair distribution function.

2.3   Experimental Set-up and Data Treatment

2.3.1   X-ray Scattering of Solutions

Figure 2.2(a) and (b) shows the principles to measure X-ray scattering from solutions.[5,6]

X-ray sources are a sealed X-ray tube or a rotating anode in the laboratory scale or intense synchrotron X-rays available in the synchrotron facilities around the world, *e.g.*, ESRF (Grenoble) and SPring-8 (Japan). Monochromatic X-rays like Mo K_α or Ag K_α radiation (the wavelength

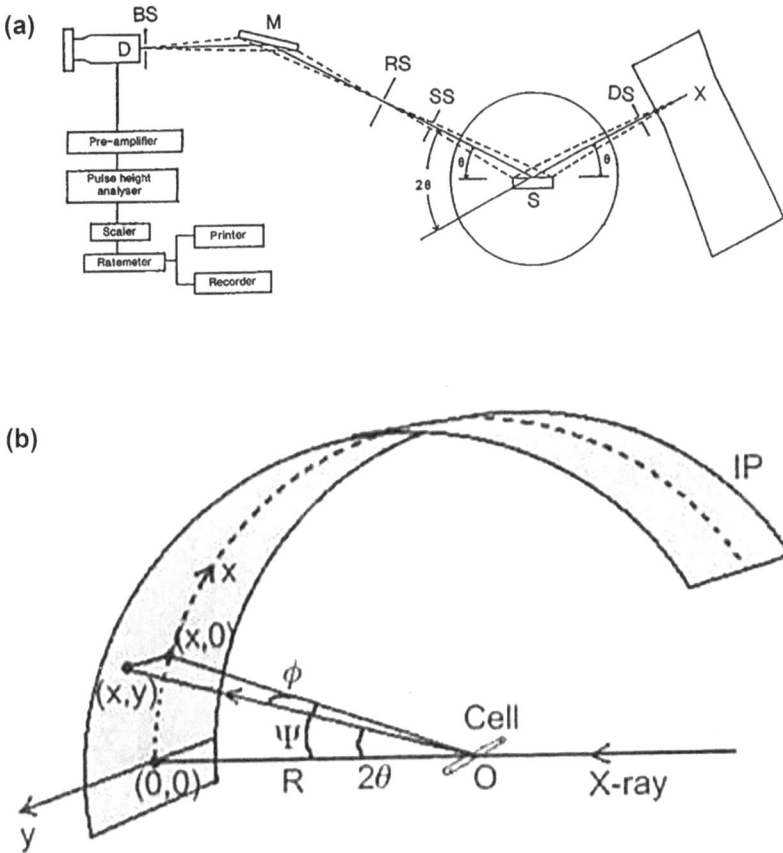

Figure 2.2 (a) A schematic diagram of a θ–θ diffractometer.[5] X: X-ray tube, S: sample, M: monochromator, D: detector, DS: divergent slit, SS: scattering slit, RS: receiving slit, BS: baffle slit. Reproduced from ref. 5 with permission from John Wiley & Sons, Copyright 2022. (b) A schematic diagram of a rapid X-ray diffractometer combined with a two-dimensional imaging plate (IP) detector. Reproduced from ref. 6 with permission from AIP Publishing, Copyright 1994.

$\lambda = 0.7107$ and 0.5609 Å for Mo K$_\alpha$ and Ag K$_\alpha$, respectively) are often used for measurements in the laboratory.[7] The Bragg–Brentano optical geometry is adopted in an X-ray diffractometer where the X-ray source and the detector move simultaneously in the opposite direction by the angle θ, and the free surface of a solution sample is kept horizontally during the measurements. The X-rays scattered from the free surface of the solutions are mono-chromatized with a curved single crystal of LiF or graphite and detected with a scintillation counter. Incoherent X-ray scattering and fluorescent scattering are removed with the monochromator. The X-ray scattering intensities are measured as a function of θ (2θ is the scattering angle). Different slits are used to make sure that X-rays illuminate only the planar part of the sample surface over a whole scattering angle range ($2\theta_{max} = \sim 140°$). Since X-ray scattering from solutions is very weak compared with that from crystalline materials, the measurements are continued until a good signal-to-noise ratio is obtained. The scattered intensities obtained with the different slits are corrected for background and absorption of a sample solution, and the individual data are merged into the data with a common slit. The normalization of experimental units to absolute electron units is made by comparing the asymptotic data with the theoretical independent scattering intensities in a high angle region or the Krogh-Moe–Norman method to integrate the experimental data over a whole scattering angle range. The structure functions $i(Q)$ are extracted from the self-scattering intensity from the normalized intensities as a function of the amplitude of scattering vector Q.

The above-mentioned θ–θ type diffractometry needs several days to complete the measurements and is thus not applied to measurements under extreme conditions, such as high-temperature and high-pressure measurements and time-resolved measurements. To overcome this difficulty, a rapid X-ray diffractometer combined with a two-dimensional imaging plate detector and a rotating anode X-ray generator was developed, as shown in Figure 2.2(b).[6] An imaging plate of a 200 mm × 400 mm size enables measurement of X-ray scattering from a solution over a scattering angle range of 0–140° simultaneously and reduces the time of measurement of several days to 30 min–1 h in the laboratory. Other advantages are a small sample needed and the ease of attaching temperature and pressure control units. A disadvantage of this method is that X-ray scattering from highly absorbing samples is unable to measure due to the transmission mode measurement and that fluorescence scattering from a sample containing heavy metal elements cannot be removed since the monochromator is placed in front of the sample. The fluorescence scattering can be removed by inserting a Zr foil (for Mo K_α radiation) before the imaging plate detector. When intense synchrotron X-rays are combined with an imaging plate detector, measurements of small amounts of solution samples in a much shorter time (several minutes) are possible. The entire data treatment including absorption and multiple scattering correction, normalization to an absolute unit, structure functions, and radial distribution functions is performed with the KURVLR program.[8]

2.3.2 Neutron Scattering of Solutions

Figure 2.3(a) and (b) illustrates the principle of neutron scattering at a reactor source and a pulsed neutron source, respectively.[5]

In the former case, the neutron spectrum corresponds to the peak in the Maxwellian, whereas in the latter case the neutron spectrum extends from the Maxwellian to the epithermal energy region. At a reactor source, incident neutrons are monochromatized with a single crystal and neutrons scattered from a solution sample are detected by a detector as a function of scattering angle 2θ by rotating the detector angle. Since the peak energy in the

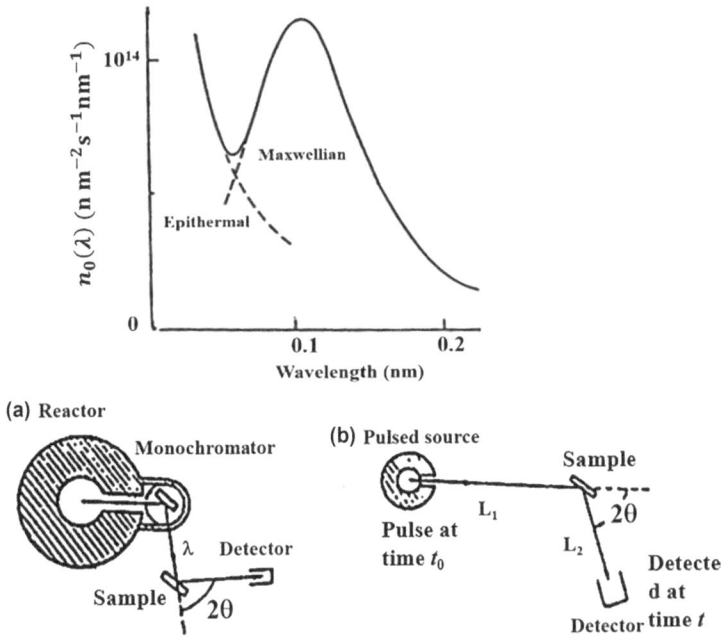

Figure 2.3 Schematic diagrams of neutron scattering of (a) the reactor source and (b) the pulsed source. The top figure shows the neutron spectrum in the Maxwellian distribution produced at the reactor source and in the Maxwellian + epithermal distribution at the pulsed source. Reproduced from ref. 5 with permission from John Wiley & Sons, Copyright 2022.

Maxwellian is at around 1 Å, the maximum amplitude of the scattering vector Q is about 10 Å^{-1}.

At a pulsed neutron source, the energy or wavelength of neutrons arriving at a diffractometer is analyzed by time-of-flight (TOF) (t) recorded at detectors. Hence, all neutrons emitted from the moderator through the flight path ($L_1 + L_2$) come to the sample, are scattered and then detected by the detectors. The neutrons created pass along the flight path L_1, are scattered at a sample and detected by detectors at a distance L_2 from the sample. The neutron count is recorded as a function of TOF with five detector banks at the scattering angles (2θ) of small-angle, 23°, 45°, 90°, and 150°. In the TOF method, the wavelength λ is given by $\lambda = ht/mL$ (m is the neutron mass, h is the Planck constant, and $L = L_1 + L_2$). The measured neutron wavelength range is available from $0.12 < \lambda/\text{Å} < 8.2$, corresponding to the amplitude of the scattering vector of $0.03 < Q/\text{Å}^{-1} < 100$. A cell used for neutron scattering measurements under ambient conditions is a cylindrical cell that is 6–10 mm in inner diameter, 0.1 mm in thickness, and 30–50 mm in height made of vanadium with a small coherent scattering length ($b_c = -0.3824$ fm) or a Ti–Zr null alloy of coherent scattering length $b_c = 0.000$ fm (a mixture of

67.6% Ti and 32.4% Zr for $b_c = -3.438$ fm for Ti and $+7.16$ fm for Zr) as shown in Table 2.1.

Neutron scattering has two major problems. One is the large incoherent scattering cross-section σ_{inc} of H which does not contain any structural information, but increases the multiple scattering and attenuation effects. To avoid these problems, fully deuterated solution samples or H/D mixtures with less than 50% H content are used for structure determination by neutron scattering. Another problem is the dynamical effect, the so-called Placzek falloff, which arises from large inelasticity effects due to the nearly-equal mass of H nuclei and incident neutrons. Figure 2.4(a) shows the raw $S(Q)$ obtained from the TOF neutron scattering experiment of 3 mol kg^{-1} aqueous ^{7}LiCl solution in D_2O. This effect is reduced by measuring deuterated samples rather than hydrogen-containing ones. There are several methods proposed to correct the dynamical falloff.[9] However, since no methods can correct the falloff completely, a polynomial fitting procedure is applied to obtain the correct structure factor $S(Q)$ as shown in Figure 2.4(b) and (c) for TOF neutron scattering of 3 mol kg^{-1} aqueous ^{7}LiCl solution in D_2O.[5]

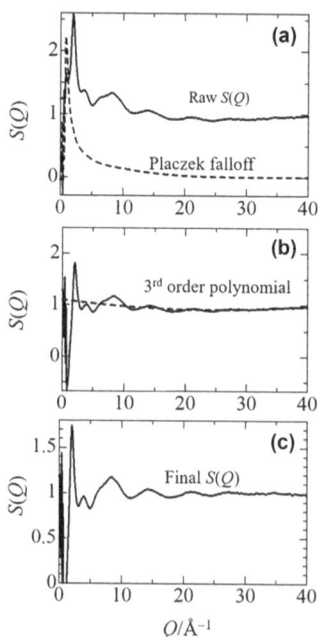

Figure 2.4 A procedure for analyzing neutron scattering data of a 3 mol kg^{-1} aqueous ^{7}LiCl solution in D_2O: (a) raw $S(Q)$ and the Placzek falloff, (b) raw $S(Q)$–the Placzek falloff and 3rd order polynomial fit, and (c) raw $S(Q)$–the Placzek falloff–3rd order polynomial +1 as the final $S(Q)$. Reproduced from ref. 5 with permission from John Wiley & Sons, Copyright 2022.

2.4 Methods of Structure Analysis

2.4.1 Radial Distribution Function Method

Figure 2.5(a) shows the experimental scattering intensity of a 2.886 mol dm^{-3} aqueous $Zn(ClO_4)_2$ solution together with the independent coherent (solid line) and incoherent (dashed line) scatterings of the atoms involved.[5,7]

Figure 2.5(b) shows the corresponding structure functions $i(Q)$ weighted by Q. Figure 2.5(c) shows the radial distribution functions obtained by the Fourier transform of the structure functions in Figure 2.5(b). The peak at 1.45 Å is assigned to the Cl–O bonds within ClO_4^{-}. The peak at 2.10 Å originates from the Zn^{2+}–$O(H_2O)$ bonds due to Zn^{2+} hydration. In addition, the O–O interactions within the tetrahedral ClO_4^{-} contribute in part to the peak. The dashed lines are the theoretical peaks calculated for the structures of ClO_4^{-} and $[Zn(H_2O)_6]^{2+}$. A quantitative analysis is performed by a least-squares fitting method of minimizing the error square sum:

$$U = \sum_{Q_{min}}^{Q_{max}} Q^2 \left\{ i(Q)_{exp} - i(Q)_{cal} \right\} \tag{2.14}$$

The theoretical structure functions $i(Q)_{cal}$ are calculated using the Debye equation:

$$i(Q)_{cal} = \sum_{p} \sum_{q} n_{pq} f_p(Q) f_q(Q) \frac{\sin(r_{pq})}{r_{pq}} \exp(-b_{pq}Q^2), \tag{2.15}$$

with the interatomic distance r_{pq}, the temperature factor b_{pq}, and the number of interactions n_{pq} for atom pair p–q. Figure 2.5(b) shows a good agreement between the experimental data (dots) and the calculated ones (solid lines) in the Q-region above 6 Å^{-1}. The structure functions in the Q-region below 6 Å^{-1} are not reproduced since the model does not consider the longer interactions, such as the solvent water structure and the second hydration of Zn^{2+}. The optimized values are obtained as $r(Zn–O_W) = 2.091(4)$ Å, $b(Zn–O_W) = 0.0046(3)$ Å^2, $n(Zn–O_W) = 5.90(4)$, $r(Cl–O) = 1.427(2)$ Å, $b(Cl–O) = 0.009(2)$ Å^2, $n(Cl–O) = 4.01(3)$. O_W and O represent the water oxygen atom and perchlorate oxygen atom, respectively.

2.4.2 Isomorphous Substitution Method

Assuming that two different solutions ML and M′L (M and M′ are isomorphous metal ions with very similar ionic radii and L is a ligand) in H_2O with the same composition have the same structure, the partial structure factors related to atoms, except for the isomorphous M and M′, should be the same and cancelled out by taking the difference of the structure

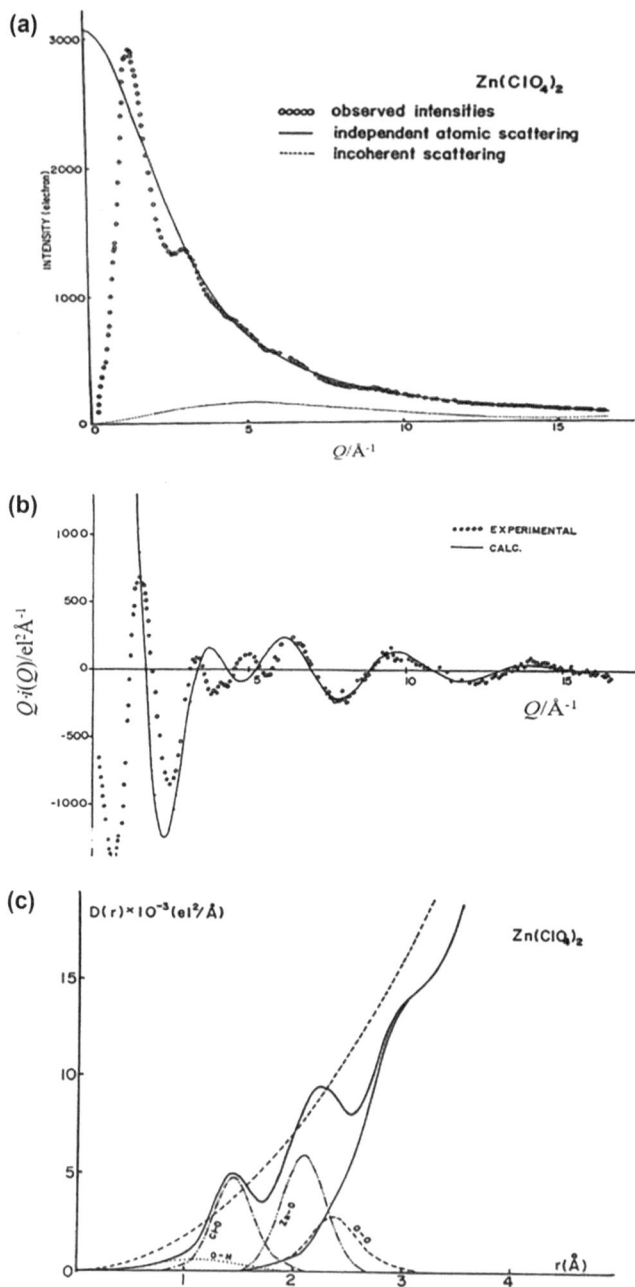

Figure 2.5 (a) X-ray scattering pattern for solutions, (b) the structure functions, and (c) radial distribution functions for an aqueous $Zn(ClO_4)_2$ solution.[5] Reproduced from ref. 7, https://doi.org/10.1246/bcsj.49.701, under the terms of the CC BY 4.0 license, https://creativecommons.org/licenses/by/4.0/.

functions of the two solutions. The differential structure function $\Delta i_M(Q)$ is expressed by

$$\Delta i_M(Q) = 2n_M n_O f_O \Delta f_M(Q)[s_{MO}(Q) - 1] + 2n_M n_L f_L \Delta f_M(Q)[s_{ML}(Q) - 1]$$
$$+ 2n_M n_H f_H \Delta f_M(Q)[s_{MH}(Q) - 1] + n_M^2 [f_M^2 - f_{M'}^2][s_{MM}(Q) - 1] \quad (2.16)$$

where $\Delta f_M = f_M - f_{M'}$ is essentially large in the isomorphous substitution method.

By Fourier transform of $\Delta i(Q)$, the RDF related only to the metal ion is obtained as

$$D^M(r) = 4\pi r^2 \rho_0^M + 2r\pi^{-1} \int_{Q_{min}}^{Q_{max}} Q\Delta i_M(Q)M(Q)f_M(Q)\Delta f_M^{-1}(Q)\sin(rQ)dQ \quad (2.17)$$

Figure 2.6 shows an example of the isomorphous substitution method applied to 0.77 mol L^{-1} aqueous $Y_2(SeO_4)_3$ (A1) and 0.78 mol L^{-1} aqueous $Er_2(SeO_4)_3$ (A2) solutions.[5,10]

Er^{3+} is chemically similar to Y^{3+} with similar ionic radii, and the hydration structures of both ions are similar to each other (see Section 9.2). In Figure 2.6(a), the peak at 1.5 Å is ascribed to the intramolecular Se–O bonds within a tetrahedral SeO_4^{2-} ion. A peak at around 2.5 Å arises from the Er–O(H$_2$O) and Y–O(H$_2$O) bonds due to the ion solvation and the

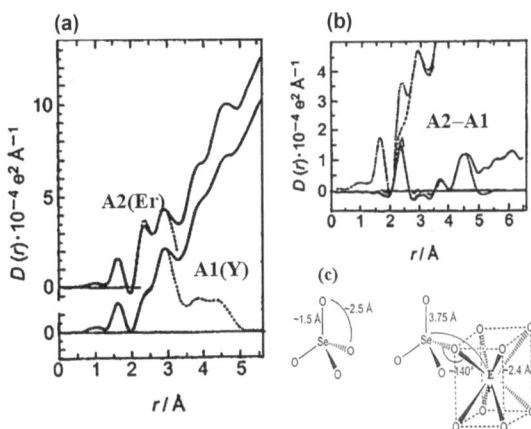

Figure 2.6 The results of the isomorphous substitution method applied to aqueous $Er_2(SeO_4)_3$ (A2) and $Y_2(SeO_4)_3$ (A1) solutions:[9] (a) the radial distribution functions of the two solutions, (b) the differential distribution functions, and (c) the most likely model of $[Er(H_2O)_7(OSeO_3)^{2+}]$ together with a tetrahedral SeO_4^{2-}. Reproduced from ref. 10 with permission from American Chemical Society, Copyright 1985.

intramolecular O–O interactions within the SeO_4^{2-} ion. A peak at around 2.9 Å arises from the first-neighbor H_2O–H_2O bonds in solvent water. A peak at 4.5 Å comes from the interactions between Er^{3+} (Y^{3+}) and water molecules in the second solvation shell. Figure 2.6(b) shows the differential radial distribution functions between the two solutions. Only the peaks due to the first-neighbor and the second-neighbor M–O interactions are revealed at 2.35 Å and 4.47 Å, respectively. A small peak at 3.75 Å is assigned to the Ln···Se interactions due to a contact ion-pair association between M^{3+} and SeO_4^{2-}. From a comparison between the experimental and theoretical $i(Q)$ using a least-squares fitting procedure, the interatomic distance (Er–O) = 2.35 Å and the coordination number = 8.0 are obtained. The most likely structure is shown in Figure 2.6(c).

2.4.3 Isotopic Substitution Method

When the coherent scattering cross-sections of isotopes have a large difference, the isotopic substitution method has an advantage in deriving a set of partial structure factors related to the substituted elements from the total scattering factors. In the case of MX in D_2O, there are ten partial structure factors (M–D, M–O, X–O, X–D, M–M, M–X, X–X, D–O, D–D, and O–O). When a cation M is replaced by its isotope M′ for the system MX–D_2O, the differential structure function $\Delta i_M(Q)$ between MX–D_2O and M′X–D_2O is given by

$$
\begin{aligned}
\Delta i_M(Q) &= 2n_M n_O b_O \Delta b_M [s_{MO}(Q) - 1] + 2n_M n_D b_D \Delta b_M [s_{MD}(Q) - 1] \\
&\quad + 2n_M n_X b_X \Delta f_M(Q) [s_{MX}(Q) - 1] + n_M^2 [b_M^2 - b_{M'}^2][s_{MM}(Q) - 1] \\
&= A_1 [s_{MO}(Q) - 1] + B_1 [s_{MD}(Q) - 1] + C_1 [s_{MX}(Q) - 1] + D_1 [s_{MM}(Q) - 1]
\end{aligned}
$$

$$(2.18)$$

since all the terms related to non-isotopically substituted atoms are cancelled out.

A similar equation is obtained when anion X is replaced by its isotope X′.

The isotopically substituted radial distribution functions $G_M(r)$ and $G_X(r)$ are obtained by Fourier transform of $\Delta_M(Q)$ and $\Delta_X(Q)$ as

$$
\begin{aligned}
G_M(r) &= A_1 [g_{MO}(Q) - 1] + B_1 [g_{MD}(Q) - 1] + C_1 [g_{MX}(Q) - 1] \\
&\quad + D_1 [g_{MM}(Q) - 1] + E_1
\end{aligned}
$$

$$(2.19)$$

where $E_1 = -(A_1 + B_1 + C_1 + D_1)$

$$
\begin{aligned}
G_M(r) &= A_2 [g_{MO}(Q) - 1] + B_2 [g_{MD}(Q) - 1] + C_2 [g_{MX}(Q) - 1] \\
&\quad + D_2 [g_{MM}(Q) - 1] + E_2
\end{aligned}
$$

$$(2.20)$$

where $E_2 = -(A_2 + B_2 + C_2 + D_2)$.

Figure 2.7(a) and (b) shows the differential scattering cross-sections of 9.3 mol kg^{-1} aqueous ^{7}Li^{35}Cl and ^{7}Li^{37}Cl solutions in D$_2$O at room temperature, respectively.[5,11]

Figure 2.7(c) and (d) shows the differential structure factor $\Delta_{Cl}(Q)$ and the corresponding radial distribution functions $G_{Cl}(r)$ obtained by Fourier transform of $\Delta_{Cl}(Q)$, respectively. In $G_{Cl}(r)$, the first peak at 2.25 Å is ascribed to the nearest-neighbor Cl–D$_1$ interactions due to the Cl$^-$ solvation. The double peaks at 3.30 Å and 3.60 Å correspond to the Cl–O and Cl–D$_2$ interactions, respectively, within the solvation shell of Cl$^-$. Thus, the hydrogen-bonded Cl···D–O is almost linear. The number of water molecules around

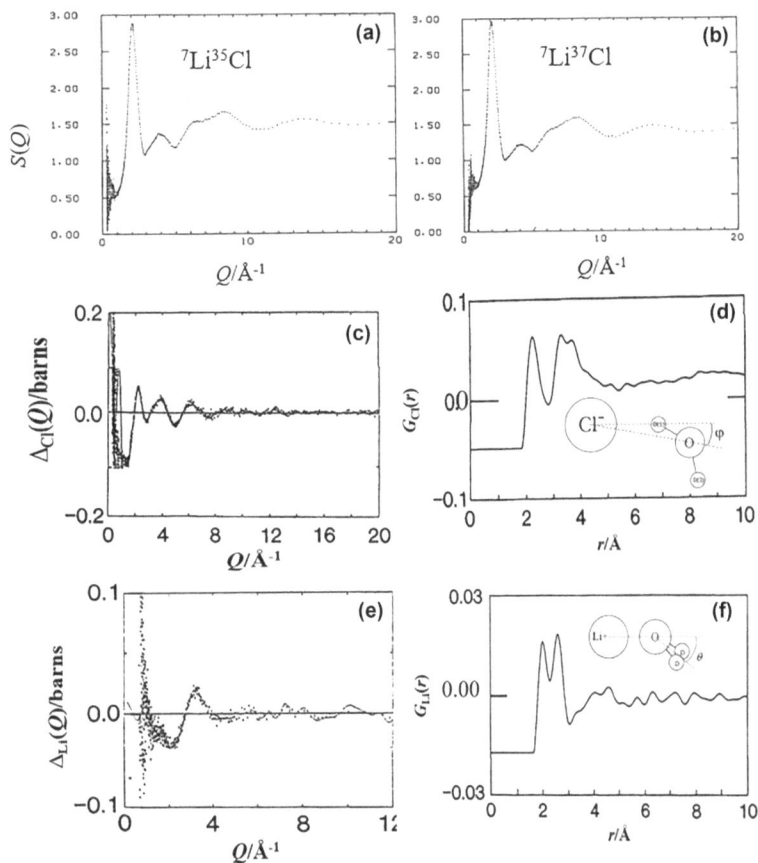

Figure 2.7 The results of the isotopic substitution method applied to aqueous isotopically substituted LiCl solutions in D$_2$O:[10] (a) and (b) $S(Q)$ of ^{7}Li^{35}Cl and ^{7}Li^{37}Cl solutions, (c) the differential structure factors between the ^{7}Li^{35}Cl and ^{7}Li^{37}Cl solutions, (d) the differential distribution function obtained by Fourier transform of (c), (e) the differential structure factors between ^{0}LiCl and ^{7}LiCl solutions, and (f) the differential distribution functions. Reproduced from ref. 11 with permission from Elsevier, Copyright 2009.

Cl^- is 5.0 from the peak area and the least-squares fitting. Figure 2.7(e) and (f) shows the differential structure factors $\Delta_{Li}(Q)$ and the radial distribution functions $G_{Li}(r)$ of 9.5 mol kg^{-1} aqueous ^{7}LiCl and ^{0}LiCl (^{0}Li is the isotopic mixture of ^{7}Li and ^{6}Li with a 44.9 : 55.1 molar ratio) solutions in D_2O. In $G_{Li}(r)$, the peaks at 2.02 Å and 2.61 Å are ascribed to the Li–O and Li···D distances, respectively, due to the Li$^+$ solvation. The number of water molecules in the primary solvation sphere is around four from an estimation of the peak area and the least-squares fitting. Using the intramolecular structure of a water molecule, the dipole of the solvated water molecules is tilted by $\theta = 40°$ from the Li–O vector. In $G_{Li}(r)$ and $G_{Cl}(r)$, there are the terms due to the ion association between M and X. Since coefficients C_1, D_1, C_2, and D_2 are smaller by about one order of magnitude than A_1, B_1, A_2, and B_2, it is practically difficult to determine the structure of the ion association. When more than three solutions with different isotopes or their mixtures are measured, the double differential structure factors $\Delta\Delta_M(Q)$ and $\Delta\Delta_X(Q)$ are extracted to obtain structural information on the M–X ion pairs.

2.4.4   EPSR Modeling Method

The EPSR modeling combined with neutron scattering data enables us to derive all the partial structure factors, partial correlation functions, the coordination number distribution, the angle distribution, and the spatial density functions in a solution. Figure 2.8(a) shows the neutron interference functions $F(Q)$ on a 2 mol dm^{-3} aqueous CaCl$_2$ solution in D_2O measured at 1 GPa and 0.1 MPa and 298 K on PLANET at the J-PARC MLF facility.[5,12]

Figure 2.8(b) shows the corresponding radial distribution functions $G(r)$ obtained by Fourier transform of the interference functions. The solid lines show the simulated values by EPSR modeling which reproduce the experimental values. Figure 2.9 shows the pair distribution functions $g(r)$ and the coordination distribution functions related to the Ca^{2+} and Cl$^-$ solvation. Upon compression to 1 GPa, Ca^{2+}, which is a structure making ion form

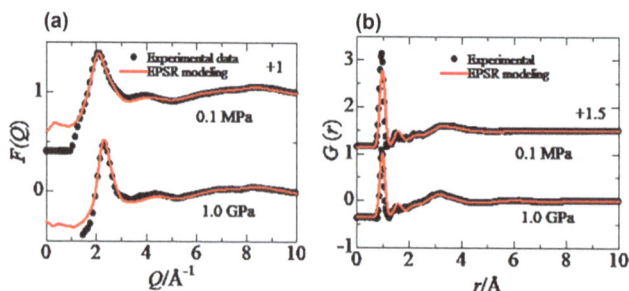

Figure 2.8   Results of EPSR modeling applied to a 2 mol kg^{-1} CaCl$_2$ aqueous solution in D_2O at pressures of 1 GPa and 0.1 MPa:[11] (a) the interference functions and (b) the radial distribution functions between the experiments and EPSR modeling. Reproduced from ref. 11 with permission from John Wiley & Sons, Copyright 2019.

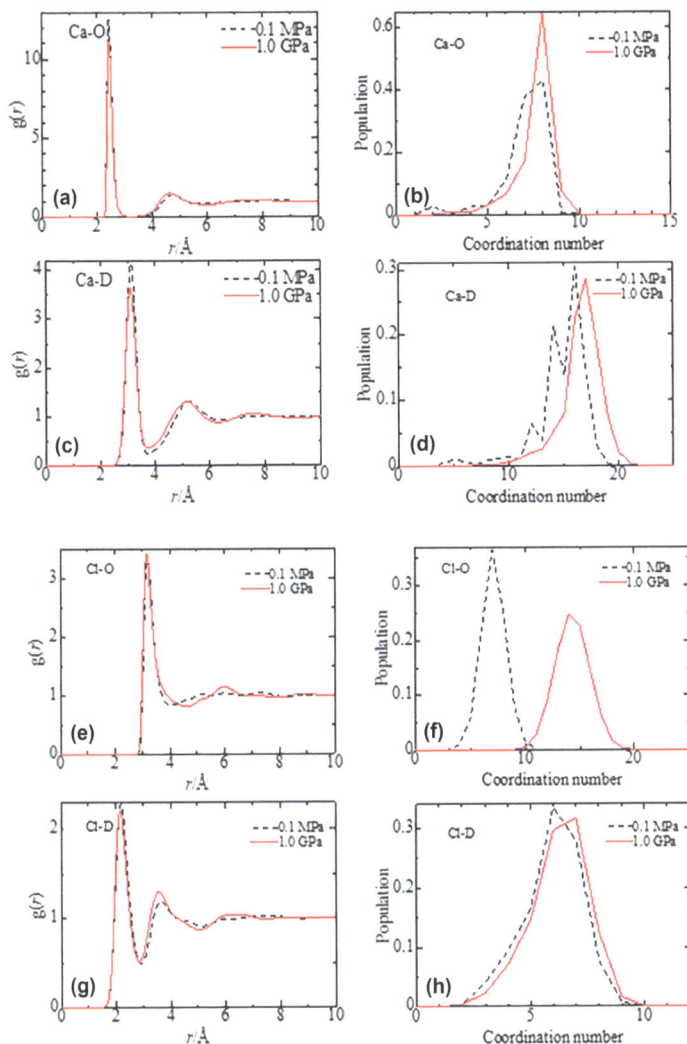

Figure 2.9 Pair distribution functions and coordination number distributions of ion solvation in a 2 mol kg^{-1} CaCl$_2$ solution in D$_2$O obtained by EPSR modeling:[11] (a and b) Ca–O (D$_2$O), (c and d) Ca–D (D$_2$O), (e and f) Cl–O, and (g and h) Cl–D. Reproduced from ref. 11 with permission from John Wiley & Sons, Copyright 2019.

rigid solvation shells around the ion, does not change the solvation structure in both the first and the second solvation shells significantly. In contrast, a large Cl$^-$ breaks the tetrahedral network structure of solvent water and does not form extended stable solvation shells around the ion under ambient conditions. However, at 1 GPa, Cl$^-$ forms the second solvation shell and the water oxygen coordination number of Cl$^-$ changes drastically from 6 at 0.1 MPa to 12 at 1 GPa, although the water hydrogen coordination number of

Cl$^-$ does not change appreciably upon compression. This is due to the structural change in solvent water as shown in Figure 2.10.

The pair correlation functions of O–O in solvent water show a structure transformation from the tetrahedral network structure to a close-packed

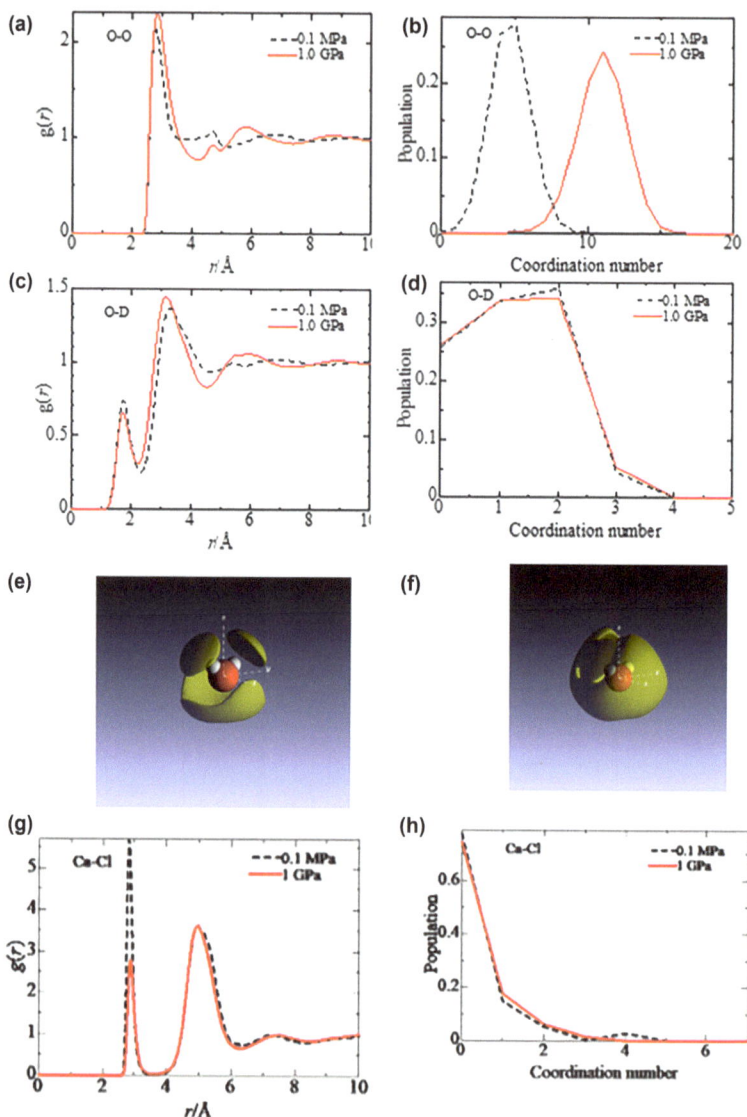

Figure 2.10 Pair distribution functions, coordination number distributions, and spatial density functions of solvent water and ion association in a 2 mol kg^{-1} CaCl$_2$ solution in D$_2$O obtained by EPSR modeling:[11] (a and b) O–O, (c and d) O–D, (e and f) SDF of neighboring water O atoms around a central water molecule at 0.1 MPa and 1 GPa, and (g and h) Ca^{2+}–Cl$^-$ ion association. Reproduced from ref. 11 with permission from John Wiley & Sons, Copyright 2019.

form of a simple liquid like Ne. The corresponding coordination number increases from 4.4 at 0.1 MPa to 12 at 1 GPa. However, the pair distribution functions and the coordination number distributions of O–H pairs are not altered very much upon compression to 1 GPa. They demonstrate the presence of hydrogen bonds among water molecules. These findings suggest the distortion of hydrogen bonds between water molecules, which are confirmed by the spatial density functions (3D structure) shown in Figure 10. Regarding the ion association, the pair distribution functions of Ca^{2+}–Cl^- do not show a significant change in the first peak due to the contact ion pairs but a shortening of the second peak position due to the solvent-separated ion pairs upon compression.

Problem 1

When you measure a 3.0 mol kg^{-1} aqueous NaCl solution by X-ray and neutron scattering, compare which atom pairs are mostly seen by each method. Use eqn (2.4) and the atomic numbers for the X-ray atomic scattering factors and the values of the coherent scattering lengths in Table 2.1 for neutron scattering.

Problem 2

Figure 2.11 shows the X-ray radial distribution function of liquid water at 298 K. Assign the peaks observed at 2.83, 4.6, and 6.8 Å. The area of the first peak at 2.83 Å amounts to 4.4 water molecules around a central water molecule. Discuss the most likely structure of liquid water, compared with the structure of hexagonal ice (I_h).

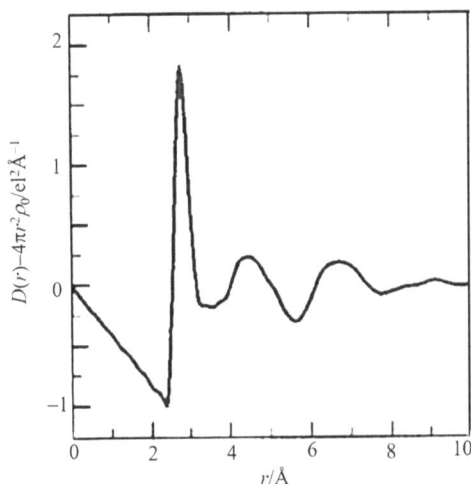

Figure 2.11   X-ray radial distribution function of liquid water at 298 K.

Answer to Problem 1

A 3.0 mol kg^{-1} aqueous NaCl solution is composed of 3.0 mol Na, 3.0 mol Cl, 55.6 mol O and 111.2 mol H (or D for neutron). The atomic fractions are 0.017, 0.017, 0.32, and 0.64 for Na, Cl, O, and H (D), respectively. The weight of atom pair p–q in eqn (2.5) is calculated as $\sum n_p n_q a_p a_q$, e.g. Na–O for X-rays as 2(Na–O and O–Na) $\times$ 0.017(Na) $\times$ 0.32(O) $\times$ 11(Na) $\times$ 8(O) = 0.96 electron^2 and for neutrons as 2 $\times$ 0.017 $\times$ 0.32 $\times$ 0.363(Na) $\times$ 0.5803(O) = 0.0023 barns.

Atom	mol/L	n_p	A.N.	b_{coh}	X-ray									
Na	3	0.0174	11	0.363	NaO	NaH	NaCl	NaNa	ClO	ClH	ClCl	OO	OH	HH
Cl	3	0.0174	17	0.9577	0.9832	0.2458	0.1127	0.0365	1.5194	0.3799	0.0871	6.6259	3.3129	0.4141
O	55.6	0.3218	8	0.5803	Neutron									
H	111.2	0.6435	1	0.6671	NaO	NaD	NaCl	NaNa	ClO	ClD	ClCl	OO	OD	DD
	172.8	1			0.00235	0.00541	0.00021	0.00004	0.00621	0.01428	0.00028	0.03486	0.16031	0.18429

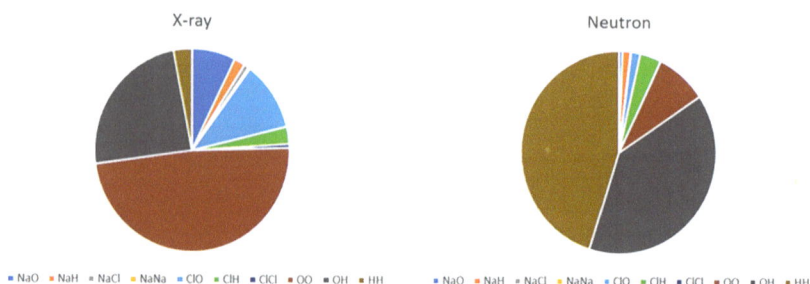

X-ray and neutron weights of the individual atom pairs for a 3 mol kg^{-1} NaCl aqueous solution.

Answer to Problem 2

The first, second, and third peaks observed at 2.83, 4.6, and 6.8 Å can be assigned to the first-neighbor, second-neighbor, and third-neighbor oxygen–oxygen interactions of water, respectively. The ratio of 4.6/2.83 is 1.6, close to $\sqrt{8/3} = 1.6$ for tetrahedral coordination. The most likely structure of liquid water is a tetrahedral hydrogen-bonded structure. The excess coordination number of 0.4 suggests that hydrogen bonds are partially disrupted and non-hydrogen bonded water molecules occupy the interstitial sites between the first- and second-neighbor coordination shells.

References

1. V. F. Sears, Neutron scattering lengths and cross sections, *Neutron News*, 1992, **3**, 26.
2. A. K. Soper, *Phys. Rev. B: Condens. Matter Mater. Phys.*, 2005, **72**, 104204.
3. A. K. Soper, *Mol. Simul.*, 2012, **38**, 1171.
4. S. W. Lovesey, *Theory of Neutron Scattering from Condensed Matter*, Clarendon Press, Oxford, 3rd edn, 1987.

5. T. Yamaguchi, *The Encyclopedia of Analytical Chemistry*, John Wiley & Sons, Ltd., 2022, DOI: 10.1002/9780470027318.a9803.
6. K. Yamanaka, T. Yamaguchi and H. Wakita, *J. Chem. Phys.*, 1994, **101**, 9830.
7. OhtakiH., T. Yamaguchi and M. Maeda, *Bull. Chem. Soc. Jpn.*, 1976, **49**, 701.
8. G. Johansson and M. Sandström, *Chem. Scr.*, 1973, **4**, 195.
9. Y. Kameda, M. Sasaki, T. Usuki, T. Othomo, K. Itoh, K. Suzuya and T. Fukunaga, *J. Neutron Res.*, 2003, **11**, 153.
10. G. Johansson and H. Wakita, *Inorg. Chem.*, 1985, **24**, 3047.
11. T. Yamaguchi, H. Ohzono, M. Yamagami, K. Yamanaka, K. Yoshida and H. Wakita, *J. Mol. Liq.*, 2010, **153**, 2.
12. T. Yamaguchi, M. Nishino, K. Yoshida, M. Takumi, K. Nagata and T. Hattori, *Eur. J. Inorg. Chem.*, 2019, 1170.

CHAPTER 3

X-ray Absorption Spectroscopy

INGMAR PERSSON

Department of Molecular Sciences, Swedish University of Agricultural Sciences, P.O. Box 7015, SE-750 07 Uppsala, Sweden
Email: ingmar.persson@slu.se

3.1 Introduction

The absorption as function of the energy/wavelength of X-rays shows sharp increases at very well-defined energies, known as absorption edges. The absorption edges of an element are directly related to the binding energies of the electrons in the different electron shells, *e.g.* 1s, 2s, 2p, 3s, 3p, 3d, *etc.*, and states, *e.g.* $^2S_{1/2}$, $^2P_{1/2}$, $^2P_{3/2}$, *etc.* The binding energy of an electron decreases with increasing principle quantum number, and an absorption edge is observed when the energy of an absorbed photon corresponds to an electronic transition or ionization potential. The notation of the absorption edges is K, L, M, *etc.* for the first, second and third electron shells of an atom/ion, respectively. The K edge has the highest energy and only one electron state. The second electron shell has two orbitals, s and p, with one ($^2S_{1/2}$) and two electron states ($^2P_{1/2}$ and $^2P_{3/2}$), respectively, and their absorption edges are denoted L_I, L_{II} and L_{III}, respectively. The energy of the three L edges decreases in the order $L_I > L_{II} > L_{III}$. Accordingly, for the third electron shell, there are three orbitals, s, p and d, and five absorption edges denoted in energy order $M_I > M_{II} > M_{III} > M_{IV} > M_V$. The absorption edge energies of all elements are summarized, *e.g.* in the X-ray Data Booklet.[1]

On the high-energy side of an absorption edge, a fine structure is observed, which can be related to the structural properties of the compounds. This fine structure is, however, not present for the noble gases as they lack

Coordination Chemistry Fundamentals Series No. 2
Metal Ions and Complexes in Solution
Edited by Toshio Yamaguchi and Ingmar Persson
© Japan Society of Coordination Chemistry 2024
Published by the Royal Society of Chemistry, www.rsc.org

neighbouring atoms. This fine structure is nowadays referred to as extended X-ray absorption fine structure (EXAFS). This phenomenon was described and explained already in 1931 by Kronig and known as the Kronig structure.[2] However, its usefulness was much later recognized and theoretically developed by Stern, Sayers and Lytle.[3] That study and the simultaneous development of synchrotron light facilities producing X-rays within a wide energy range with unprecedented intensity became the starting point for X-ray absorption spectroscopy (XAS) as a very powerful and versatile method for structural studies of matter in all states of aggregation.

3.2 Theory of EXAFS

When the incoming energy, E, is larger than the threshold energy, the energy required to excite an inner-core electron to continuum, E_0, photoelectrons are ejected with the kinetic energy E_k, $E_k = E - E_0$. The wavelength of the ejected photoelectron is given by the de Broglie expression: $\lambda_e = h/(2m_e \cdot (E - E_0))^{\frac{1}{2}}$, where h is Planck's constant and m_e is the mass of the electron. The photoelectron wave from the absorbing atom can be back-scattered by the surrounding atoms. The outgoing and back-scattered waves will then interfere. As the kinetic energy of the outgoing wave increases, the wavelength becomes shorter, with increasing E. This interference will induce variations in the electron density on the absorbing atom in such a way that a constructive interference will increase the electron density on the absorbing atom and thereby increase the probability of the photon to be absorbed. Consequently, a destructive interference will decrease the electron density on the absorbing atom and thereby decrease the probability of the photon to be absorbed (Figure 3.1). As the wavelength of the formed photoelectron wave will decrease with increasing E, while the distances between the absorbing atom and its neighbouring atoms are constant, a change in interference from constructive to destructive will cause changes in the absorption from maximum to minimum. This will give rise to oscillations in the absorption after the absorption edge for all compounds with neighbouring atoms around the absorbing atom. This means that a sinusoidal wave in the absorption will be formed for every unique distance in the studied sample. The goal of data treatment of EXAFS data is therefore to extract the different sinusoidal waves and determine their phases and amplitudes to extract the structural parameters of the sample under study.

3.3 Experimental Set-up

A standard experimental set-up for an XAS experiment consists of two–three ion chambers for detection in the transmission mode, an X-ray fluorescence detector, an ion chamber as the Lytle detector,[4] or modern solid state detectors[5,6] (Figure 3.2). The first ion chamber, I_0, determines the intensity of the incoming X-ray beam. The second ion chamber, I_1, determines the

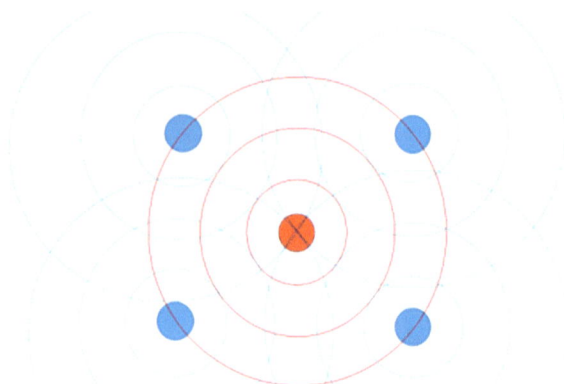

(a)

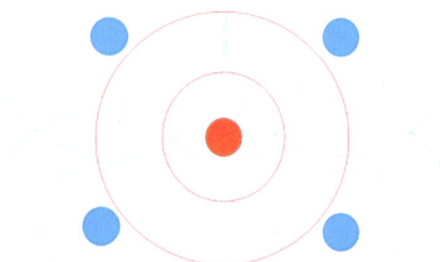

(b)

Figure 3.1   (a) Constructive interference of the outgoing photoelectron wave (red wave) and the back-scattered one (blue wave). (b) Destructive interference of the outgoing photoelectron wave (red wave) and the back-scattered one (blue wave).

intensity of the X-ray beam after absorption in the sample, and the third ion chamber, I_2, determines the intensity after the absorption in a reference sample, which is normally the pure element of the same element under study. Perpendicular to the X-ray beam, an X-ray fluorescence detector, IF, detects the fluorescence signal from the sample ideally placed 45° to the incoming X-ray beam (Figure 3.2).

3.4   Data Collection

Data collection of an XAS spectrum is made in the energy range of -200 eV to $+1000$ eV or more relative to the threshold energy, E_0. The XAS spectrum is divided into three sections: pre-edge, -200 to -20 eV relative to E_0; edge-

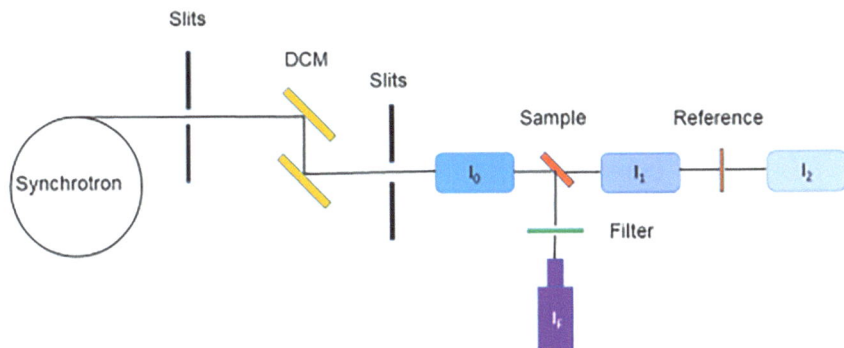

Figure 3.2 EXAFS experimental set-up for simultaneous detection of transmission and fluorescence data; DCM – double crystal monochromator, I_0, I_1 and I_2 – first, second and third ion chambers, respectively, and I_F – fluorescence detector.

region or XANES (X-ray absorption near edge structure), -20 to $+200$ eV relative to E_0; and the EXAFS (extended X-ray absorption fine structure) region, from *ca.* $+50$ eV to as high energies as possible without too much noise in the data, normally 600–1200 eV above E_0 (Figure 3.3). Data collection can be performed according to two principles: step scans, in which a pre-determined step in energy is applied in the different regions with large steps, 5–10 eV, in the pre-edge region, small steps, 0.1–0.2 eV, in the edge region, and with increasing step size in the EXAFS region to obtain equal step size in the wave vector *k* (see below), and continuous scans, in which the monochromator motor changes the angle in the double crystal monochromator with a constant change in theta/second, theta being the scattering angle in Bragg's law to obtain a certain wavelength/energy from the monochromator. Examples of experimental XAS data are shown in Figure 3.3. The negative slope of the transmission XAS spectrum depends on the decreasing total absorption of the sample with increasing photon energy. The positive slope of the fluorescence XAS spectrum depends on increasing fluorescence quantum yield with increasing photon energy.

3.5 EXAFS Data Treatment

The data treatment is in principle the same regardless of whether the data have been collected in the transmission or fluorescence mode. Only the principles of the EXAFS data treatment will be shown here. The first step of subtracting the pre-edge slope of the data set is shown in Figure 3.4a, and the resulting spectrum is shown in Figure 3.4b after normalizing the absorption edge step to unity. The next and by far the most crucial step in the EXAFS data treatment is to subtract the spline, the red function in Figure 3.4b; the spline is how the spectrum should have looked like if the absorbing atom has not had neighbouring atoms. Thereafter, the spline is

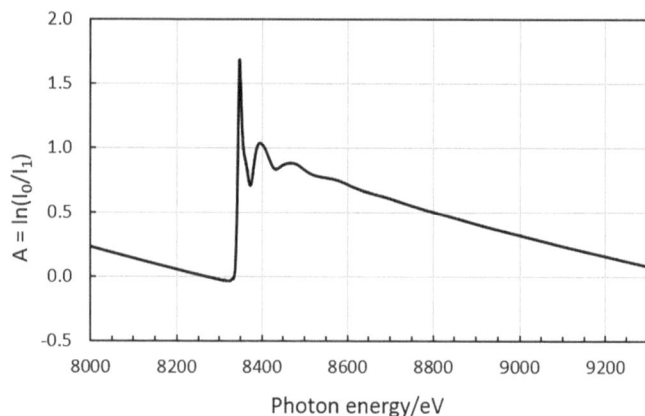

(a)

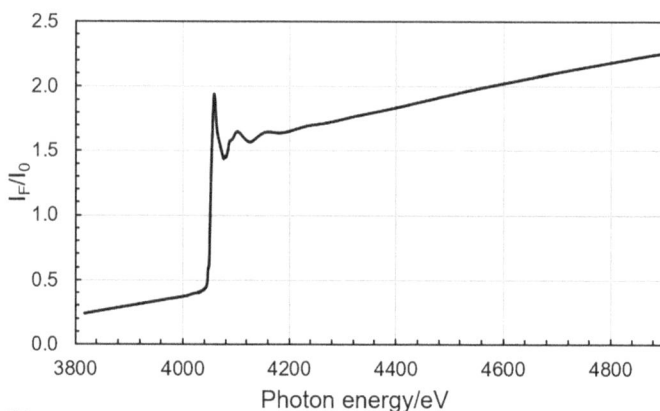

(b)

Figure 3.3   (a) Raw X-ray absorption spectrum of an aqueous solution of nickel(II) perchlorate recorded in the transmission mode. (b) Raw X-ray absorption spectrum of an aqueous solution of calcium(II) perchlorate recorded in the fluorescence mode.

subtracted from the experimental data, and the resulting function is shown in Figure 3.4c. The energy scale in eV is now converted to the photoelectron wavenumber or wave vector, k, according to the following equation:

$$k = \sqrt{(2m_e \cdot (E - E_0)/h^2)} = \sqrt{(0.2625 \cdot (E - E_0))} \qquad (3.1)$$

where m_e is the mass of the electron and h is Planck's constant; $k = 0$ at E_0. The EXAFS function, $\chi(k)$, is now plotted in the form $\chi(k) \times k^3$ *versus* k to make the high-energy data more visible as shown in Figure 3.4d. The final step in the data treatment is to convert data from inverse to real space by Fourier transformation as shown in Figure 3.4e.

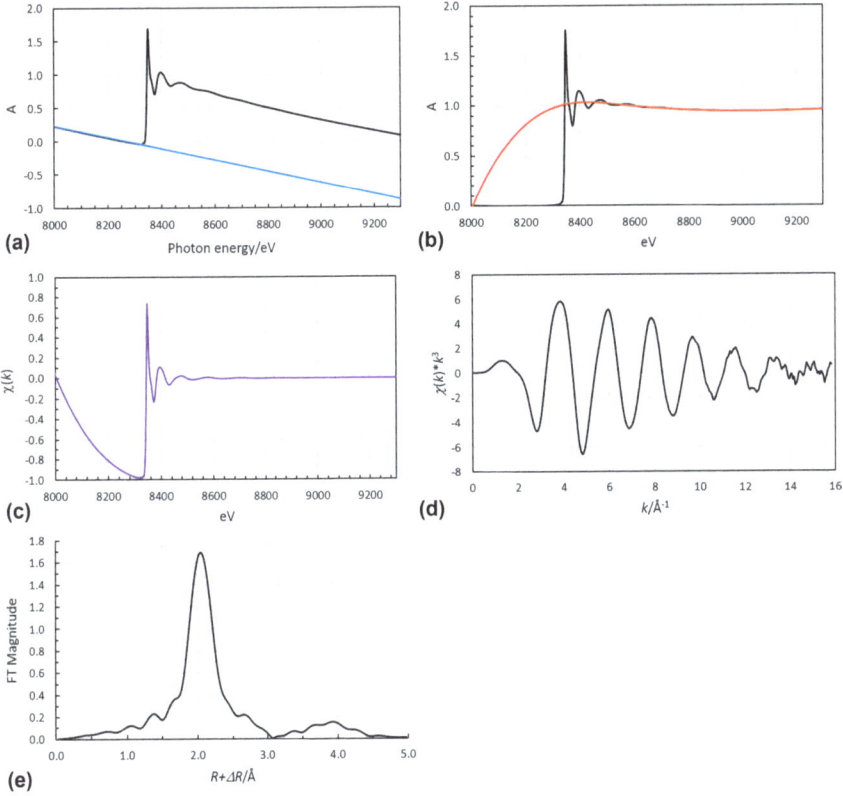

Figure 3.4 (a) Raw X-ray absorption spectrum of an aqueous solution of nickel(II) perchlorate recorded in the transmission mode (black line) and the extended pre-edge (blue line). (b) Raw X-ray absorption spectrum of an aqueous solution of nickel(II) perchlorate after pre-edge subtraction and normalization (black line) and the spline function (red line). (c) EXAFS function ($\chi(k)$) of an aqueous solution of nickel(II) perchlorate after the subtraction of the spline function from the normalized raw data (purple line). (d) Raw EXAFS function of an aqueous solution of nickel(II) perchlorate. (e) Fourier transform of the raw EXAFS function of an aqueous solution of nickel(II) perchlorate.

To evaluate the structural parameters, the experimental data are fitted with the EXAFS equation:

$$\chi_i(k) = \sum_j \frac{N_j \cdot S_o^2(k)}{k \cdot R_j^2} \cdot |f_{\text{eff}}(k)| \cdot e^{-\left(2k^2 \cdot \sigma_j^2\right)} \cdot e^{-\left(\frac{2R_j}{\Lambda(k)}\right)} \cdot \sin\left(2kR_j + \phi_{i,j}(k)\right) \quad (3.2)$$

where N_j is the number of scatterers in the jth shell, R_j is the distance between the absorbing atom and the back-scatterers in the jth shell in single back-scattering and for multiple-scattering it is the half of the total path length, S_o^2 is the amplitude reduction factor, f_{eff} is the effective amplitude function for each scattering path, $\exp[-2\sigma_j^2 \cdot k^2]$ is the Debye–Waller factor in the harmonic approximation, the Debye–Waller coefficient σ_j accounts for thermal and configurational disorder, $\Lambda(k)$ is the photoelectron mean free path, $\exp[-2R_j/\Lambda(k)]$ is the mean path factor, and $\exp[2R_j + \Phi_{ij}(k)]$ is the

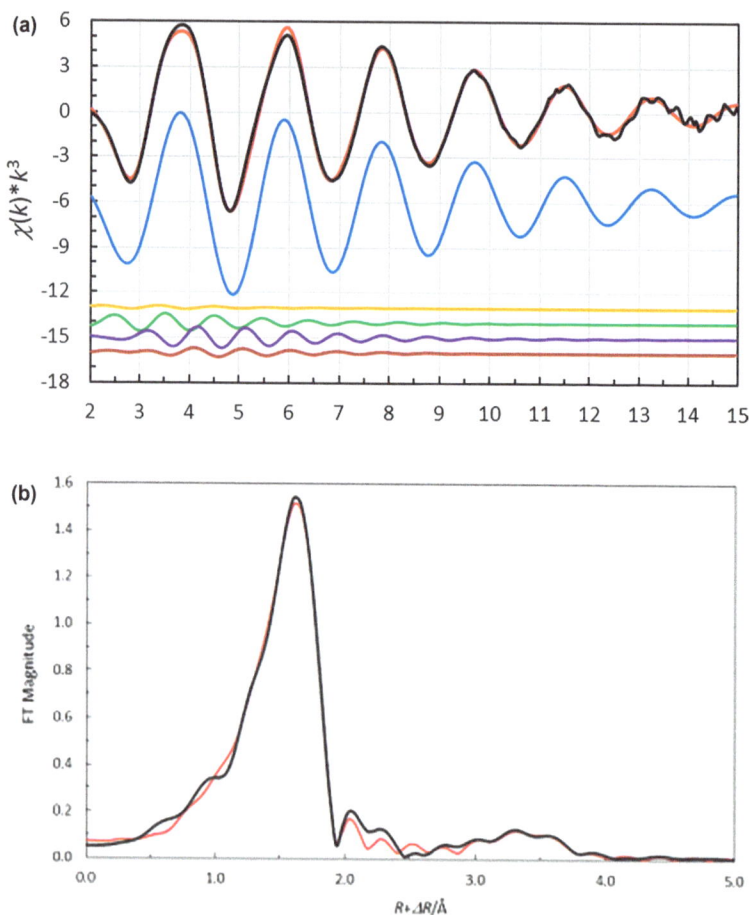

Figure 3.5 (a) Fit of the raw EXAFS function of an aqueous solution of nickel(II) perchlorate; raw EXAFS data (black line), calculated EXAFS function (red line), and the individual contributions in the model hexaaquanickel(II) complex, Ni–O single-scattering (light blue line, offset: −6), Ni–O–O non-linear multiple-scattering (yellow line, offset: −13), Ni–O–O linear multiple-scattering (green line, offset: −14), Ni–O–O–O linear multiple-scattering of different oxygens (purple line, offset: −15) and Ni–O–O–O linear multiple-scattering of the same oxygen (brown line, offset: −16). (b) Fit of the Fourier transform of the raw EXAFS data of an aqueous solution of nickel(II) perchlorate; the black line represents the raw data and the red line represents the calculated data.

total phase, where $\Phi_{ij}(k)$ is the phase shift due to the Coulomb potential of the central atom i and the back-scattering atom j.

The experimental data are fitted with the EXAFS equation by refining distance from the absorbing atom, R_j, amplitude reduction factor, S_o^2, Debye–Waller coefficient, σ_j, and the number of distances, N_j. It is important to note that R_j and S_o^2 relate to the phase of the EXAFS signal, and σ_j and N_j to the amplitude. They are pairwise strongly correlated and should not be refined simultaneously until the final refinement. The fitting of the EXAFS data of the hydrated nickel(II) ion in aqueous solution is shown in Figure 3.5 together with the individual contributions from the individual single- and multiple-scattering paths.

3.6 Usability of EXAFS

EXAFS can be applied on all states of aggregation. Only distances around the absorbing atom can be detected, which in most cases is an advantage. The term $\exp[-2R_j/\Lambda(k)]$ in the EXAFS equation shows that the damping of the EXAFS signal increases sharply with increasing distance. Furthermore, the term $\exp[-2\sigma_j{}^2 \cdot k^2]$ damps the EXAFS signal sharply with increasing bond distance distribution. This means that only distances up to 4–5 Å from the absorbing atom can be detected, and long diffuse distances as a second hydration sphere are hardly possible to detect with EXAFS. All chemical species radiated in an experiment will contribute to the EXAFS signal, which means it is an advantage to have as few chemical species containing the absorbing atom in the sample as possible. The very high intensities possible to obtain in modern synchrotron light allow us to study very dilute samples, submillimolar, with good accuracy. Furthermore, a full EXAFS scan can be performed in seconds, which makes it possible to determine structural changes with a time resolution of seconds. Detailed information can therefore be obtained on, *e.g.*, catalytic pathways.

Problem 1

Why does the absorption of a compound vary sinusoidally with increasing energy of X-rays after the absorption edge?

Problem 2

In the EXAFS function, which parameters will affect the phase and the amplitude, respectively?

Answer to Problem 1

As the energy increases in an X-ray absorption scan, the wavelength of the outgoing photoelectron wave will also continuously increase. When the wavelength of the photoelectron wave is equal to the length of a single- or multiple-scattering event or a multiple of it, there will be a positive interference between the outgoing and back-scattered waves (see Figure 3.1a), which will result in a maximum electron density on the absorption atom and thereby also maximum absorbance. When the wavelength of the photoelectron wave is equal to half the length of a single- or multiple-scattering event or a multiple of it, there will be a negative interference between the outgoing and back-scattered waves (see Figure 3.1b), which will result in a minimum electron density on the absorption atom and thereby also minimum absorbance. With increasing energy of the incoming X-rays, there will be sequence maxima and minima in the absorbance causing a sinusoidal absorbance.

Answer to Problem 2

The parameters in the EXAFS function affecting the phase are distance and mean pathway, and the amplitude is affected by the number of distances and the Debye–Waller coefficient (see eqn (3.2)). Parameters that are strongly correlated should not be refined simultaneously.

References

1. A. Thompson, D. Attwood, E. Gullikson, M. Howells, K.-J. Kim, K. Kirz, J. Kortright, I. Lindau, Y. Liu, P. Pianetta, A. Robinson, J. Scofield, J. Underwood, G. Williams and H. Winick, *X-ray Data Booklet*, Lawrence Berkley National Laboratory, 2009.
2. R. de L. Kronig, *Z. Phys.*, 1931, **70**, 317.
3. E. A. Stern, D. E. Sayers and F. W. Lytle, *Phys. Rev. B: Condens. Matter Mater. Phys.*, 1975, **11**, 4836.
4. F. W. Lytle, R. B. Greegor, D. R. Sandstrom, E. C. Marques, J. Wong, C. L. Spiro, G. P. Huffman and F. E. Huggins, *Nucl. Instrum. Methods Phys. Res.*, 1984, **226**, 542.
5. https://www.mirion.com/products/passivated-implanted-planar-silicon-pips-detectors (downloaded July 4, 2022).
6. https://www.ketek.net/sdd/technology/working-principle/ (downloaded July 4, 2022).

CHAPTER 4

Nuclear Magnetic Resonance

TATSUYA UMECKY

Faculty of Science and Engineering, Saga University, Honjo-machi, Saga, 840-8502, Japan
Email: umecky@cc.saga-u.ac.jp

4.1 Introduction

Nuclear magnetic resonance (NMR) is a phenomenon in which the magnetic nuclei of a substance placed in a magnetic field resonantly absorb radio-frequency energy. NMR spectroscopy can be applied to a wide range of substances because several ions and complexes generally contain at least one NMR-active nucleus.[1] Even for the same nucleus, the NMR chemical shifts can differ depending on the structures of ion solvation and the complex in a solution because the radiofrequency energy absorption is affected by the electron density around the nucleus. Therefore, one-dimensional NMR spectra provide valuable information about the structures of ion solvation and the complex in a solution. Additionally, using the Fourier-transform (FT) technique, multiple radiofrequency pulses can be applied during NMR measurements. Relaxation times, self-diffusion coefficients, and two- (or more) dimensional spectra can also be measured with FT-NMR spectroscopy. FT-NMR spectroscopy is a unique tool that provides a microscopic picture of the structure and dynamics of solvated ions and complexes in different solutions. Using specialized NMR cells and probes,[2–4] *in situ* solution-state NMR spectra can be measured under specific conditions, that is, high pressure, supercritical state, and electric fields. In this section, the experimental measurements of one-dimensional spectra, relaxation times, self-diffusion coefficients, and two-dimensional spectra are discussed for solutions containing ions and complexes.

Coordination Chemistry Fundamentals Series No. 2
Metal Ions and Complexes in Solution
Edited by Toshio Yamaguchi and Ingmar Persson
© Japan Society of Coordination Chemistry 2024
Published by the Royal Society of Chemistry, www.rsc.org

4.2   One-dimensional Spectra

One-dimensional NMR spectra can be obtained using the simplest single-pulse sequence.[5] To precisely determine the NMR chemical shifts in a spectrum, suitable reference compounds for the observed nucleus are used.[1] Most of the NMR spectra are measured using internal and external referencing methods. The internal referencing method is the technique of adding the reference compound directly to the solution under investigation and often used to identify unknown compounds. On the other hand, in the external referencing method, the liquid solution containing the reference compound is sealed in a capillary or a cylindrical tube and separated from the solution under investigation. When the exact value of the chemical shift in one-dimensional spectra is required, the external referencing method is a better choice than the internal referencing method. This is because, in the internal referencing method, the reference compounds in the studied solution are either impurities or, in some cases, reactants. The value of the chemical shift determined by the external referencing method includes the difference between the volume magnetic susceptibilities of the sample and reference solutions. Hence, correcting the volume magnetic susceptibilities is essential in the external referencing method. In the external referencing measurements using coaxial cylindrical double tubes, the volume magnetic susceptibilities can be corrected as follows:[6]

$$\delta_{cor} = \delta_{obs} - \frac{4\pi}{3}\left(\chi_{v,sam} - \chi_{v,ref}\right) \times 10^6, \tag{4.1}$$

where δ_{cor} and δ_{obs} are the corrected and observed NMR chemical shifts, respectively. Moreover, $\chi_{v,sam}$ and $\chi_{v,ref}$ are the volume magnetic susceptibilities of the sample and reference solutions, respectively. The values of $\chi_{v,sam}$ and $\chi_{v,ref}$ can be determined using the bulbed-capillary external double referencing method[6] and other techniques, *e.g.*, magnetic balance (Gouy method). Solvation- and complexation-induced chemical shifts for the solutions under investigation offer relevant information about solvation and complexation of ions, respectively. Furthermore, solvation-induced chemical shifts of probe molecules in the solutions are suggested as indirect methods for determining the Gutmann donor number (D_N) and Gutmann–Mayer acceptor number (A_N). D_N and A_N are determined from the ^{23}Na NMR spectra of $NaClO_4$ and the ^{31}P NMR spectra of triethyl phosphine oxide, respectively.[7]

In a 1:1 metal complex formation reaction, two different metal ions are involved, that is, uncomplexed (but solvated) (M) and complexed (ML) metal ions. When the ligand exchange between M and ML (M $\rightleftharpoons$ ML) is faster than the NMR timescale, the δ_{cor} of metal ions can be expressed using the weighted chemical shifts of M (δ_M) and ML (δ_{ML}) using the following equation:

$$\delta_{cor} = x_M\delta_M + x_{ML}\delta_{ML}, \tag{4.2}$$

where x_M and x_{ML} are the mole fractions of M and ML in the complex formation equilibrium, respectively. The complex formation constant (K_f) is represented as follows:

$$\delta_{cor} = \delta_{ML} + \{(\delta_M - \delta_{ML})/2K_f C_M\}\left(A + \sqrt{A^2 + 4K_f C_M}\right) \tag{4.3}$$

where C_M is the total concentration of the metal ion. Moreover, A can be expressed as follows:

$$A = K_f C_M - K_f C_L - 1 \tag{4.4}$$

where C_L is the total concentration of the ligand. The numerical value of K_f can be determined from the dependence of C_L on the chemical shifts of metal ions. Figure 4.1(a) shows the ^{7}Li NMR spectra of lithium bis(trifluoromethanesulfonyl)amide (Li[NTf$_2$]) in acetonitrile solutions containing 1,4,7,10,13-pentaoxacyclopentadecane (15C5) at different 15C5 : Li$^+$ molar ratios ([15C5]/[Li]). At [15C5]/[Li] < 1.0, the ^{7}Li peak of the acetonitrile solution shifts toward the lower field with increasing 15C5 concentration. At [15C5]/[Li] > 1.0, the ^{7}Li peak is independent of the molar ratio [15C5]/[Li], suggesting that the stoichiometric proportion of the Li$^+$–15C5 complex is 1. Figure 4.1(b) shows the δ_{cor}(^{7}Li) of Li[NTf$_2$] in acetonitrile and 1-ethyl-3-methylimidazolium bis(trifluoromethanesulfonyl)amide ([C$_2$C$_1$im][NTf$_2$]) solutions as a function of [15C5]/[Li] at 298 K. The K_f values determined using eqn (4.3) were 3.9×10^4 for acetonitrile and 2.2×10^4 for [C$_2$C$_1$im][NTf$_2$]. The ^{7}Li peak of Li[NTf$_2$] in [C$_2$C$_1$im][NTf$_2$] shifts toward the

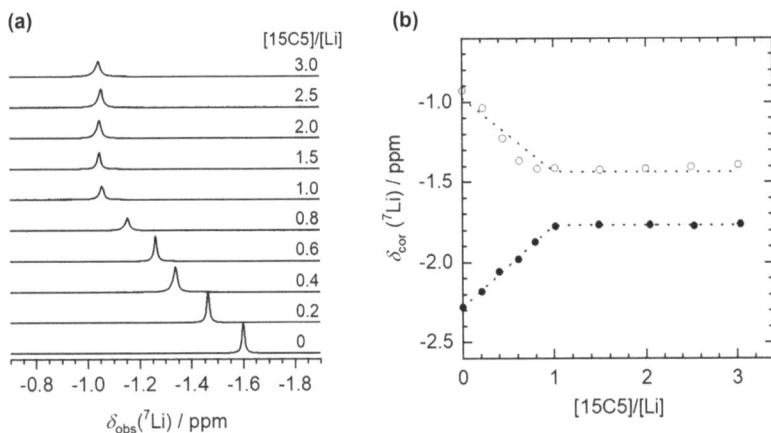

Figure 4.1 (a) ^{7}Li NMR spectra of Li[NTf$_2$] in acetonitrile solutions containing 15C5 at different molar ratios of 15C5 : Li$^+$ ([15C5]/[Li]) at 298 K. (b) ^{7}Li chemical shifts (δ_{cor}) of Li[NTf$_2$] in the acetonitrile (closed circles) and [C$_2$C$_1$im][NTf$_2$] (open circles) solutions as a function of the molar ratio ([15C5]/[Li]).

higher field. The difference between the shift directions of the ^{7}Li peaks in the cases of acetonitrile and [C$_2$C$_1$im][NTf$_2$] is attributed to the electron density difference between the solvated and complexed Li$^+$. As Li$^+$ is complexed with 15C5, the electron density of Li$^+$ in acetonitrile decreases, whereas that of Li$^+$ in [C$_2$C$_1$im][NTf$_2$] increases. In addition, Figure 4.1(b) shows that the δ_{cor}(^{7}Li) values of acetonitrile are always smaller than those of [C$_2$C$_1$im][NTf$_2$], regardless of the complexation, implying that the electron densities of solvated and complexed Li$^+$ in acetonitrile are higher than those in [C$_2$C$_1$im][NTf$_2$]. If detecting the NMR signals of metal elements is difficult owing to low frequency and low sensitivity, K_f can also be determined from the dependence of C_M on the chemical shifts of elemental species, such as ^{1}H and ^{13}C, in ligands. For the complexation of metal ions with macrocyclic ligands, K_f, the thermodynamic quantities (ΔH and ΔS) and the rate constants determined by the NMR methodology have been explored in a previous study.[8]

The peak area in one-dimensional spectra set at a sufficiently longer pulse delay than longitudinal relaxation time is proportional to the abundance of the chemical species. Quantitative NMR spectroscopy that utilizes the peak area in one-dimensional spectra is one of the quantification methods. The tautomerization of 2,4-pentandione (Hacac), which is one of the typical bidentate ligands, has been quantitatively investigated in various solutions and extended nano-spaces of size up to 100 nm through one-dimensional spectra.[9,10] Most recently, Maki *et al.* applied quantitative ^{27}Al NMR spectroscopy to investigate the hydrolysis equilibrium of aluminum solutions.[11] They identified the aluminum complexes in aqueous solutions and determined their abundances.

4.3   Relaxation Times

Longitudinal (T_1) and transverse (T_2) relaxation times can be obtained with multi-pulse sequences.[12] Standard double-pulse sequences of T_1 and T_2 are inversion-recovery and spin-echo pulse sequences, respectively. T_1 and T_2 of the ions and the complexes in solutions can be related to the correlation times (τ_c) for rotational and jump motions, if the extreme narrowing condition is satisfied.[13] At lower viscosity and higher temperature, τ_c is relatively short, indicating that the extreme narrowing condition is almost satisfied. A dipolar nucleus, *e.g.*, ^{1}H, ^{13}C, ^{15}N, ^{19}F, and ^{31}P, with a spin quantum number of $\frac{1}{2}$ relaxes through the intra-molecular dipole–dipole interaction with another dipolar nucleus (S) that is directly attached to the observed nucleus (I). In this condition, τ_c can be determined from the dipole relaxation times ($T_{1,dd}$ and $T_{2,dd}$) as follows:

$$\frac{1}{T_{1,dd}} = \frac{1}{T_{2,dd}} = N\left(\frac{\mu_0}{4\pi}\right)^2 \gamma_I^{\,2}\gamma_S^{\,2}r^{-6}\left(\frac{h}{2\pi}\right)^2\tau_c, \qquad (4.5)$$

where N is the number of dipolar nuclei directly attached to the observed nucleus, μ_0 is the permeability of vacuum, γ is the magnetogyric ratio, and h is the Planck constant. The subscripts I and S indicate nuclei I and S, respectively, and r is the internuclear distance between I and S. It is noteworthy that the observed longitudinal and transverse relaxation times $(T_{1,obs}, T_{2,obs})$ are not completely equal to $T_{1,dd}$ and $T_{2,dd}$, respectively, because the dipolar nucleus relaxes through not only the dipole–dipole relaxation mechanism but other mechanisms as well. Meanwhile, the quadrupolar nucleus, *e.g.*, ^{2}H, $^{6/7}$Li, ^{9}Be, ^{14}N, ^{18}O, and ^{27}Al, with spin quantum numbers larger than $\frac{1}{2}$ relaxes through the quadrupole interaction with the electric field gradient at the nucleus. In this condition, τ_c can be determined from the quadrupole relaxation times $(T_{1,q}$ and $T_{2,q})$ as follows:

$$\frac{1}{T_{1,q}} = \frac{1}{T_{2,q}} = \frac{3\pi^2}{10} \frac{2I+3}{I^2(2I-1)} \left(1 + \frac{\eta_q^2}{3}\right) \left(\frac{eQq}{h}\right)^2 \tau_c, \tag{4.6}$$

where I is the spin quantum number of the observed nucleus, η_q is the asymmetry parameter, and eQq/h is the quadrupole coupling constant with the principal axis. Furthermore, when the quadrupolar nucleus even slightly relaxes through any mechanism other than quadrupole relaxation, the $T_{1,obs}$ and $T_{2,obs}$ of the quadrupolar nucleus are not equal to $T_{1,q}$ and $T_{2,q}$, respectively. Kanakubo *et al.* measured the T_1 values of ^{9}Be of the central metal ion and ^{13}C of the methine group from the ligand of bis(acetylacetonato)-beryllium ([Be(acac)$_2$]) in acetonitrile at eight different temperatures.[14] They successfully determined the τ_c of [Be(acac)$_2$] in acetonitrile solutions. Thereafter, the τ_c values of tris(acetylacetonato)aluminum ([Al(acac)$_3$]) and tris(1,1,1,5,5,5-hexafluoroacetylacetonato)aluminum ([Al(hfa)$_3$]) in acetonitrile and supercritical CO$_2$ solutions were also reported.[15] Table 4.1 lists the $T_{1,obs}$ values of ^{13}C and ^{27}Al, nuclear Overhauser effect factor (η_{nOef}), τ_c, and eQq/h for ~ 0.30 mol dm^{-3} [Al(acac)$_3$] and ~ 0.080 mol dm^{-3} [Al(hfa)$_3$] in the acetonitrile-d_3 solution at 313 K. The η_{nOef} values can be determined from the ^{13}C signal heights of one-dimensional spectra using the completed and inverse-gated decoupling methods:

$$\eta_{nOef} = \frac{M_c}{M_{ig}} - 1 \tag{4.7}$$

Table 4.1 Experimental data of $T_{1,obs}(^{13}$C$)$, η_{nOef}, τ_c, $T_{1,obs}(^{27}$Al$)$, and eQq/h for ~ 0.30 mol dm^{-3} [Al(acac)$_3$] and ~ 0.080 mol dm^{-3} [Al(hfa)$_3$] in the acetonitrile-d_3 solution at 313 K.

	$T_{1,obs}(^{13}$C$)$/s	η_{nOef}	τ_c/ps	$T_{1,obs}(^{27}$Al$)$/ms	(eQq/h)/MHz
[Al(acac)$_3$]	2.24	1.94	19.3	5.71	3.09
[Al(hfa)$_3$]	2.08	1.89	20.2	5.04	3.22

where M_c and M_{ig} are the signal heights with the completed and inverse-gated decoupling sequences, respectively. $T_{1,dd}(^{13}C)$ is determined from $T_{1,obs}(^{13}C)$ and η_{nOef} using the following equation:

$$\frac{1}{T_{1,dd}} = \frac{1}{T_{1,obs}} \times \frac{\eta_{nOef}}{\eta_{max}} \tag{4.8}$$

where η_{max} is the maximum value of the nuclear Overhauser effect. The η_{max} value is expressed as the product of the constant φ and the ratio of γ_S and γ_I, which is 0.5 under the extreme narrowing condition:

$$\eta_{max} = \varphi \times (\gamma_S / \gamma_I) \tag{4.9}$$

When the ^{13}C nucleus ($\gamma = 6.73 \times 10^7$ $T^{-1}s^{-1}$) is irradiated by the ^1H nucleus ($\gamma = 26.75 \times 10^7$ $T^{-1}s^{-1}$), the η_{max} value is 1.99, and τ_c is determined from the experimental data of $T_{1,obs}(^{13}C)$ and η_{nOef} using eqn (4.5) and (4.8). Furthermore, eQq/h can be obtained from τ_c and the experimental data of $T_{1,obs}(^{27}Al)$ assuming that the ^{27}Al nucleus relaxes only through the quadrupole relaxation mechanism ($T_{1,obs}(^{27}Al) \approx T_{1,q}(^{27}Al)$). The τ_c values in Table 4.1 indicate that, in acetonitrile, the rotational motion of [Al(acac)$_3$] is slightly faster than that of [Al(hfa)$_3$]. Hayamizu *et al.* measured the T_1 of ^1H, ^7Li, and ^{19}F of [C$_2$C$_1$im][NTf$_2$] and 1-ethyl-3-methylimidazolium bis(fluorosulfonyl)amide ([C$_2$C$_1$im][FSA]) with and without lithium electrolytes.[16] In the case of viscous solutions such as ionic liquids, the extreme narrowing condition is often not satisfied. Hence, τ_c cannot be determined using the above equations. Hayamizu *et al.* successfully determined the τ_c of the 1-ethyl-3-methylimidazolium cation (C$_2$C$_1$im$^+$), bis(trifluoromethanesulfonyl)-amide anion (NTf$_2^-$), bis(fluorosulfonyl)amide anion (FSA$^-$), and Li$^+$ for ionic liquid solutions using the temperature dependence of T_1 of ^1H, ^7Li, and ^{19}F.

4.4 Self-diffusion Coefficients

Self-diffusion coefficients (D) can be obtained from the multi-pulse sequences of field gradient pulses.[17] A pulsed-gradient spin-echo (PGSE) sequence is often used to determine the D of metal ions and complexes in solutions. When the NMR peak with a PGSE sequence is very small and undetectable owing to the shorter T_2 of the observed nucleus, a pulsed-gradient stimulated-echo (PGSTE) sequence, where the π pulse of a PGSE sequence splits into two $\pi/2$ pulses, can be used. When performing D measurements, the height of the liquid sample should be approximately 5 mm. This is because heat convection in the NMR tube significantly affects the D values. Moreover, the correction of the magnetic gradient strength using the accepted literature data, *e.g.*, 2.30×10^{-9} $m^2 s^{-1}$ at 298 K for H$_2$O,[18] is absolutely imperative in D measurements. The D value is a quantity that represents translational motion; however, it provides useful information about the structure of the solvated metal ions and the complexes in solutions. The values of D for metal ions and complexes in various solutions have been reported. More recently, Hayamizu *et al.* determined D, T_1, and T_2 values for alkali metal ions in aqueous solutions at 303 K using ^{7}Li, ^{23}Na, ^{87}Rb, and ^{133}Cs NMR spectroscopies.[19] They evaluated the dynamic ionic radius of alkali metal ions on the basis of the Stokes–Einstein relation. They found that the dynamic ionic radii of alkali metal ions are in the order of $Cs^+ < Rb^+ < Na^+ < Li^+$, in contrast to the corresponding static ionic radii.

Apart from the PGSE and PGSTE sequences, diffusion-ordered spectroscopy (DOSY) is also occasionally used to determine the D of ions and complexes in various solutions.[20] A DOSY spectrum with D on the vertical axis and δ_{obs} on the horizontal axis is a pseudo-two-dimensional NMR spectrum. The value of D obtained from the abovementioned PGSE and PGSTE sequences represents the diffusivity of a single component. On the other hand, the D obtained from DOSY represents the diffusivity of multiple components with the distribution of D. For one nucleus, a one-dimensional NMR spectrum is the overlapping of the NMR peaks of all compounds in sample solutions. A DOSY spectrum of a multi-component solution facilitates the assignment of the NMR peaks because these peaks can be separated based on the difference in the D values of the multiple components. Carper *et al.* reported the ^{1}H DOSY spectra of the mixture of 1-ethyl-3-methylimidazolium chloride and aluminum chloride at a mixing ratio of 1:2.[21] Figure 4.2 shows the ^{1}H DOSY spectrum of 1-ethyl-3-methylimidazolium acetate ([C$_2$C$_1$im][AcO]) at 298 K. The diffusion time and gradient pulse width were fixed at 150 ms and 6.5 ms, respectively. Magnetic gradient strengths were varied up to 0.6 T m^{-1}. In the ^{1}H DOSY spectrum, the C$_2$C$_1$im$^+$ cation and AcO$^-$ anion are clearly separated. The D of AcO$^-$ is smaller than that of C$_2$C$_1$im$^+$. Recently, Byers *et al.* reported the ^{1}H DOSY

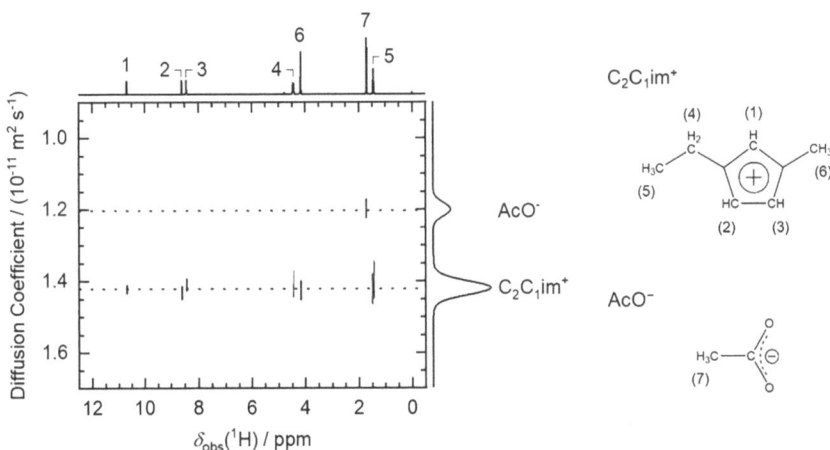

Figure 4.2 ^{1}H DOSY spectra of the [C$_2$C$_1$im][AcO] ionic liquid at 298 K.

spectra of paramagnetic complexes in organic liquid solutions.[22] From the ^{1}H DOSY spectrum, the monomer–dimer equilibrium of a zirconium–cobalt complex was suggested.

The mobility (μ) of ions in solutions can be determined when the D measurements are conducted under electric field application conditions. Because specialized NMR cells and probes are required to measure μ, the experimental data of μ of ions in solutions are scarcely available in the literature. A previous study has reported the μ values of the cations and anions of $C_2C_1im^+$- and 1-butyl-3-methylimidazolium-based ionic liquids.[23] Despite the liquids being highly viscous, the μ values of the cations and anions were almost equal and relatively large.

4.5 Two-dimensional Spectra

Two-dimensional NMR spectra of heteronuclear single quantum coherence (HSQC) and heteronuclear multiple quantum coherence (HMQC) are useful when two nuclei are directly linked by a covalent bond. A heteronuclear multiple bond correlation (HMBC) spectrum, which demonstrates correlations between two nuclei that are separated by several covalent bonds, is useful when it is difficult to obtain HSQC and HMQC spectra. Two-dimensional spectra of homonuclear Overhauser effect spectroscopy (NOESY) and heteronuclear Overhauser effect spectroscopy (HOESY) provided information about the internuclear correlation. Correlation peaks can be detected when the two nuclei are spatially close. Therefore, NOESY and HOESY spectra provide information on the intra- and intermolecular correlations, whereas HSQC, HMQC, and HMBC spectra are limited to the intramolecular correlation. The conformational studies of cyclodextrin complexes have been reported.[24–26] Figure 4.3 shows the [7]Li-detected [7]Li–[1]H HOESY spectra of 1-octyl-4-aza-1-azoniabicyclo[2.2.2]octane bis(trifluoromethanesulfonyl)amide

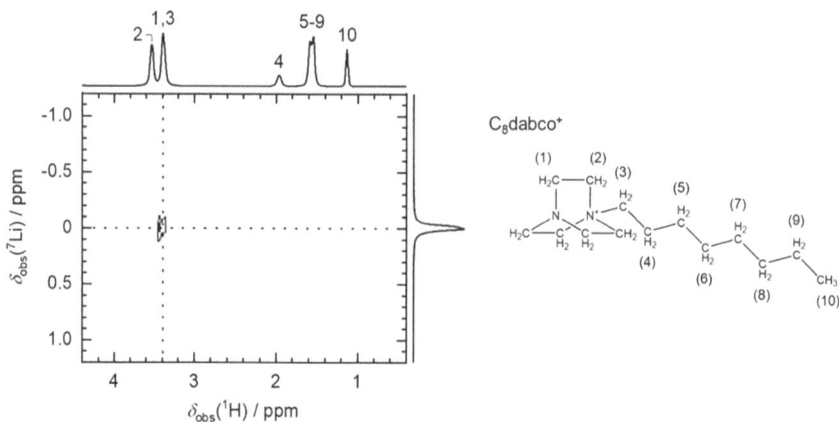

Figure 4.3   [7]Li–[1]H HOESY spectra of the [C$_8$dabco][NTf$_2$] solution with Li[NTf$_2$] at 313 K.

([C$_8$dabco][NTf$_2$]) with Li[NTf$_2$] at 313 K. The mixing time and accumulation number were set to 0.6 s and 8, respectively. Only one correlation peak in Figure 4.3 is observed between the Li$^+$ and hydrogen atoms of the hetero-cyclic ring of the C$_8$dabco$^+$ cation, which proves that Li$^+$ interacts with the nitrogen atom of the tertiary amino group of the C$_8$dabco$^+$ cation.[27]

Problem 1

Explain the dependence of the DN of solvents on the ^{23}Na NMR peak of Na$^+$ in solution.

Problem 2

Explain the dependence of the AN of solvents on the ^{31}P NMR peak of triethyl phosphine oxide.

Problem 3

Calculate the η_{max} of the ^{15}N nucleus under the extreme narrowing condition when the ^{15}N nucleus ($\gamma = -2.71 \times 10^7$ T^{-1} s^{-1}) is irradiated by the ^{1}H nucleus. Draw a diagrammatic illustration of ^{1}H-decoupled ^{15}N NMR spectra using each of the completed and inverse-gated decoupling methods.

Problem 4

For sodium formate in deuterium oxide solution at 298 K, $T_{1,obs}(^{13}C)$, M_c, and M_{ig} are 21.0 s, 85.4, and 35.8, respectively. Calculate the τ_c of the formate anion in deuterium oxide solution at 298 K on the assumption that the ^{13}C nucleus relaxes under the extreme narrowing condition.

Hint: $N = 1$, $\mu_0/4\pi = 10^{-7}$, γ (^{13}C) $= 6.73 \times 10^7$ T^{-1} s^{-1}, γ (^{1}H) $= 26.75 \times 10^7$ T^{-1} s^{-1}, $r_{CH} = 1.09 \times 10^{-10}$ m, $h = 6.626 \times 10^{-34}$ J s.

Answer to Problem 1

When the donor site of the solvent donates electrons to Na$^+$, the electron density around the donor atom nucleus decreases. The interaction of Na$^+$ with the electron-donating site of the solvent gives an occasion to a decrease in the electron density of the sodium atom induced by the interelectronic repulsion. The decrease in the electron density of the Na$^+$ ion becomes remarkable as the DN of the solvent increases. Therefore, the ^{23}Na peak shifts to the lower field as the DN of the solvent increases.[7]

Answer to Problem 2

When the oxygen atom of triethyl phosphine oxide donates electrons to the electron-accepting site of the solvent molecule, the electron density around

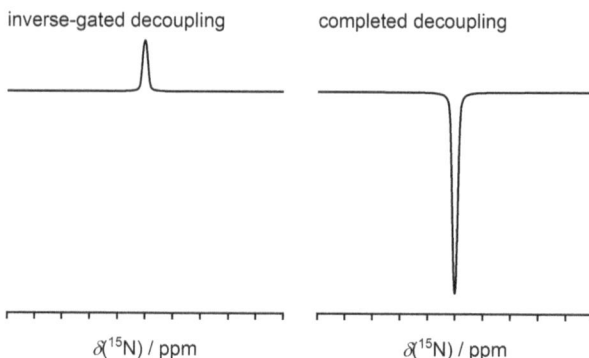

inverse-gated decoupling   completed decoupling

$\delta(^{15}N)$ / ppm $\delta(^{15}N)$ / ppm

Figure 4.4 ^{1}H-decoupled ^{15}N NMR spectra.

the oxygen nucleus decreases. The interaction of the electron-accepting site of the solvent with the oxygen atom of triethyl phosphine oxide gives an occasion to a decrease in the electron density of the phosphorus atom induced by repolarization of the P–O bond. The decrease in the electron density of the phosphorus atom becomes remarkable as the AN of the solvent increases. Therefore, the ^{31}P peak shifts to the lower field as the AN of the solvent increases.[7]

Answer to Problem 3

Substituting $\varphi = 0.5$, $\gamma_S = 26.75 \times 10^7$ T^{-1}s^{-1}, and $\gamma_I = -2.71 \times 10^7$ T^{-1}s^{-1} into eqn (4.9) yields the following equation:

$$\eta_{max} = 0.5 \times \frac{26.75 \times 10^7 \text{ T}^{-1}\text{s}^{-1}}{-2.71 \times 10^7 \text{ T}^{-1}\text{s}^{-1}} = -4.94$$

The negative η_{max} means that the ^{1}H-decoupling ^{15}N peak using the completed decoupling method is observed in the opposite direction to that using the inverse-gated decoupling method. The absolute height of the former ^{1}H-decoupling ^{15}N peak is, at most, 3.94 times ($\eta_{max} + 1$) higher than that of the latter ^{1}H-decoupling ^{15}N peak (Figure 4.4).

Answer to Problem 4

Substituting $M_c = 85.4$ and $M_{ig} = 35.8$ into eqn (4.7) yields the following equation:

$$\eta_{nOef} = \frac{M_c}{M_{ig}} - 1 = \frac{85.4}{35.8} - 1 = 1.39$$

Next, substituting $T_{1,obs}(^{13}C) = 21.0$ s, $\eta_{nOef} = 1.39$, and $\eta_{max} = 1.99$ into eqn (4.8) yields the following equation:

$$\frac{1}{T_{1,dd}} = \frac{1}{T_{1,obs}} \times \frac{\eta_{nOef}}{\eta_{max}} = \frac{1}{21.0} \times \frac{1.39}{1.99} = 0.0333 \text{ s}^{-1}$$

Finally, substituting $1/T_{1,dd} = 0.0333$ s^{-1}, $N = 1$, $r = 1.09 \times 10^{-10}$ m into eqn (4.5) yields $\tau_c = 1.55 \times 10^{-12}$ s (1.55 ps).

Abbreviations

15C5	1,4,7,10,13-Pentaoxacyclopentadecane
AcO$^-$	Acetate anion
[Al(acac)$_3$]	Tris(acetylacetonato)aluminum(III)
[Al(hfa)$_3$]	Tris(1,1,1,5,5,5-hexafluoroacetylacetonato)aluminum(III)
AN	Gutmann–Mayer acceptor number
[Be(acac)$_2$]	Bis(acetylacetonato)beryllium(II)
C$_2$C$_1$im$^+$	1-Ethyl-3-methylimidazolium cation
C$_8$dabco$^+$	1-Octyl-4-aza-1-azoniabicyclo[2.2.2]octane cation
C_L	Total concentration of the ligand
C_M	Total concentration of the metal ion
D	Self-diffusion coefficient
DN	Gutmann donor number
DOSY	Diffusion-ordered spectroscopy
eQq/h	Quadrupole coupling constant
FSA$^-$	Bis(fluorosulfonyl)amide anion
FT	Fourier transform
Hacac	2,4-Pentandione
HMBC	Heteronuclear multiple bond correlation
HMQC	Heteronuclear multiple quantum coherence
HOESY	Heteronuclear Overhauser effect spectroscopy
HSQC	Heteronuclear single-quantum coherence
h	Planck constant
K_f	Complex formation constant
M	Metal ion
ML	1 : 1 complex
N	Number of dipolar nuclei directly attached to the observed nucleus
NMR	Nuclear magnetic resonance
NOESY	Nuclear Overhauser effect spectroscopy
NTf$_2^-$	Bis(trifluoromethanesulfonyl)amide anion
PGSE	Pulsed gradient spin-echo
PGSTE	Pulsed gradient stimulated echo
r	Internuclear distance
T_1	Longitudinal relaxation time
T_2	Transverse relaxation time
$T_{1,obs}$	Observed longitudinal relaxation time
$T_{2,obs}$	Observed transverse relaxation time
$T_{1,dd}$	Dipole longitudinal relaxation time
$T_{2,dd}$	Dipole transverse relaxation time
$T_{1,q}$	Quadrupole longitudinal relaxation time
$T_{2,q}$	Quadrupole transverse relaxation time

x_M	Mole fraction of uncomplexed M
x_{ML}	Mole fraction of complexed ML
γ	Magnetogyric ratio
δ_{cor}	Corrected chemical shift
δ_M	Chemical shift of uncomplexed M
δ_{ML}	Chemical shift of complexed ML
δ_{obs}	Observed chemical shift
η_{max}	Maximum of the nuclear Overhauser effect
η_{nOef}	Nuclear Overhauser effect factor
μ	Mobility
μ_0	Permeability of vacuum
τ_c	Correlation time
$\chi_{v,ref}$	Volume magnetic susceptibility of the reference
$\chi_{v,sam}$	Volume magnetic susceptibility of the sample

Acknowledgements

Two-dimensional NMR measurements were conducted at the Analytical Research Center for Experimental Sciences, Saga University.

References

1. R. K. Harris, E. D. Becker, S. M. C. de Menezes, R. Goodfellow and P. Granger, *Pure Appl. Chem.*, 2001, **73**, 1795.
2. I. T. Horváth and J. M. Millar, *Chem. Rev.*, 1991, **91**, 1339.
3. N. Matubayasi, C. Wakai and M. Nakahara, *J. Chem. Phys.*, 1997, **107**, 9133.
4. M. Holz, *Chem. Soc. Rev.*, 1994, **23**, 165.
5. S. Berger and S. Braun, in *200 and More NMR Experiments*, Wiley-VCH, Weinheim, 2011, pp. 44–53.
6. K. Mizuno, Y. Tamiya and M. Mekata, *Pure Appl. Chem.*, 2004, **76**, 105.
7. V. Gutmann, in *The Donor–Acceptor Approach to Molecular Interactions*, Plenum Press, New York, 1978, pp. 17–33.
8. R. M. Izatt, K. Pawlak, J. S. Bradshaw and R. Bruening, *Chem. Rev.*, 1991, **91**, 1721.
9. J. N. Spencer, E. S. Holmboe, M. R. Kirshenbaum, D. W. Firth and P. B. Pinto, *Can. J. Chem.*, 1982, **60**, 1178.
10. T. Tsukahara, K. Nagaoka, K. Morikawa, K. Mawatari and T. Kitamori, *J. Phys. Chem. B*, 2015, **119**, 14750.
11. H. Maki, G. Sakata and M. Mizuhata, *Analyst*, 2017, **142**, 1790.
12. S. Berger and S. Braun, in *200 and More NMR Experiments*, Wiley-VCH, Weinheim, 2011, pp. 160–166.
13. A. Abragam, in *Principles of Nuclear Magnetism*, Oxford University Press, New York, 1961, pp. 264–353.
14. M. Kanakubo, H. Ikeuchi and G. P. Sato, *J. Chem. Soc., Faraday Trans.*, 1998, **94**, 3237.

15. T. Umecky, M. Kanakubo and Y. Ikushima, *J. Phys. Chem. B*, 2011, **115**, 10622.
16. K. Hayamizu, S. Tsuzuki, S. Seki and Y. Umebayashi, *J. Chem. Phys.*, 2011, **135**, 084505.
17. S. Berger and S. Braun, in *200 and More NMR Experiments*, Wiley-VCH, Weinheim, 2011, pp. 467–469.
18. M. Holz, S. R. Heil and A. Sacco, *Phys. Chem. Chem. Phys.*, 2000, **2**, 4740.
19. K. Hayamizu, Y. Chiba and T. Haishi, *RSC Adv.*, 2021, **11**, 20252.
20. S. Berger and S. Braun, in *200 and More NMR Experiments*, Wiley-VCH, Weinheim, 2011, pp. 515–517.
21. W. R. Carper, G. J. Mains, B. J. Piersma, S. L. Mansfield and C. K. Larive, *J. Phys. Chem.*, 1996, **100**, 4724.
22. M. P. Crockett, H. Zhang, C. M. Thomas and J. A. Byers, *Chem. Commun.*, 2019, **55**, 14426.
23. T. Umecky, Y. Saito and H. Matsumoto, *J. Phys. Chem. B*, 2009, **113**, 8466.
24. H.-J. Schneider, F. Hacket, V. Rüdiger and H. Ikeda, *Chem. Rev.*, 1998, **98**, 1755.
25. T. Akita, K. Yoshikiyo and T. Yamamoto, *J. Mol. Struct.*, 2014, **1074**, 43.
26. S. Hazra and G. S. Kumar, *RSC Adv.*, 2015, **5**, 1873.
27. T. Umecky, K. Suga, E. Masaki, T. Takamuku, T. Makino and M. Kanakubo, *J. Mol. Liq.*, 2015, **209**, 557.

CHAPTER 5

Vibrational Spectroscopy

HAJIME TORII

Department of Applied Chemistry and Biochemical Engineering, Faculty of Engineering, and Department of Optoelectronics and Nanostructure Science, Graduate School of Science and Technology, Shizuoka University, 3-5-1 Johoku, Naka-ku, Hamamatsu 432-8561, Japan
Email: torii.hajime@shizuoka.ac.jp

5.1 General

In typical vibrational spectroscopy, we observe molecular vibrations *via* absorption or emission of infrared (IR) light or *via* scattering of light (visible, ultraviolet, or near-IR) with a different frequency from that of the incident light (called the Raman effect). Many vibrational modes are more or less delocalized over a molecule or a group of molecules, but some of them are known to be sufficiently localized to particular functional groups, and those modes appear at their own characteristic frequencies, or in their characteristic frequency regions, in the spectra. As a result, if we observe IR or Raman spectra in a wide frequency range (*e.g.*, 400–4000 cm^{-1}), we can get information on identification of molecular species in the sample. However, this is not the only way we can utilize vibrational spectroscopic techniques. Some vibrational modes are known to be sufficiently sensitive to intermolecular interactions, so that if we can correctly analyse the vibrational band profiles of each of those modes, *i.e.*, frequency shift, intensity change, and/or band shape (intensity distribution within a band as well as broadening due to relaxation processes), we can get information on intermolecular interactions and/or mutual molecular configurations in condensed phases or in clusters of molecules.

Coordination Chemistry Fundamentals Series No. 2
Metal Ions and Complexes in Solution
Edited by Toshio Yamaguchi and Ingmar Persson
© Japan Society of Coordination Chemistry 2024
Published by the Royal Society of Chemistry, www.rsc.org

The requirements for realization of this are fourfold: (1) the vibrational mode should be localized to a particular functional group or to a rather narrow spatial region (in a molecular sense) around it, so that the spectral signal is sufficiently site specific; (2) the vibrational band should be spectrally localized to avoid complication of analysis arising from spectral congestion; (3) the frequency, intensity, and/or intra-band intensity distribution should be sensitive to intermolecular interactions; and (4) the mechanism giving rise to this sensitivity should (preferably) be made clear so that we can get a transparent relationship between the spectral features and the mutual molecular configurations.

With regard to monoatomic metal ions themselves, there is no internal vibrational mode. As a result, according to the abovementioned scenario, interactions with ligands or solvents may be probed by perturbation on the vibrations of the ligands or solvents. In addition, another conceivable way to achieve this goal would be to measure and analyse the bands arising directly from intermolecular vibrational modes, such as the metal–ligand or metal–solvent stretching mode, which usually appear in a rather low-frequency region. In the following, some important aspects of these will be given with some typical example cases.

5.2 Mechanism of Frequency Shifts – The Simplest Cases and Beyond

How are vibrational frequencies affected by intermolecular interactions? In the simplest cases, it is explained by the mechanical anharmonicity of the vibrational mode and the interaction with the intermolecular electric field.[1,2]

The potential energy V may be expanded by the vibrational coordinate Q (with the potential energy minimum of the isolated molecule being taken as its origin) and the electric field E as

$$
\begin{aligned}
V &= V_0 + \frac{1}{2}\left(\frac{\partial^2 V}{\partial Q^2}\right)_0 Q^2 + \frac{1}{6}\left(\frac{\partial^3 V}{\partial Q^3}\right)_0 Q^3 + \cdots + \left(\frac{\partial V}{\partial E}\right)_0 E + \left(\frac{\partial^2 V}{\partial E \partial Q}\right)_0 EQ + \cdots \\
&= V_0 + \frac{1}{2}k_0 Q^2 + \frac{1}{6}\left(\frac{\partial^3 V}{\partial Q^3}\right)_0 Q^3 + \cdots - \mu_0 E - \left(\frac{\partial \mu}{\partial Q}\right)_0 EQ + \cdots
\end{aligned}
$$

$$(5.1)$$

where subscript 0 indicates that the quantity is evaluated at the origin, k_0 is defined as $k_0 = (\partial^2 V/\partial Q^2)_0$, and μ $(\equiv -\partial V/\partial E)$ is the dipole moment. For a small displacement of Q under electric field E, we have

$$
-\frac{\partial V}{\partial Q} \cong \left(\frac{\partial \mu}{\partial Q}\right)_0 E
$$

$$(5.2)$$

As a result, the field-induced displacement ΔQ is given as

$$\Delta Q \cong \frac{1}{k_0} \left(\frac{\partial \mu}{\partial Q} \right)_0 E \tag{5.3}$$

From eqn (5.1), we have

$$\frac{\partial^2 V}{\partial Q^2} = k_0 + \left(\frac{\partial^3 V}{\partial Q^3} \right)_0 Q + \cdots \tag{5.4}$$

so that the vibrational force constant (and hence the vibrational frequency) is affected rather linearly by the field-induced displacement ΔQ (if it is small) and, combining with eqn (5.3), also by the electric field itself.

This type of linear dependence of the vibrational frequency shift on the electric field E is seen in the relationship between the observed stretching frequency of the C=O bond of acetone and the estimated electric field operating on it in some aqueous electrolyte solutions.[3] The frequency becomes lower upon increasing the concentration of the salt (NaCl, LiCl, and $ZnCl_2$) as shown in Figure 5.1(A), and this low-frequency shift is linearly related to the estimated increase in the electric field as shown in Figure 5.1(B) and (C). In fact, the electric field E that induces a frequency shift is not limited to the one operating mutually among the species constituting a solution, but may also originate externally. The frequency shift occurring in the latter case is called the vibrational Stark effect[4] and has been observed for the stretching modes of many polar functional groups.

However, it is also clear from Figure 5.1 that the electric field is not the sole factor that controls the frequency shift; a separate linear dependence of the frequency on the field is seen for each salt. It is claimed that a non-electrostatic mechanism also plays an important role in this case.[3]

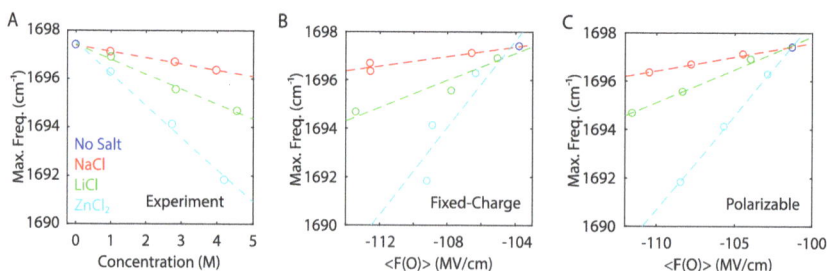

Figure 5.1   Variation of the peak frequency of the observed C=O stretching IR band of acetone dissolved in aqueous solutions of electrolytes (red: NaCl, green: LiCl, light blue: $ZnCl_2$) as a function of (A) electrolyte concentration and of the mean electric fields operating on the carbonyl O atom predicted from MD simulations using (B) fixed-charge and (C) polarizable potential functions. Adapted from ref. 3 with permission from American Chemical Society, Copyright 2020.

Even if the electrostatic interaction plays a major role in the frequency shift, the treatment expressed with eqn (5.1)–(5.4) may sometimes be too simple because the electric field operating in a condensed-phase system is generally non-uniform. To take this aspect into account, vibrational frequency maps have been developed for various vibrational modes of polar functional groups.[5] Here, each frequency map consists of the parameters that should be multiplied to the electrostatic potentials and/or electric fields operating on some atomic sites (in some cases also on some additional off-atom sites), so that we can estimate the frequency shift by evaluating those electrostatic potentials and/or electric fields. This strategy has been recognized to be effective for describing hydrogen-bond-induced frequency shifts of various vibrational modes,[5] even in the case of the C≡N stretching mode of acetonitrile,[6] where dipolar solvation induces a low-frequency shift, while hydrogen-bond formation on the N atom induces a high-frequency shift. In this sense, application of the strategy of vibrational frequency map would be promising for interpreting the vibrational spectra of electrolyte solutions.

5.3 More Specific Example Cases

Even when there is no abovementioned type of physical mechanism yet clarified, thorough experimental investigations may provide us some empirical rules on the relationship between the vibrational and structural properties. They may also be supported by first-principles theoretical calculations that are carefully performed on model systems.

In the case of the C–O symmetric and antisymmetric stretching modes of the carboxylate (COO^-) group interacting with one or two metal ions, it is known that the frequencies depend greatly on the type of coordination,[7] which is classified as[8] (a) the unidentate (or monodentate) form, where a metal ion interacts primarily with only one of the two oxygen atoms of the COO^- group, (b) the bidentate (or chelating) form, where a metal ion interacts rather equally with both oxygen atoms, (c) the bridging (or bridging bidentate) form, where a metal ion interacts with one of those oxygen atoms, and another metal ion interacts with the other oxygen atom, (d) the pseudo-bridging form, where one of these metal ions is replaced by a hydrogen atom of a water molecule, and (e) the ionic form, where the metal···carboxylate interactions are regarded as sufficiently weak. An empirical rule with regard to the frequency difference between the two modes (denoted as Δ_{a-s}) has been derived[7] as Δ_{a-s}(unidentate) > Δ_{a-s}(ionic) ~ Δ_{a-s}(bridging) > Δ_{a-s}(bidentate). More quantitatively, this rule relates the value of $\Delta_{a-s} > 200$ cm^{-1} to the unidentate form and the value of $\Delta_{a-s} < 105$ cm^{-1} to the bidentate form.

This is supported by first-principles theoretical calculations carried out for the deuterated acetate ion (CD_3COO^-) interacting with Mg^{2+}, Ca^{2+}, Na^+, and/or water.[8] For the $CD_3COO^- \cdots Mg^{2+}$ and $CD_3COO^- \cdots Ca^{2+}$ complexes in the unidentate form, Δ_{a-s} is calculated as 346 and 260 cm^{-1}, respectively, while for the same complexes in the bidentate form it is calculated as 26 and 102 cm^{-1}, respectively. For the $CD_3COO^- \cdots Na^+$ complex, it is calculated in the range of 150–200 cm^{-1} regardless of the type of coordination, which is close to the value of 208 cm^{-1} obtained for the free ion, because the metal$\cdots$carboxylate interactions are sufficiently weak in this case (regarded as in the ionic form). This dependence of the value of Δ_{a-s} on the type of coordination is correlated to the changes in the OCO angle (θ_{OCO}) and the asymmetry of the two C–O bond lengths (δr_{CO}), expressed to a good approximation as[8]

$$\Delta_{a-s}/cm^{-1} = 1818.1 \left(\delta r_{CO}/\text{Å}\right) + 16.47 \left[(\theta_{OCO}/\text{deg}) - 120\right] + 66.8 \qquad (5.5)$$

The correlation to δr_{CO} arises from the changes in the nature of the two modes. For the ionic form, they are truly the symmetric and antisymmetric stretching modes, and the same holds for the bidentate form because the two C–O bonds interact equally with the metal ion. However, for the unidentate form, the C–O bond interacting with the metal ion is longer than the other, and the "symmetric" and "antisymmetric" stretching modes tend to be close to rather independent stretches of the longer and shorter C–O bonds, respectively, as shown in Figure 5.2. In other words, they tend to be

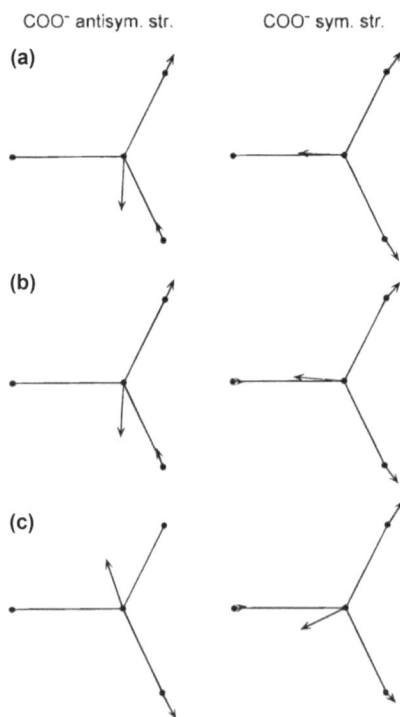

Figure 5.2 Vibrational patterns of the COO$^-$ antisymmetric and symmetric stretches of the carboxylate group in the (a) ionic, (b) bidentate, and (c) unidentate forms. Adapted from ref. 8 with permission from American Chemical Society, Copyright 1996.

single- and double-bond stretching modes, leading to a large frequency separation, Δ_{a-s}. Meanwhile, the correlation to θ_{OCO} arises from the changes in the interaction force constant between the stretches of the two C–O bonds and in the mixing of the OCO bend and the CC stretch with the COO⁻ symmetric stretch. The value of θ_{OCO} tends to be smaller for the bidentate form, leading to a smaller value of Δ_{a-s} because of these two factors.

This sensitivity of the COO⁻ symmetric and antisymmetric stretching modes to the type of coordination is used to examine the structural characteristics of the binding sites in Ca^{2+}-binding proteins, such as parvalbumins, calmodulin, and troponin C.[9] At each of those binding sites, a few side chains of Asp and Glu residues are involved in the interactions with the Ca^{2+} ion. IR spectroscopy is applicable to samples in aqueous solution. As a result, it is possible to detect the involvement of a water molecule on the COO⁻ group at the Ca^{2+} binding site, *i.e.*, a change from the unidentate form in the crystal to the pseudo-bridging form in aqueous solution, by measuring the IR spectrum of an aqueous sample and comparing it with the result of X-ray analysis of the crystal.

In some cases, intensity analysis may provide us important information on the solvation structures of metal ions in solution. *N,N*-Dimethylformamide (DMF) and *N,N*-dimethylacetamide (DMA) have a spectrally separated band at 660 and 740 cm^{-1}, respectively, in their Raman spectra. In either case, the band width is sufficiently narrow, so that a distinct band or a clear shoulder band arising from the interaction with a metal ion is observed upon solvation of an electrolyte, as shown in Figure 5.3.[10] By analysing the changes in the intensities of the newly generated band and the original solvent band as a function of the concentration, it is possible to estimate the coordination number of the metal ion in each solvent. In this way, it has

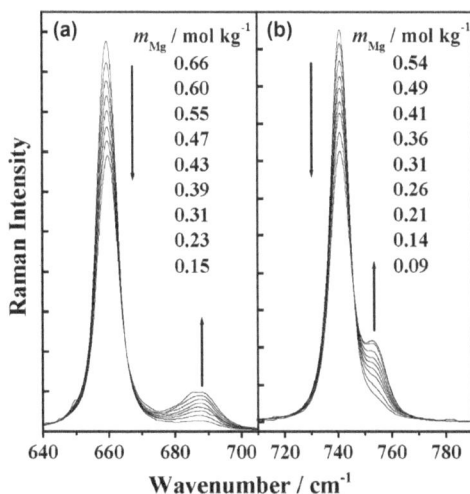

Figure 5.3 Raman spectra of the solutions of $Mg(ClO_4)_2$ dissolved in (a) DMF and (b) DMA with varying concentrations. Reproduced from ref. 10 with permission from John Wiley & Sons, Ltd, Copyright 2006.

been clarified that the coordination number is 6, 7, 8, and 8 for Mg^{2+}, Ca^{2+}, Sr^{2+}, and Ba^{2+}, respectively, in either solvent.[10] Because the two bands are spectrally separated from each other, in the case of the DMF–DMA liquid mixtures, it is possible to estimate the coordination number of each solvent species to examine whether there is any preferential solvation occurring around a metal ion. It has been shown that practically no preferential solvation occurs around Mg^{2+} and Ba^{2+}, while Ca^{2+} significantly prefers DMA in the DMF–DMA liquid mixtures.[10]

5.4 Noncoincidence Effect of Polarized Raman Spectra

It is known that, for the stretching modes of some polar functional groups, resonant vibrational coupling tends to occur among neighbouring identical molecules in condensed phases. When an electrolyte is dissolved, negatively charged atoms of those molecules are likely to cluster around a cation, resulting in a change in the spatial configurations of the solvent molecules and, hence, in the characteristics of the resonant vibrational coupling. Consequently, the spectral phenomena arising from the resonant vibrational coupling should be properly taken into account for a correct understanding of the spectral features observed for electrolyte solutions.

Resonant vibrational coupling occurs mainly for the vibrational modes with strong IR intensities. This is because the IR intensity is proportional to the square of the magnitude of the dipole derivative $\partial\mu/\partial Q$, which is the derivative of the dipole moment μ with respect to the vibrational coordinate Q, and the vibrational coupling constant according to the transition dipole coupling (TDC) mechanism[11–13] also varies with the dipole derivatives $\partial\mu/\partial Q$ of the two coupled modes. The latter is derived as follows. The potential energy arising from the dipole–dipole interaction is expressed as

$$V = \frac{\boldsymbol{\mu}_j \cdot \boldsymbol{\mu}_k - 3(\boldsymbol{n}_{jk} \cdot \boldsymbol{\mu}_j)(\boldsymbol{n}_{jk} \cdot \boldsymbol{\mu}_k)}{r_{jk}^3} \tag{5.6}$$

where $\boldsymbol{\mu}_j$ and $\boldsymbol{\mu}_k$ are the dipole moments of the jth and kth molecules, r_{jk} is the distance between them, and $\boldsymbol{n}_{jk}$ is the unit vector along the line connecting them. The coupling constant is obtained by taking the derivative once each with respect to Q_j and Q_k, which are the coordinates of the coupled vibrational modes of the two molecules. It is expressed as

$$\frac{\partial^2 V}{\partial Q_j \partial Q_k} = \frac{(\partial\boldsymbol{\mu}_j/\partial Q_j) \cdot (\partial\boldsymbol{\mu}_k/\partial Q_k) - 3[\boldsymbol{n}_{jk} \cdot (\partial\boldsymbol{\mu}_j/\partial Q_j)][\boldsymbol{n}_{jk} \cdot (\partial\boldsymbol{\mu}_k/\partial Q_k)]}{r_{jk}^3} \tag{5.7}$$

When the liquid structure is mainly determined by the dipole–dipole interaction of the solvent molecules, such as in the case of neat dipolar liquids, and if the dipole derivative $\partial\mu/\partial Q$ is (exactly or nearly) parallel to the permanent dipole μ of the molecule, this coupling constant is negative on thermal average. However, when the liquid structure is mainly determined by other factors, there is a possibility that the coupling constant is positive on thermal average. The latter cases include hydrogen-bonding liquids and also the spatial regions around the ions in electrolyte solutions, where the ion–solvent interactions are stronger than the solvent–solvent interactions and lead to solvent clustering around the ions.

The most conspicuous spectral phenomenon arising from the resonant vibrational coupling would be the noncoincidence effect[14] of polarized Raman spectra (often called the Raman noncoincidence effect). When we measure a polarized Raman spectrum of a macroscopically isotropic (even if locally ordered) sample, such as a bulk liquid or a solution, in a way that we detect the vertically (V) and horizontally (H) polarized scattered light with the vertically polarized exciting light, we have two components: one is the isotropic component (I_{iso}), which arises from the trace part of the Raman scattering tensor, and the other is the anisotropic component (I_{aniso}), which arises from the other part of the (symmetric) Raman scattering tensor. These are related to the VV and VH spectral profiles as $I_{\mathrm{iso}} = I_{\mathrm{VV}} - (4/3)I_{\mathrm{VH}}$ and $I_{\mathrm{aniso}} = I_{\mathrm{VH}}$. When the resonant vibrational coupling is sufficiently strong, it is possible to detect a frequency separation between them (usually represented by the difference in their first moments as $\tilde{\nu}_{\mathrm{NCE}} = \tilde{\nu}_{\mathrm{aniso}} - \tilde{\nu}_{\mathrm{iso}}$), and this phenomenon is called the Raman noncoincidence effect. It occurs because the isotropic component arises from the in-phase combination of the coupled vibrations and, hence, its frequency is particularly sensitive to the resonant vibrational coupling (a lower $\tilde{\nu}_{\mathrm{iso}}$ for a negative vibrational coupling). As a result, the value of $\tilde{\nu}_{\mathrm{NCE}}$ is positive when the resonant vibrational coupling is negative on thermal average, and the reverse also holds. Combined with the above-mentioned dependence of the resonant vibrational coupling on the structural configurations of the coupled modes, it is possible to analyse the liquid structures, including the solvation structures around the ions, by referring to the sign and magnitude of the noncoincidence effect.

As an example, the polarized Raman spectra in the C=O stretching band region observed for neat liquid acetone and the solution of $Mg(ClO_4)_2$ in acetone (mole fraction $x_{\mathrm{salt}} = 0.045$)[15] are shown in Figure 5.4. In the case of neat liquid acetone shown in Figure 5.4(b), the isotropic and anisotropic Raman bands are observed at 1708.9 and 1714.1 cm^{-1}, respectively, indicating that $\tilde{\nu}_{\mathrm{NCE}} = 5.2$ cm^{-1}, in accord with the expectation that the resonant vibrational coupling is negative on thermal average for a vibrational mode of a neat dipolar liquid with the dipole derivative being parallel to the molecular dipole. (It should be mentioned in passing that the band at ~ 1750 cm^{-1} arises from Fermi resonance with the combination level of the antisymmetric C–C stretching and the in-plane C=O bending, and the band at ~ 1676 cm^{-1} arises from the ^{13}C=O stretching present at natural abundance.) Upon solvation of $Mg(ClO_4)_2$, a new band appears in each spectrum, at 1729.0 cm^{-1} in the isotropic Raman spectrum and at ~ 1695 cm^{-1} in the anisotropic Raman spectrum, resulting in a negative noncoincidence ($\tilde{\nu}_{NCE} \cong -34$ cm^{-1}), in accord with

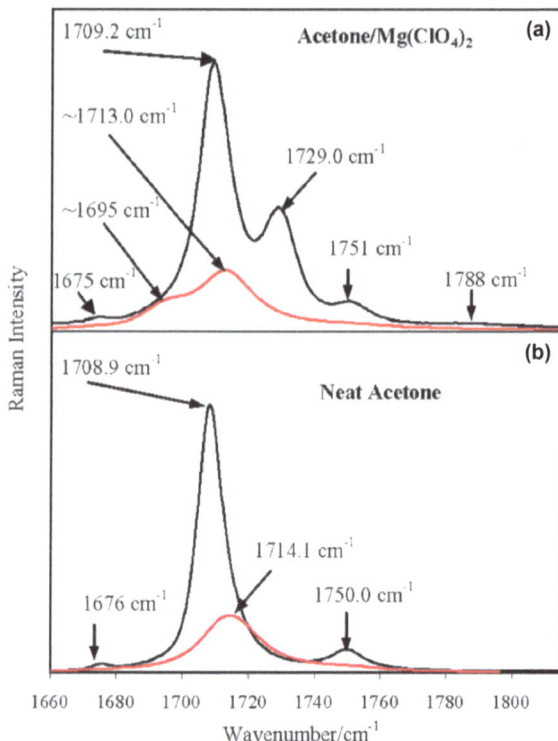

Figure 5.4   Isotropic (black line) and anisotropic (red line) Raman spectra of (a) the solution of $Mg(ClO_4)_2$ dissolved in acetone ($x_{salt} = 0.045$) and (b) neat liquid acetone in the frequency region of the C=O stretching mode. Adapted from ref. 15 with permission from the PCCP Owner Societies, Copyright 2010.

the expectation that the supposed liquid structure with many solvent C=O bonds pointing toward each Mg^{2+} ion gives rise to positive vibrational coupling based on eqn (5.7). From comparison with the results of first-principles theoretical calculations carried out for the (acetone)$_n$Mg^{2+} clusters with $n = 3$, 4, and 6,[15] it is derived that the value of $\tilde{\nu}_{NCE} \cong -34$ cm^{-1} is consistent with the coordination number of 6 ($n = 6$). Similar spectral features are observed for the solutions of the other alkaline earth and alkali metal salts (the perchlorates of Ca^{2+}, Sr^{2+}, Ba^{2+}, Li^+, and Na^+) in acetone, with the values of $\tilde{\nu}_{NCE}$ in the range from -14.7 to -23.0 cm^{-1}.[15,16] It is possible to estimate the coordination number around each metal ion by comparing the value of $\tilde{\nu}_{NCE}$ with those calculated for the cluster species with varying numbers of solvent molecules.

This technique may also be applicable to a solution in a mixed solvent if the bands of the constituent solvents appear at different frequencies. For example, in the case of the equimolar liquid mixture of propylene carbonate and diethyl carbonate, the C=O stretching bands appear at 1790 and 1745 cm^{-1} in the isotropic Raman spectrum and at 1798 and 1746 cm^{-1} in the anisotropic Raman spectrum,[17] as shown in Figure 5.5(c) and (d).

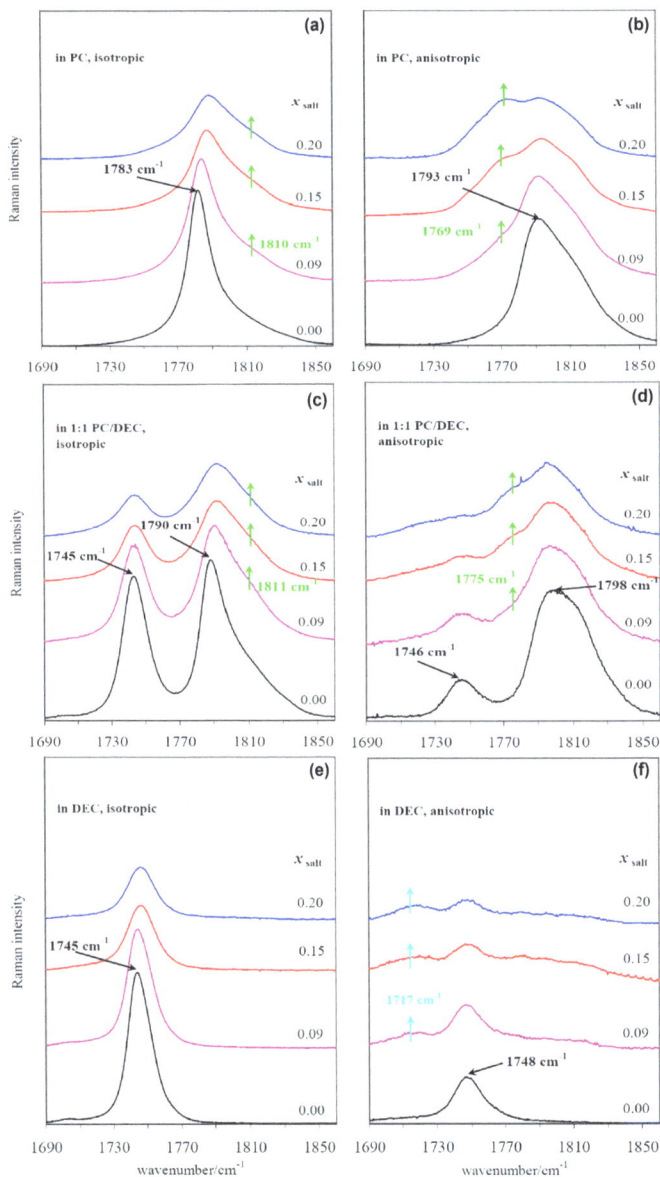

Figure 5.5 (a, c and e) Isotropic and (b, d and f) anisotropic Raman spectra observed in the spectral region around the C=O stretching mode for the solutions of LiClO$_4$ in (a and b) liquid propylene carbonate (PC), (e and f) liquid diethyl carbonate (DEC), and (c and d) an equimolar PC/DEC liquid mixture, with mole fractions x_{salt} = 0.00, 0.09, 0.15, and 0.20. The light green and light blue upward arrows indicate the frequency positions of the PC and DEC cluster component bands, respectively. Reproduced from ref. 17 with permission from American Chemical Society, Copyright 2015.

The higher-frequency band in each spectrum originates from propylene carbonate and the other from diethyl carbonate because of the hybridization effect on the carbonyl group of a five-membered ring (propylene carbonate) that makes the C=O stretching mode appear at a higher frequency position.[18] As LiClO$_4$ is dissolved in this liquid mixture, a newly generated band appears at 1811 cm^{-1} in the isotropic Raman spectrum and at 1775 cm^{-1} in the anisotropic Raman spectrum, indicating that $\tilde{\nu}_{NCE} = -36$ cm^{-1}. These frequency positions are close to those observed for the solution of LiClO$_4$ in propylene carbonate (1810 and 1769 cm^{-1}, *i.e.*, $\tilde{\nu}_{NCE} = -41$ cm^{-1}) shown in Figure 5.5(a) and (b). In contrast, in a lower-frequency region, where a newly generated band appears at 1717 cm^{-1} in the anisotropic Raman spectrum of the solution in diethyl carbonate as shown in Figure 5.5(f), there is no new band clearly recognized for the solution in the mixed solvent as shown in Figure 5.5(c) and (d). These results indicate that, in the solution of LiClO$_4$ in the equimolar mixed solvent, preferential solvation occurs around the Li$^+$ ion, with a higher population of propylene carbonate around the ion than in the bulk. From further analyses conducted on the observed spectra in the 880–960 cm^{-1} region, where the OCO bending mode of diethyl carbonate and the Cl–O symmetric stretching mode of the ClO$_4^-$ ion appear, and, from comparison with first-principles theoretical calculations on cluster species, it is suggested that the solvation structure with three propylene carbonate molecules and one diethyl carbonate molecule around a Li$^+$ ion is the most probable solvation structure in the solution in the equimolar mixed solvent.

5.5   Intermolecular Low-frequency Modes

Clustering of solvent molecules around a metal ion gives rise to spectral changes also in the low-frequency region. These spectral changes are clearly seen for the solutions of alkali metal salts in aprotic polar solvents like dimethyl sulfoxide (DMSO), *N*-methyl-2-pyrrolidone (NMP), and tetrahydrofuran (THF).[19–21] From the results of the measurements carried out for the solutions of alkali metal ions with various combinations of anions and solvents summarized in Table 5.1, it is recognized that the frequency positions of the bands that are newly generated in the far-IR spectra upon solvation of the salts are determined mainly by the metal ion, and at most only secondarily by the anion and the solvent. More specifically, for the solutions of salts dissolved in DMSO or NMP, there is essentially no dependence of the frequency position on the anion, indicating that this spectral feature arises primarily from the solvation structure consisting only of the metal ion and the solvent, without participation of the anion. In contrast, in the case of the solutions of salts dissolved in THF, there is a weak dependence of the frequency position on the anion. As an example, the spectra observed for salts dissolved in NMP are shown in Figure 5.6.

It is considered that the vibrations of a metal ion in a solvent cage give rise to this spectral feature.[19–21] For example, in the case of the solution of the

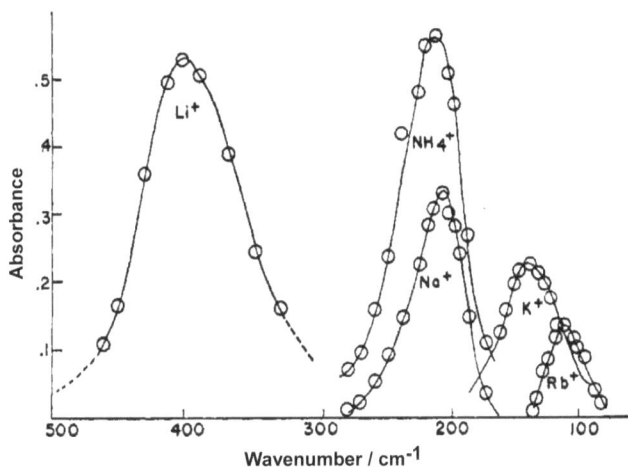

Figure 5.6 Far-IR spectra of the solutions of alkali metal and ammonium salts dissolved in *N*-methyl-2-pyrrolidone. Adapted from ref. 20 with permission from American Chemical Society, Copyright 1969.

Li$^+$ ion in NMP, a detailed analysis[20] shows that each Li$^+$ ion is surrounded by four solvent molecules and vibrates in the cage formed by those solvent molecules. In this sense, it is natural that the frequency goes down as the metal ion becomes heavier, with minor dependence on the solvent species. These modes are strongly IR active because a metal ion, as a charged particle, goes back and forth in a solvent cage, inducing a large modulation of the system's dipole moment, *i.e.*, a large dipole derivative. It should also be noted that the existence of this spectral feature itself indicates that the solvent cage is sufficiently long-lived compared with the vibrational period of this mode[21] (*e.g.*, 83 fs for the mode at 400 cm^{-1}).

In the solutions in protic solvents also, metal ions (if sufficiently dilute) are considered to be surrounded by the electronegative atoms of the solvent and, as a result, observation of a similar spectral feature is expected. Indeed, for the solution of LiI in methanol, a newly generated band is observed at around 460 cm^{-1},[22] which is close to the frequencies observed for the same salt in DMSO (429 cm^{-1}),[19] NMP (398 cm^{-1}),[20] and THF (373 cm^{-1})[21] as well as for LiCl in methanol ($\sim$460 cm^{-1}).[22] In methanol, this frequency position is close to that of the libration band of the solvent ($\sim$570 cm^{-1}), but the existence of this newly generated band is clearly visible. The effect of hydration is recognized from the spectral measurement on the solution of LiI in the mixture of 5% H$_2$O in DMSO,[22] where the bands at $\sim$400 cm^{-1} (observed in DMSO) nearly disappear and a new band is observed at 550 cm^{-1}.

In methanol, the effects of salt solvation are observed also in a lower-frequency region,[22] at 82 and 240 cm^{-1} for LiI, and at 135 and 220 cm^{-1} for LiCl. These are assigned to the vibrations of the contact ion pairs (240 and 220 cm^{-1}) and to the vibrations of the solvent-separated ion pairs and of the anions (82 or 135 cm^{-1}). With regard to the latter, the band is observed at almost the same frequency position among the solutions of LiI, NaI (86 cm^{-1}), and RbI (83 cm^{-1}).

The existence of similar vibrational modes, called rattling modes, have also been discussed for aqueous solutions of alkali and alkaline earth metal salts.[23-25] On the basis of the analysis over the spectra observed for the aqueous solutions of various combinations of metal cations and anions, it is derived that cooperativity effects between the cations and the anions are small, so that the rattling modes of the cations and anions appear at their own separate frequencies. The frequency positions of the cation rattling modes are shown in Figure 5.7. For example, the rattling modes for the alkali metal ions are observed at ~400, ~170, ~155, ~120 and ~100 cm^{-1} for Li$^+$, Na$^+$, K$^+$, Rb$^+$, and Cs$^+$, respectively. These values are close to those observed for the DMSO solutions of these ions summarized in Table 5.1. According to the theoretical analysis shown in ref. 24, the rattling modes observed at 175 cm^{-1} for Mg^{2+}, for example, is called the Θ mode, where the hexa-coordinated Mg^{2+} ion moves back and forth in one direction (assumed z), in-phase with the two water molecules located in the $\pm z$ direction, and out-of-phase with the other four water molecules located in the $\pm x$ and $\pm y$ directions. [Here, x, y, and z may be permutated so that this mode is three-fold degenerate.] Another IR active mode (called the Ψ mode) is predicted at a higher

Table 5.1 Frequencies (in cm^{-1}) of the far-IR bands that are newly generated upon solvation of alkali metal salts in various aprotic polar solvents.

Solvent	Anion	Cation				
		Li$^+$	Na$^+$	K$^+$	Rb$^+$	Cs$^+$
DMSO[a]	Cl$^-$	429	199			
	Br$^-$	429	199	153	125	
	I$^-$	429	198	153	123	110
	NO$_3^-$	429	206	154	125	
	ClO$_4^-$	429	200		122	109
	SCN$^-$		200	153		
	BPh$_4^-$		198			
NMP[b]	Br$^-$	398	204			
	I$^-$	398	204	140	106	
	NO$_3^-$	398	206			
	ClO$_4^-$	398	204	140	106	
	SCN$^-$	398	207	138		
	BPh$_4^-$		205	138	106	
THF[c]	Cl$^-$	387				
	Br$^-$	378				
	I$^-$	373	184			
	NO$_3^-$	407				
	BPh$_4^-$	412	198			
	Co(CO)$_4^-$	413	192	142		

[a] Ref. 19.
[b] Ref. 20.
[c] Ref. 21.

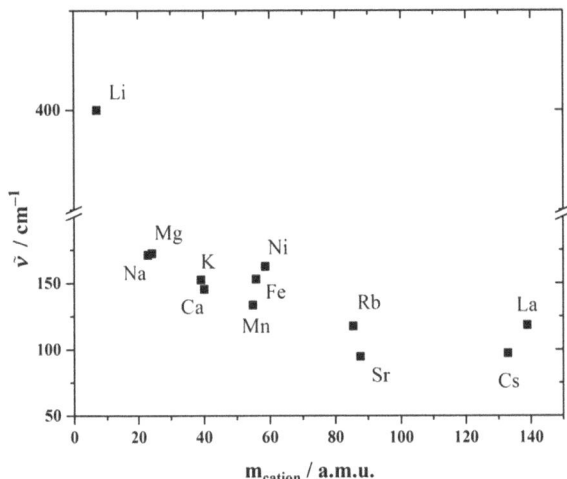

Figure 5.7 Frequency positions of the cation rattling modes in the THz (far-IR) spectra of the aqueous solutions of alkali and alkaline earth metal chloride salts plotted as a function of the atomic mass of the cation. Reproduced from ref. 25 with permission from Wiley-VCH Verlag GmbH & Co. KGaA, Copyright 2019.

frequency, ~ 600 cm^{-1} for Mg^{2+}, but it was out of the range of the THz (far-IR) spectroscopic measurement in the study described there.[24]

Problem 1

According to eqn (5.3), in the case of $k_0 = 2.7 \times 10^{-4}$ $E_h a_0^{-1} m_e^{-1}$ (corresponding to the vibrational frequency of ~ 3600 cm^{-1}), $|\partial\mu/\partial Q| = 0.020$ $e\, m_e^{-1/2}$ ($\cong 4.1$ D Å^{-1} amu$^{-1/2}$, corresponding to the IR intensity of 7.1×10^2 km mol^{-1}), and $|E| = 0.040$ $E_h e^{-1} a_0^{-1}$ ($\cong 2.1 \times 10^8$ V cm^{-1}), with E being parallel to $\partial\mu/\partial Q$, how large is the field-induced displacement ΔQ along the normal mode? [E_h: Hartree energy, e: elementary charge, a_0: Bohr radius, m_e: electron rest mass].

Problem 2

According to eqn (5.5), if we assume $\theta_{OCO} = 120°$, how large δr_{CO} can we estimate when we observe $\Delta_{a-s} = 250$ cm^{-1}?

Problem 3

Calculate the vibrational coupling constants by using eqn (5.7) for the following three cases:

(1) Two molecules in a dipolar parallel head-to-tail arrangement: molecule j located at $r_j = (0, 0, -a)$ with $\partial\mu_j/\partial Q_j = (0, 0, d)$, and molecule k located at $r_k = (0, 0, a)$ with $\partial\mu_k/\partial Q_k = (0, 0, d)$.

(2) Two molecules in a dipolar antiparallel side-by-side arrangement: molecule j located at $r_j = (0, -a, 0)$ with $\partial\mu_j/\partial Q_j = (0, 0, -d)$, and molecule k located at $r_k = (0, a, 0)$ with $\partial\mu_k/\partial Q_k = (0, 0, d)$.

(3) Two molecules in octahedral clustering around a metal ion: molecule j located at $r_j = (0, 0, a)$ with $\partial\mu_j/\partial Q_j = (0, 0, d)$, and molecule k located at $r_k = (0, a, 0)$ with $\partial\mu_k/\partial Q_k = (0, d, 0)$.

Answer to Problem 1

ΔQ is calculated as approximately $3.0 \, a_0 \, m_e^{1/2}$ ($\cong 3.7 \times 10^{-2}$ Å amu$^{1/2}$). For a vibrational mode with a reduced mass of ~ 1 amu (such as an OH stretching mode), this corresponds to the change in the bond length of 3.7×10^{-2} Å.

Answer to Problem 2

A simple calculus shows $\delta r_{CO} \cong 0.10$ Å.

Answer to Problem 3

(1) $r_{jk} = 2a$ and $n_{jk} = (0, 0, 1)$, so that $\partial^2 V/\partial Q_j \partial Q_k = -d^2/(4a^3)$.

(2) $r_{jk} = 2a$ and $n_{jk} = (0, 1, 0)$, so that $\partial^2 V/\partial Q_j \partial Q_k = -d^2/(8a^3)$.

(3) $r_{jk} = \sqrt{2}a$ and $n_{jk} = (0, 1, -1)/\sqrt{2}$, so that $\partial^2 V/\partial Q_j \partial Q_k = 3\sqrt{2}d^2/(8a^3)$.

Note that the vibrational coupling constant is negative for cases 1 and 2, and is positive for case 3.

References

1. H. Torii, in *Atoms, Molecules and Clusters in Electric Fields: Theoretical Approaches to the Calculation of Electric Polarizability*, ed. G. Maroulis, Imperial College Press, London, 2006, pp. 179–214.
2. H. Torii and R. Ukawa, *J. Phys. Chem. B*, 2021, **125**, 1468.
3. N. H. C. Lewis, A. Iscen, A. Felts, B. Dereka, G. C. Schatz and A. Tokmakoff, *J. Phys. Chem. B*, 2020, **124**, 7013.
4. S. D. Fried and S. G. Boxer, *Acc. Chem. Res.*, 2015, **48**, 998.
5. C. R. Baiz, B. Błasiak, J. Bredenbeck, M. Cho, J.-H. Choi, S. A. Corcelli, A. G. Dijkstra, C.-J. Feng, S. Garrett-Roe, N.-H. Ge, M. W. D. Hanson-Heine, J. D. Hirst, T. L. C. Jansen, K. Kwac, K. J. Kubarych, C. H. Londergan, H. Maekawa, M. Reppert, S. Saito, S. Roy, J. L. Skinner, G. Stock, J. E. Straub, M. C. Thielges, K. Tominaga, A. Tokmakoff, H. Torii, L. Wang, L. J. Webb and M. T. Zanni, *Chem. Rev.*, 2020, **120**, 7152.
6. H. Torii, *J. Phys. Chem. A*, 2016, **120**, 7137.
7. G. B. Deacon and R. J. Phillips, *Coord. Chem. Rev.*, 1980, **33**, 227.
8. M. Nara, H. Torii and M. Tasumi, *J. Phys. Chem.*, 1996, **100**, 19812.

9. M. Nara, H. Morii and M. Tanokura, *Biochim. Biophys. Acta*, 2013, **1828**, 2319.

10. M. Asada, T. Fujimori, K. Fujii, R. Kanzaki, Y. Umebayashi and S. Ishiguro, *J. Raman Spectrosc.*, 2007, **38**, 417.

11. D. E. Logan, *Chem. Phys.*, 1986, **103**, 215.

12. H. Torii and M. Tasumi, *J. Chem. Phys.*, 1993, **99**, 8459.

13. H. Torii, in *Novel Approaches to the Structure and Dynamics of Liquids: Experiments, Theories and Simulations*, ed. J. Samios and V. A. Durov, Kluwer, Dordrecht, 2004, pp. 343–360.

14. G. Fini, P. Mirone and B. Fortunato, *J. Chem. Soc., Faraday Trans. 2*, 1973, **69**, 1243.

15. M. G. Giorgini, H. Torii and M. Musso, *Phys. Chem. Chem. Phys.*, 2010, **12**, 183.

16. M. G. Giorgini, H. Torii, M. Musso and G. Venditti, *J. Phys. Chem. B*, 2008, **112**, 7506.

17. M. G. Giorgini, K. Futamatagawa, H. Torii, M. Musso and S. Cerini, *J. Phys. Chem. Lett.*, 2015, **6**, 3296.

18. H. K. Hall Jr and R. Zbinden, *J. Am. Chem. Soc.*, 1958, **80**, 6428.

19. B. W. Maxey and A. I. Popov, *J. Am. Chem. Soc.*, 1969, **91**, 20.

20. J. L. Wuepper and A. I. Popov, *J. Am. Chem. Soc.*, 1969, **91**, 4352.

21. W. F. Edgell, J. Lyford, IV, R. Wright, W. Risen Jr and A. Watts, *J. Am. Chem. Soc.*, 1970, **92**, 2240.

22. B. Guillot, P. Marteau and J. Obriot, *J. Chem. Phys.*, 1990, **93**, 6148.

23. D. A. Schmidt, O. Birer, S. Funkner, B. P. Born, R. Gnanasekaran, G. W. Schwaab, D. M. Leitner and M. Havenith, *J. Am. Chem. Soc.*, 2009, **131**, 18512.

24. S. Funkner, G. Niehues, D. A. Schmidt, M. Heyden, G. Schwaab, K. M. Callahan, D. J. Tobias and M. Havenith, *J. Am. Chem. Soc.*, 2012, **134**, 1030.

25. G. Schwaab, F. Sebastiani and M. Havenith, *Angew. Chem., Int. Ed.*, 2019, **58**, 3000.

CHAPTER 6

Computational Approach

SATORU IUCHI[a] AND HIROFUMI SATO*[b]

[a] Graduate School of Informatics, Nagoya University, Japan; [b] Department of Molecular Engineering & Fukui Institute for Fundamental Chemistry, Kyoto University, Japan
*Email: hirofumi@moleng.kyoto-u.ac.jp

6.1 Computational Approach

Computer technology has rapidly developed recently: the computational power has increased quickly, and powerful computational resources have become widely available. Consequently, we can now analyse experimental data and simulate chemical phenomena based on the principle of physical chemistry. In this chapter, such computer simulations are outlined. For coordination complexes in solutions, two computational methods are generally required. One is quantum chemistry which provides information on the properties of individual molecules. Another is computational methods such as statistical mechanics and molecular simulations, which provide information on the properties of solutions as a molecular ensemble. Chemical phenomena in solutions reflect the nature of individual molecules and molecular ensembles. Due to space limitations, the theoretical background of computational methods is mainly introduced here. For a deeper understanding of methods, we recommend readers consult ref. 1 and 2.

6.1.1 Quantum Chemistry

Chemical properties such as the reactivities and properties of individual molecules are governed by the electron and its motions. The electron motions obey the principle of quantum mechanics, and quantum chemistry is a

Coordination Chemistry Fundamentals Series No. 2
Metal Ions and Complexes in Solution
Edited by Toshio Yamaguchi and Ingmar Persson
© Japan Society of Coordination Chemistry 2024
Published by the Royal Society of Chemistry, www.rsc.org

theoretical framework for molecular properties. The time-independent Schrödinger equation, which describes the stationary state of electrons, is usually the starting point:

$$H\Psi = E\Psi \tag{6.1}$$

The Hamiltonian H represents all the interactions involving nuclei and electrons composed of a molecule. The electronic wavefunction Ψ represents a quantum mechanical state of the molecule. For example, the wavefunction of a system including N electrons is a function of all the coordinates of the N electrons (multi-electron wavefunction). This electronic wavefunction must satisfy an anti-symmetric property concerning an exchange of two arbitral coordinates as

$$\Psi(r_1, r_2 \cdots r_i \cdots r_j \cdots r_N) = -\Psi(r_1, r_2 \cdots r_j \cdots r_i \cdots r_N) \tag{6.2}$$

In the Hartree–Fock (HF) method, known as one of the fundamental methods for quantum chemistry, the electronic wavefunction for a closed system is represented by a single Slater determinant, which satisfies the anti-symmetric property in eqn (6.2). A single determinant is a good approximation to represent the electronic structure of various molecules composed of typical elements. In other words, the treatment only with a single Slater determinant corresponding to Figure 6.1(a) offers a sufficiently accurate description of the electronic wavefunction in many molecules. However, this treatment is generally not sufficient enough to describe electronic states for coordination-metal complexes because the electronic structures for co-ordination-metal complexes are characterised by a mixture of various electronic configurations (*e.g.* Figure 6.1(a)–(d)). Thus, the electronic wavefunction is accurately described by a linear combination of the Slater determinants corresponding to these electronic configurations. This complicated treatment is attributed to molecular orbitals for d-electrons, which characterises the electronic structures of coordination-metal complexes. The d-electrons are relatively close in energy, and the electronic energies corresponding to these electronic configurations are relatively close (static electron correlation). In addition, because d-electrons are distributed

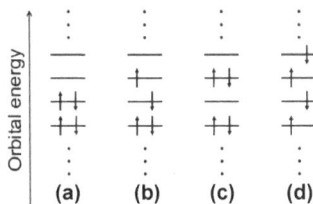

Figure 6.1 Schematic pictures of electronic configurations, where α and β electrons are occupying various orbitals: (a) the ground state configuration and (b)–(d) examples of excited state configurations.

around the metal centres, effects arising from electron collisions (dynamic electron correlation) become essential. The HF method cannot describe these electron correlation effects. Therefore, based on the HF method, various electronic structure theories have been proposed to include these effects so far, for example, Møller–Plesset perturbation methods such as MP2, the multi-configuration based wavefunction methods such as CASSCF, their perturbation methods such as MCQDPT and CASPT, and coupled-cluster methods such as CCSD. These electronic structure methods can be readily performed using various program packages. Although these methods, including the HF, are derived under some approximations, they solve the electronic Schrödinger equation in a first-principles manner. In this regard, these methods are called *ab initio* electronic structure theories. While the accuracy of these methods is high, the computational cost is generally highly demanding, making applications to large molecular systems challenging. In the case of transition metal systems, applications have been limited to systems containing several metal atoms in practice. However, it can be expected that applications will be expanded steadily by a further increase in computer power and the development of more efficient theories.

The electronic wave-function Ψ in eqn (6.2) contains information on all the coordinates of N electrons. One might consider that having a more simplified description would be preferable. The square of the wavefunction $|\Psi|^2$ represents the probability density of the particle (electron in the present case). In practice, it would be enough to obtain the probability density at a spatial position, $r = (x,y,z)$, *i.e.*, the probability distribution function of electrons $\rho(r)$. This idea is fundamental of density functional theory (DFT) calculations. In the wavefunction-based methods mentioned above, various electronic configurations (*e.g.* Figure 6.1(a)–(d)) are explicitly considered. Then the quantum mechanical state is described by summing up these contributions appropriately. By contrast, the DFT methods directly determine the quantum mechanical state from information on $\rho(r)$.

However, the functional that is rigorously determined by $\rho(r)$ is unknown, and thus approximate functionals for electron correlation and exchange interactions are utilized. Besides, hybrid methods which combine the pure DFT and HF methods have been widely used. The B3LYP (Becke 3-parameter Lee–Yang–Parr) functional is an example. In the B3LYP, the exchange–correlation functional is a mixture of the HF exchange and other functionals. The three parameters for the mixture were determined so that computational results agree well with experimental data on the properties of various molecules. As expected from these theoretical backgrounds, DFT methods generally provide accurate results with lower computational costs than *ab initio* electronic structure theory methods. DFT methods have been widely used since the mid-1990s, when the theories and program codes were well developed. Their applications to transition metal complexes have made remarkable progress, and computational methods have become popular in research on transition metal complexes. The DFT methods are powerful tools, and computations can be relatively easily carried out. Nevertheless, the development of basic foundations is still in progress, and we always need careful examination as to whether calculations are performed using appropriate theories for computed properties.

6.1.2 Molecular Simulations and Statistical Mechanics

The solution is an ensemble composed of an enormous number of molecules. Thermodynamics is a theoretical framework that explains energy balance in such a molecular ensemble. The macroscopic properties of a molecular ensemble originate from the microscopic properties of individual molecules. Thus, theories to characterise the molecular ensemble are necessary to connect the macroscopic and microscopic properties. Molecular simulation is a representative technique, a computational experiment that enables understanding of the macroscopic properties by simulating a molecular system. Specifically, an appropriate size "box" is prepared. Then, an appropriate number of molecular models (Figure 6.2) are put in that box to satisfy the condition of the realistic system, such as density. Positions and orientations of individual molecules take various values due to the thermal motions of the molecules, which is particularly important to dealing with a solution system. Hence, it is necessary to compute an appropriate statistical ensemble. The widely used molecular simulation methods are roughly classified into molecular dynamics and Monte Carlo simulations, as explained below.

6.1.2.1 *Molecular Dynamics (MD) Simulation*

According to Newton's equations of motion, we can compute the positions and momenta of particles at a future time from a given set of their positions and momenta at an arbitral time. In most cases, all the particles in the box are affected by each other. This interaction is described with appropriate functions (potential energy functions). The total potential energy of the system (U) is often calculated as the sum of interactions between two particles, the so-called pair potentials. In solving Newton's equations of motion numerically, the forces acting on particles are calculated at each time step as the derivative of the potential function $(-\nabla_{r_i} U)$.

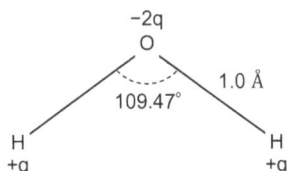

Figure 6.2 A typical water model, known as SPC. Appropriate charges are put onto oxygen and hydrogen atoms. A water molecule is represented by further adding parameters of the Lennard-Jones potential on oxygen atoms.

In MD simulations, we pay attention to a microscopic variable $x(t)$ at time t and then compute its time average X by following the motions of individual molecules based on Newton's equations of motion:

$$X = \lim_{\tau \to \infty} \frac{1}{\tau} \int_0^\tau x(t) \mathrm{d}t \tag{6.3}$$

This concept forms a fundamental of the MD simulation. In practice, sufficiently long-time simulation is performed because an infinite time simulation is impossible. However, the appropriate simulation length depends on the target physical properties and system. In addition, the above equation is applicable to physical quantities characterised by an equilibrium system. For the computation of dynamical properties, many trajectories (positions and momenta of individual molecules along time) should be accumulated and then averaged over time.

6.1.2.2   Monte Carlo Method (Metropolis Method)

Thermodynamic equilibrium is characterised as a statistical ensemble of positions and orientations of molecules, described by the partition function or the configuration integral. For the canonical ensemble composed of N identical kind particles, the configurational integral is given as

$$Q_N(V, T) = \iint \cdots \int \exp(-U/k_B T) \mathrm{d}r^N \tag{6.4}$$

where T and k_B are the temperature and the Boltzmann constant, respectively. Computer experiments simulating practical phenomena by utilizing random processes are generally called Monte Carlo methods. For molecular systems, physical properties are computed as ensemble averages of the particle's positions by generating random numbers. However, such a simple procedure is inefficient and not practical. The Metropolis algorithm is usually used to compute the ensemble averages efficiently, where various configurations are generated utilizing the Markov chain such that their probabilities satisfy the Boltzmann distribution. In the Monte Carlo method, a macroscopic physical variable is computed by using microscopic physical variable x_i at each configuration as

$$X = \langle x_i \rangle \tag{6.5}$$

where $\langle \cdots \rangle$ denotes the statistical ensemble average.

Various numerical techniques have been developed in these two molecular simulation methods (MD and MC), and many potential energy functions that adequately represent the interaction between molecules have been vigorously developed. A proper understanding of statistical ensembles is crucial to appropriately utilizing these molecular simulation methods.

6.1.2.3 Liquid State Theory (Integral Equation Theory of Liquids)

Because the configuration integral, eqn (6.4), contains multiple integrals corresponding to the number of the system's degrees of freedom, the direct integration is impossible. The aforementioned molecular simulation methods, instead, numerically deal with the integral. However, it would be desirable to have analytical equations that can be evaluated directly, even if those include approximations. One of the purposes of the statistical mechanics is to provide a mathematical connection between macroscopic and microscopic variables, and various approaches have been developed over the past two centuries. In particular, several methods to treat liquid states and solutions have been proposed and established, which have been applied to various solution systems. The reference interaction site model (RISM) theory is an example, and the extended RISM theory development enables it to apply to various solution systems, including polar molecules. By inputting temperature, density, and the information on molecular models widely used for MD simulations, we can obtain the correlation functions that characterise the system's statistical mechanical properties, such as the radial distribution function (RDF). From the correlation functions, all the thermodynamic properties can be also computed. Although the equations are derived with some approximations, the statistical mechanical equations can be directly evaluated. As a result, the computational cost is generally much smaller than those of the molecular simulations. In addition, the solvation free energies and the reaction free energy profiles can be computed by avoiding statistical sampling problems which should be paid attention in molecular simulations. In other words, all the computed properties are based on the correct statistical ensemble.

6.1.3 Computational Methods for Metal Complexes in Solutions

While quantum mechanics describes the microscopic world where electrons play a primary role, our daily world is characterised by Newtonian mechanics (classical mechanics). One might wonder where their boundary is. The thermal de Broglie wavelength is an indicator given as

$$\varLambda = \frac{h}{\sqrt{2\pi m k_B T}} \approx \frac{h}{p} \tag{6.6}$$

where h is the Planck constant, and m and p are the particle mass and momentum, respectively. In the framework of quantum mechanics, all substances show aspects of a wave. According to the above equation, a substance with a small wavelength can be regarded as a classical particle. A computed value of $\varLambda$ for a cryogenic system is close to the interparticle distance, meaning that quantum effects are important for describing the system. By contrast, Λ values for a molecule in many systems around the room temperature are sufficiently short, and thus molecular motions are classical and can be distinguished from the quantum mechanical behavior of individual molecules in the system. This fact implies the following two issues.

(1) Unless quantum mechanical processes such as chemical reactions are not considered, we can describe molecular motions by classical mechanics and understand macroscopic properties based on the classical statistical ensemble.

(2) However, under certain conditions, systems composed of atoms and molecules correspond to the boundary between classical and quantum mechanics. Therefore, there could be many cases where both classical and quantum mechanical descriptions are needed.

For metal complexes in solutions, the liquid structure is investigated using classical mechanical methods. As mentioned in (1), appropriate molecular models, *i.e.*, interaction potential energy functions, are crucial. Concerning (2), more advanced methods are required by combining quantum and statistical mechanics. A quantum mechanical calculation utilizing the dielectric continuum model (PCM method) is one of the most straightforward related approaches in this direction. Polar solvents such as water create large electric fields, which often significantly affect the electronic states of solute molecules. In most transition metal complexes, electronic states attributed to the various configurations of d-electrons are close in energy, making the solvation effects essential. A method combining quantum and statistical mechanics becomes indispensable.

6.1.4 Examples of Advanced Theories and Computations

In MD simulations, molecular motions are directly dealt with, and dynamical properties such as vibrational motions of metal complexes and solvent behaviors around them are computed. An appropriate interaction potential energy function is the key to obtaining the physical properties with sufficient accuracy. When a system is composed of the typical elements only, employing simple potential functions such as the Lennard-Jones plus the electrostatic (Coulomb) interaction functions is enough to describe the system adequately. However, this is not always the case for the systems containing metal complexes that have various electronic states. Therefore, various potential energy functions representing the interactions involving metal ions have been vigorously developed, including the empirical type of potentials.

Alternatively, approaches to directly utilize quantum chemical calculations have been developed instead of employing such potential energy functions. For example, in the Born–Oppenheimer dynamics, the time-independent Schrödinger equation (eqn (6.1)) is solved at every MD timestep to compute the total potential energy of the system and forces acting on atoms. Therefore, the electronic structure of target systems incorporating solvent effects is expected to be described accurately, as far as the quantum chemical method employed is appropriate for the target properties. The Car–Parrinello method is a widely used, similar approach, but the orbitals are treated as variables. These *ab initio* MD simulations are useful, though the drawback is enormous computational costs, meaning that great care must be taken for the appropriate sampling.

The hybrid approach utilizing both quantum chemical calculations and potential energy functions reduces the computational costs while keeping the advantage of the quantum chemical calculations. In this method, called QM/MM, metal complexes are described by the quantum mechanical (QM) method. In contrast, solvent molecules are described using classical mechanics (MM: molecular mechanics): quantum chemical calculations are repeated at various solvent configurations generated from MD simulations. Although the method still requires huge computational costs, solvent effects can be incorporated into the description of the metal complexes.[3] By combining with MD simulations (QM/MM-MD), the vibrational spectroscopies of the metal complexes in solutions and the solvent motions around the metal complexes are analysed.

Similarly, the liquid state theory can be combined with quantum chemical calculations (RISM-SCF method).[4] Figure 6.3 shows the solvation structures

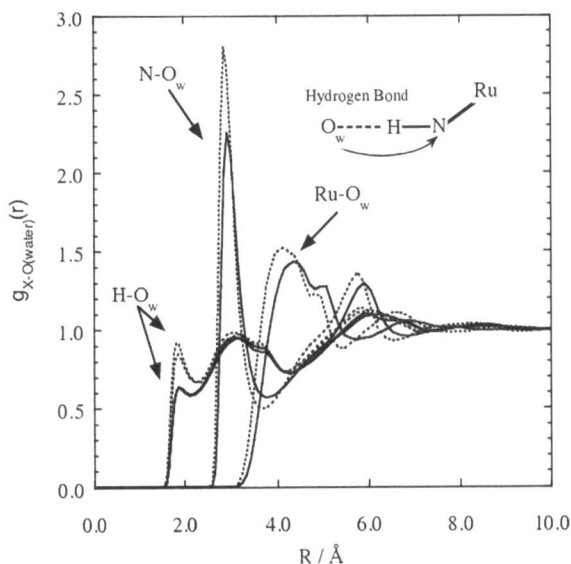

Figure 6.3 Radial distribution functions of $[Ru(NH_3)_6]^{2+/3+}$ in aqueous solution computed by RISM-SCF.[5] Solid and dashed lines represent +2 and +3 complexes, respectively. It is evident that the ammonia ligand is strongly coordinated by water molecules. Reproduced from ref. 5 with permission from American Chemical Society, Copyright 2002.

(represented with radial distribution functions) of $[Ru(NH_3)_6]^{2+/3+}$ in an aqueous solution computed by RISM-SCF.[5] It is seen that the solvation structure changes concerning the change of the oxidation in the metal center. This change is understood by the linear response regime (Marcus theory) by detailed analysis. In addition, the quantum chemical calculation allows us to analyse the electronic states of the metal complexes. In this case, both the central metals and ligands were stabilized by solvation compared to the same metal systems in the gas phase. The method has been further extended to RISM-SCF-SEDD or RISM-SCF-cSED.[6] With these methods, the metal complexes are described by quantum chemistry and solvations by RISM, and the reaction free energy changes can be computed. Figure 6.4 shows an example where the free energy changes of the hydration reaction of

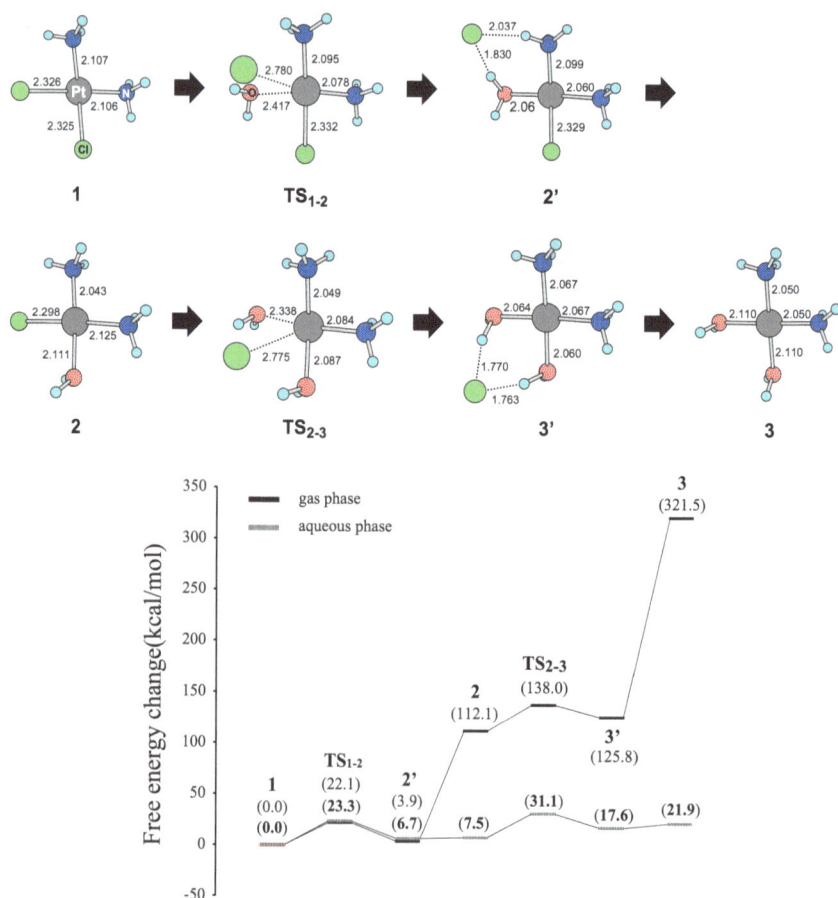

Figure 6.4   Free energy changes of the hydration reaction of cisplatin in aqueous solution computed by RISM-SCF-SEDD.[7] Reproduced from ref. 7 with permission from the Royal Society of Chemistry.

cisplatin in aqueous solution were computed by RISM-SCF-SEDD.[7] In this process, the chloride anions attached to the central Pt are sequentially replaced with two water molecules. The lower panel exhibits the energy change along with the substitution. For the isolated molecular system (the gas phase), the free energy becomes higher and higher to reach more than 300 kcal mol^{-1}. Of course, the process never happens because there is no reason to separate the cation and anion from each other to eliminate the significant coulombic stabilization. The profile is dramatically changed in aqueous solution, and the reaction can occur with a reasonable barrier height. The key to this dramatic change is the solvation. Each ion is individually solvated by the surrounding water molecules, compensating for the ion's attractions. The bond energy is typically a few hundred kcal mol^{-1}, and the solvation energy of an ion is also the order of a few hundred kcal mol^{-1} with the opposite sign, depending on the ionic valence. Hence, the total energy change is determined by the balance of these contributions, a typical scenario of bond alternation in a polar solvent. It would be meaningful to draw attention to the charge change during the process. The total change of the metal complex is altered from zero to two with the substitution, leading to a significant change in the solvation structure. This situation means that the evaluation of the free energy change is not readily accomplished in the standard sampling technique in molecular simulations. Thanks to the analytical nature of the RISM employed here, in which the solvation free energy is given in a closed form of the equation, all the charge states can be treated with equivalent accuracy, and the free energy change is readily computed.

For transition metal complexes, various electronic states such as d–d excited states and charge-transfer excited states draw remarkable attention. The MD simulations for excited states provide the dynamics of solvent and excited-state relaxation. In fact, MD simulations in this direction have been performed by using quantum chemical calculations and QM/MM methods. However, the electronic states of transition metal complexes are generally complicated, often requiring advanced quantum chemical calculations such as MCQDPT and CASPT methods. MD simulations with these advanced methods require high computational costs and are still limited. A study on the d–d excited states of Ni^{2+} in an aqueous solution overcomes this problem by constructing advanced interaction potential functions.[8] In this example, the model electronic Hamiltonian was constructed to reproduce quantum chemical calculation results. The adequate functions to represent electrostatic, charge transfer, and exchange interactions between a water molecule and a transition metal ion were constructed based on quantum mechanics. This model Hamiltonian is easily applied to MD simulations, and all the triplet d–d excited states are computed through the diagonalization of a small Hamiltonian matrix. Figures 6.5 and 6.6 show the computed electronic absorption spectrum and RDFs, respectively. The computed spectrum agrees well with the experimental one. In addition, the computations of the transition lifetimes and non-adiabatic MD simulations have revealed the

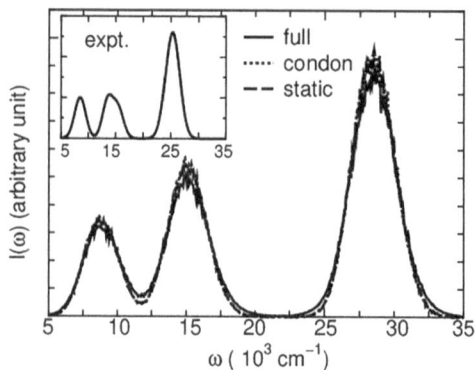

Figure 6.5   Electronic absorption spectrum of Ni^{2+} in aqueous solution computed by MD simulations using the model Hamiltonian with three different conditions.[8a] The inset graph shows an experimental spectrum reproduced from four Gaussian functions reported by C. K. Jørgensen (*Acta Chem. Scand.*, 1955, **9**, 1362). Reproduced from ref. 8(a) with permission from AIP Publishing, Copyright 2004.

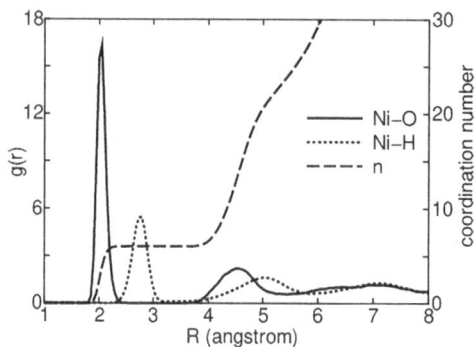

Figure 6.6   Radial distribution functions for Ni^{2+}-O and Ni^{2+}-H and coordination number (denoted as n in the graph) in aqueous solution computed by MD simulations using the model Hamiltonian.[8a] Reproduced from ref. 8(a) with permission from AIP Publishing, Copyright 2004.

details of the electronic relaxation process after photoexcitation to the d–d excited state.

The model Hamiltonian has been further applied to a different transition metal complex[9] and self-assembly systems.[10] The process by which molecules spontaneously assemble to form highly ordered structures is called self-assembly. An example is coordination self-assembly, in which the transition metal is responsible for the framework formation. In many cases, the assembled systems contain several tens of transition metals, making the use of general-purpose electronic structure calculations (DFT calculations) impractical. Therefore, model Hamiltonians have been used to reveal the details of the formation process and the relative stability of different-sized complexes in the solution phase.[10] Such investigations require the exploration of many structures of complexes, and the ability to rapidly perform a large number of calculations provided by the model Hamiltonian is crucial.

Metal complexes in solutions have a very complicated electronic structure. However, by computing such complicated electronic structures based on appropriate quantum chemical methods, we can understand complicated properties systematically. Nowadays, we can quickly obtain numerical results by using computers. However, we may easily misunderstand them unless we correctly understand theoretical foundations. We should avoid using computational methods like a black box to obtain only self-serving numerical results and keep interested in theoretical background, caring about the limitations of computational methods.

Problem 1

By analysing the RDFs in Figure 6.6, we can understand the first solvation structure of Ni^{2+} in an aqueous solution at the molecular level. Analyse the first peaks for Ni–O and Ni–H and describe the orientation of the water molecules to the nickel ion. From the graph, how many water molecules directly coordinate with the nickel ion?

Problem 2

As shown in Figure 6.5, the calculated absorption spectra are consistent with the experimental result, providing details of the electronic structure of the hydrated nickel ion in reality. The *ab initio* MCQDPT method showed that each of the three bands (8436 cm^{-1}, 13 904 cm^{-1}, and 27 323 cm^{-1}) comprises three excited states (*i.e.*, a total of nine excited states). Explain the physical background of three excited states that compose a band. Also, explain why each band has a certain width.

Problem 3

What are the advantages and disadvantages of *ab initio* electronic structure theory calculations and DFT calculations for coordination-metal complexes?

Problem 4

What are the advantages and disadvantages of MD simulations, MC simulations, and RISM theory for the solution system of coordination-metal complexes?

Answer to Problem 1

A sharp peak at ~ 2.05 Å in the Ni–O RDF corresponds to the first solvation shell. In this region, the coordination number computed by integrating the

Ni–O RDF (dashed line in Figure 6.6) is 6. These indicate that six water molecules are strongly coordinated to Ni^{2+} in the first solvation shell. A first peak position in the Ni–H RDF is shifted to a larger distance (~ 2.8 Å), indicating that the water molecules in the first solvation shell orient with the oxygen atom pointing toward Ni^{2+}.

Answer to Problem 2

The first, second, and third bands in the absorption spectrum are assigned to the excitations from the ground state to the 1st–3rd, 4–6th, and 7–9th excited states, respectively. In the stable $[Ni(H_2O)_6]^{2+}$ structure, each excitation energy corresponds to three excited states (triply degenerate) due to the symmetric structure (O_h, more precisely T_h symmetry) in the ground state. All the excitation energies fluctuate in aqueous solution due to the structural fluctuation of the metal complex and the effects of solvents. These fluctuations were taken into account through MD simulations, and thus the three computed bands show certain widths.

Answer and Additional Information to Problem 3

The accuracy of *ab initio* electronic structure methods is generally high because they solve the electronic Schrödinger equation in a first-principles manner. However, applications to large coordination-metal complexes are limited because of their high computational costs. DFT methods generally provide accurate results with smaller computational costs than those of *ab initio* methods and thus have an advantage in calculating the ground state properties of large coordination-metal complexes. However, appropriate DFT methods must be carefully chosen for target properties. For example, although time-dependent DFT (TD-DFT) calculations have been widely utilized to compute excited states, it should be checked whether TD-DFT is appropriate for target excited states.

Answer and Additional Information to Problem 4

These methods can compute various properties of coordination-metal complexes in solutions, for example, the RDFs between the metal ion and solvents. MD simulations also have an advantage in computing dynamical properties, such as solvent dynamics around the metal complex, because the motions of molecules are traced by solving Newton's equations of motion. However, in MD and MC simulations, a physical property is computed as a statistical ensemble average over microscopic events, requiring sufficiently long simulations in general. In contrast, the RISM requires a much smaller computational cost than these simulations because it directly deals with the mathematical equations in statistical mechanical theory to calculate RDFs, thermodynamic properties, *etc.* Although it is free from the so-called 'sampling error', the accuracy depends on the applied approximations.

The RISM-SCF method combines the quantum chemical calculation and liquid state theory, allowing analysis of the electronic states of the metal complexes. The MD (or MC) simulations can be combined with the electronic structure theory, as described in the nickel example. QM/MM-MD simulations also combine MD simulations with widely-used, standard quantum chemical methods.

References

1. F. Jensen, *Introduction to Computational Chemistry*, Wiley, 3rd edn, 2017.
2. M. P. Allen and D. J. Tildesley, *Computer Simulation of Liquids*, Oxford University Press, 2nd edn, 2017.
3. T. S. Hofer, B. R. Randolf and B. M. Rode, in *Solvation Effects on Molecules and Biomolecules*, ed. S. Canuto, Springer, 2008, ch. 10, pp. 247–278.
4. H. Sato, in *Molecular Theory of Solvation*, ed. F. Hirata, Kluwer, 2003, ch. 2, pp. 61–99.
5. H. Sato and F. Hirata, *J. Phys. Chem. A*, 2002, **106**, 2300.
6. D. Yokogawa, *Chem. Phys. Lett.*, 2013, **587**, 113.
7. D. Yokogawa, K. Ono, H. Sato and S. Sakaki, *Dalton Trans.*, 2011, **40**, 11125.
8. (a) S. Iuchi, A. Morita and S. Kato, *J. Chem. Phys.*, 2004, **121**, 8446; (b) S. Iuchi, A. Morita and S. Kato, *J. Chem. Phys.*, 2005, **123**, 024505.
9. S. Iuchi and N. Koga, *J. Comput. Chem.*, 2021, **42**, 166.
10. (a) Y. Matsumura, S. Iuchi, S. Hiraoka and H. Sato, *Phys. Chem. Chem. Phys.*, 2018, **20**, 7383; (b) Y. Yoshida, S. Iuchi and H. Sato, *Phys. Chem. Chem. Phys.*, 2021, **23**, 866.

CHAPTER 7

Solvent Properties

TOSHIO YAMAGUCHI[a,b]

[a] Key Laboratory of Salt Lake Resources Chemistry of Qinghai Province, Qinghai Institute of Salt Lakes, Chinese Academy of Sciences, Xining, Qinghai 810008, P. R. China; [b] Professor Emeritus, Department of Chemistry Faculty of Science, Fukuoka University, 8-19-1 Nanakuma, Jonan, Fukuoka 814-0180, Japan
Email: yamaguch@fukuoka-u.ac.jp

7.1 Introduction

Solvents are the main substances indispensable to solutions. The knowledge of solvents and solutes is essential for understanding the properties, structure, and function of metal ions and complexes in solution. In coordination chemistry, solvents are widely used as media for synthesis, reactions, and equilibria of metal complexes. However, the outcome of a chemical reaction or equilibrium depends to a large extent on the solvent and its properties. It is therefore desirable to know the macroscopic and microscopic properties of solvents. Solvents have their unique structures, which are responsible for their properties. This chapter explains the various properties of the solvents used in coordination chemistry. For commonly used solvents, their liquid structures and characteristics are described.

7.2 Physicochemical Properties of Solvents

A solution consists of a solvent and a solute. Therefore, interactions in a solution can be divided into solvent–solvent interactions, solute–solvent interactions (solvation), and solute–solute interactions (ion association, complex formation). The solvent–solvent interaction is reflected in the the

Coordination Chemistry Fundamentals Series No. 2
Metal Ions and Complexes in Solution
Edited by Toshio Yamaguchi and Ingmar Persson
© Japan Society of Coordination Chemistry 2024
Published by the Royal Society of Chemistry, www.rsc.org

physical properties of the pure solvent. The physical properties that represent the characteristics of the solvent include boiling and melting points, density, viscosity, refractive index, dielectric permittivity, dipole moment, electron-pair donor ability, specific heat, the heat of vaporization, the heat of solidification, evaporation entropy, and light absorption (visible/ultraviolet/infrared). Furthermore, the important properties of solvents in experiments, which are not directly related to the reaction, are smell, toxicity, volatility, flammability, purity, uniformity, degradability, price, and biodegradability required to be environmentally friendly at disposal.

7.3 Types of Solvents

Except for special cases such as solid solutions, a solvent is usually a liquid. Generally, a liquid medium is often referred to as a solvent in a temperature range (0 to 100 °C, 1 atm (0.1013 MRI)) in which water is in a liquid state. In this chapter, solutions for this category are described. However, in actual reactions, a wider range of substances are employed. SO_2 and NH_3, gases under ambient conditions, can be liquefied at low temperature or high pressure and used as a solvent for different purposes. In analytical chemistry, ionic crystals, such as sodium carbonate, are melted at high temperatures to form a molten salt (melting). It may be used as a solvent by setting the temperature to 851 °C. In particular, molten aluminum chloride (melting point: 170.9 °C) or a low melting point hydrated salt is used as a solvent in a chemical reaction. In addition, supercritical carbon dioxide (ambient temperature and a pressure above 100 atm) and supercritical water (the critical point is 374 °C and 218 atm) are used as a medium. Supercritical argon and low melting point metals are also used as a medium for unusual reactions. Most of the solvents used at room temperature are molecular liquids with a very small degree of ionization of the solvent itself. For example, the water dissociates electrolytes, but its self-dissociation is very small $(K_w = [H^+][OH^-] = 10^{-14} \ mol^2 \ dm^{-6}$ at 25 °C). On the other hand, recently, room temperature ionic liquids, which consist of a cation and an anion, are liquids at ambient temperature and have drawn much attention in solvent properties and applications.

7.4 Water Characteristics and Anomalies

General complex formation, including biological reactions, occurs in an aqueous solution. Apart from synthesis, equilibrium and reactions in non-aqueous solvents are often considered specific, but water is a particular solvent in terms of solvent characteristics. Before explaining the properties of various solvents, the anomalous properties of water are described. The melting point (0 °C) and boiling point (100 °C) of water are abnormally higher than those of similar compounds. For example, the boiling point of H_2O estimated from the change in the boiling point of group 16 hydrides (H_2O, H_2Se, H_2S) would be about −70 °C. Furthermore, the boiling point of water could be $-162\ °C$, estimated from that of methane which has a similar molecular weight to water. *N*-Pentane $(C_2H_5-CH_2-C_2H_5)$ and its methylene group $(-CH_2-)$ replaced with oxygen $(-O-)$ diethyl ether $(C_2H_5-O-C_2H_5)$ have similar boiling points of $36\ °C$ and $35\ °C$, respectively. Therefore, the boiling points of water and CH_4 should not differ greatly from each other without specific interaction.

In addition, water has a very high heat of vaporization $(H_2O$: about $44\ kJ\,mol^{-1}$, CH_4: about $8\ kJ\,mol^{-1})$ and a very high specific dielectric constant $(H_2O$: $\sim 78)$. The density of liquid water (about $1.00\ g\ cm^{-3}$) is larger than that of ice (about $0.92\ g\ cm^{-3}$) and becomes the largest at $4\ °C$. The evaporation entropy and specific heat are also abnormally large. These properties are presumed to be due to the hydrogen bonds between water molecules and the higher-order association structure of water formed by them. D_2O (heavy water) is sometimes used instead of H_2O (light water). For example, in NMR measurements, ^1H included in a solute is the most important probe to be measured, but most solvents, including water, contain hydrogen ^1H, which is a major obstacle to solute measurement. Therefore, in the case of an aqueous solution, D_2O is used as the solvent; however, even if ^2D is an isotope of ^1H, the physical properties of D_2O, such as pK_a and pH, are very different from those of H_2O, and thus care should be taken in rigorous equilibrium and reaction analyses.

7.5   Ionic Liquids

As mentioned in Section 7.3, a solvent, like water and organic solvents, is hardly ionized at ambient temperature. On the other hand, a room-temperature ionic liquid (RTIL), which is a solvent composed only of ions and has a low melting point ($100\ °C$ or less), is different from conventional solvents; therefore, their characteristics are briefly explained here. An RTIL exhibits high ionic conductivity and has a wide redox window (potential window) because it is an ionic dissociation solvent. In addition, the vapor pressure is very low and the flammability is also low. Due to these properties, RTILs are often used in electrochemical and physico-chemical studies and applications.

Furthermore, because of their low vapor pressure, they are also considered as environmentally friendly materials. Many RTILs are difficult to mix with water, ethers, and esters. However, because they have the property of dissolving benzene, alkanes, alkylated aromatic compounds, *etc.*, they are widely used in analytical chemistry, such as solvent extraction, coordination chemistry, and synthetic organic chemistry. Various kinds of organic salts become ionic liquids. Figure 7.1 shows the structure of the cation of 1-ethyl-3-methylimidazolium tetrafluoroborate $[C_2min][BF_4]$, which is the first RTIL developed. Imidazolium salts $[C_nmim][X]$ with various chain lengths are often used. In recent years, those with longer and more branched alkyl chains and additional functional groups have been developed. Anions as halides, BF_4^-, PF_6^- and $(CF_3SO_2)_2N^-$ are often used (see Chapter 17).

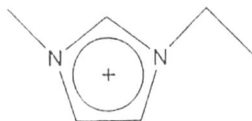

Figure 7.1 Structure of the C_2min ion.

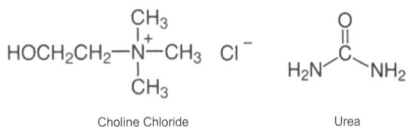

Figure 7.2 Molecular structure of DES (choline chloride and urea).

7.6 Deep Eutectic Solvents (DESs)

DESs have attracted much attention as an emerging class of green solvents related to ionic liquids (ILs). DESs are the eutectic composition of certain hydrogen bond donors and acceptors, showing abnormally deep melting point depression (Figure 7.2). A typical DES is the $1:2$ mole fraction combination of powdered choline chloride ((2-hydroxyethyl)-trimethylammonium chloride) (melting point $T_m \sim 302\ ^\circ C$) with crystalline urea ($T_m \sim 133\ ^\circ C$), which results in a liquid ($T_{eu} = 12\ ^\circ C$). DESs are regarded as a class of ILs since they show similar characteristics, such as high thermal stability, low volatility, and low vapor pressure. Unlike RTILs, DESs are inexpensive, biodegradable, non-toxic, and easier to prepare compared to RTILs. DESs have been used in various application fields, such as metallurgy and electrodeposition, separation and gas capture, power system and battery technologies, biocatalysis and organic chemistry, biomass processing, biomolecular structures, folding, and stability, pharmaceutical and medical research, and nanomaterials synthesis. The details of the applications are summarized in ref. 1.

7.7 Solvent Parameters

Although the abovementioned properties of a solvent itself, such as the dielectric constant and viscosity, due to the solvent–solvent interaction, are related to the chemical equilibrium and the chemical reaction, it is the solute–solvent interaction (ion solvation) that dominates complex formation in solution. In a solution, complex formation proceeds between solvated reactants and products; thus, ion solvation is heavily involved in the reaction. The solute–solvent interaction is complex, depending on the structure and physicochemical properties of the solvent. Solvents are classified based on various characteristics. The classification methods differ depending on researchers, and different solvent parameters are proposed for the same interaction and aspects depending on the measurement and calculation methods. The following shows the typical solvent classification methods and solvent parameters that are relatively often used.[2–6]

7.7.1 Functional Groups

Solvents are classified based on their functional groups and constituent elements, *e.g.*, aliphatic hydrocarbons, aromatic hydrocarbons, solvents containing OH groups, and other groups containing oxygen atoms, solvents containing halogens, nitrogen, and sulphur donor solvents, *etc.*

7.7.2 Dielectric Constant

The dielectric constant is the most fundamental parameter for estimating the electrostatic interaction between the solute and solvent. Thermodynamics in solution, such as the activity coefficient, ion-pair formation (ion association) and other ionic interactions, and solvation energy of ions, can be treated fairly well theoretically with the dielectric constant.

7.7.3 Coordination Ability

Complex formation, including a metal ion in solution, is always a competition between the solvent and potential ligands to bind to the metal ion. The ligand must be able to form stronger bonds to the metal ion than the ligand(s), especially as the concentration of the solvent in most cases is much larger than that of the ligand. The strength of the chemical bonds in a metal complex is dependent on the bonding characteristics of the metal ion, regarded as electron-pair acceptors, and of the ligands, regarded as electron-pair donors. A first attempt to systemise and postulate possible complex formation and the stability of the formed complexes was made by Chatt and coworkers by comparing reported stability constants.[7] They found that class (a) metal ions form more stable complexes in aqueous solution with ligands with donor atoms first in groups 15–17, N, O, and F, than in aqueous solution with ligands with other donor atoms. On the other hand, class (b) metal ions form the most stable complexes with ligands with P, S, and I. Class (a) metal ions form mainly electrostatic interactions and typical class (a) metal ions include alkali, alkaline earth, lanthanoid and actinoid ions. Class (b) metal ions form bonds with a higher degree of cavalency as, *e.g.* copper(I), silver(I), mercury(II), palladium(II) and platinum(II). This concept was further developed by Pearson who proposed the hard–soft acid-base (HSAB) concept.[8–10] In this concept, "hard" applies to species that are small, have high charge states, and are weakly polarizable, and 'soft' applies to species that are large, have low charge states and are strongly polarizable. The "hard" and "soft" metal ions are equivalent to the class (a) and class (b) metal ions (Lewis acids). The "hard" ligands (Lewis bases) are those with O and F donor atoms, and the "soft" ligands are those with P, S donor atoms and the heavier halide ions. The most stable metal complexes are formed between hard acids and bases, and soft acids and bases, while the complexes between hard acids and soft bases and soft acids and hard base are expected to be weak.

A large number of concepts have been proposed to qualitatively evaluate the electron-pair acceptor and donor properties of solvents.[11,12] The first widely applied of these concepts, the acceptor and donor numbers, A_N and D_N, respectively, were proposed Gutmann. The acceptor number is an empirical parameter to quantitatively evaluate the electrophilic properties of solvents and probed by the ^{31}P shift of triethylphosphine oxide in the solvent under study.[11] The donor number is a quantitative measure of the nucleophilic property (Lewis basicity) of solvents and is defined as the negative heat of formation, in $kcal\,mol^{-1}$, of the adduct formation between antimony(v) chloride and a solvent in dilute 1,2-dichloromethane solution.[12] The concept of the D_N scale was originally proposed by Lindqvist and Zackrisson.[13] A number of qualitative scales to estimate Lewis basicity were proposed afterwards as summarized in ref. 9. In this respect, the usefulness of the D_N scale has been questioned.[15] The donor strength scale, D_S,[14] uses the soft electron-pair acceptor, mercury(II) bromide as the probe, which forms solvate bonds with the highest degree of covalency of the proposed electron-pair donor scales. However, it is very important to keep in mind that the bonding properties of the chosen probe in the Lewis basicity scale will strongly affect the outcome of the scale and for which kind of metal ions, Lewis acids, it can successfully be applied on. If one wants to get information about the solvation of different solvents to a certain metal, one shall choose the scale with a probe as similar as possible to the metal ion one is interested in. In this respect, the D_S scale gives the most accurate information about the solvents with the highest electron-pair donor capability.

7.7.4 Polarity

The term "polar" is often used, such as polar solvent or non-polar solvent. Polar solvents have the property of dissolving electrolytes and many substances that are insoluble in non-polar solvents. Strong ion–dipole and dipole–dipole interactions, hydrogen bonding, *etc.* work between a solute and a solvent. "Polarity" must be considered comprehensively, including the dipole moment, dielectric constant, donor property, acceptor property, *etc.* Still, in organic reactions, parameters such as the E_T-value, Z-value, Y-value, and X-value are often proposed and used as solvent parameters to show the effect of "polarity". These parameters are highly correlated with each other. The E_T value is often used to measure solvent polarity or acceptability in complex formation.

7.7.5 Intermolecular Forces

In a system where relatively strong interactions such as coordination bonds, electrostatic interactions, and hydrogen bonds can be ignored, the dispersion force (London force) between molecules is the main interaction. The regular solution theory explains this intermolecular force quantitatively, and the proposed dissolution parameter can explain the distribution equilibrium fairly quantitatively.

7.7.6 Protic/Aprotic Solvents and Acidic/Basic Solvents

A solvent able to release a hydrogen atom as a proton is a protic solvent. A solvent without a hydrogen atom or only alkyl hydrogens, as they do not dissociate, is an aprotic solvent, *e.g.* benzene, toluene, chloroform and hexane. An acidic solvent is a solvent that donates a proton or has a hydrogen atom for hydrogen-bonding or accepts a lone pair of electrons, *e.g.* acetic acid and 2,2,2-trifluoroethanol. A solvent that accepts a proton or a lone pair of electrons having a proton-accepting or lone-electron pair-donating group is a basic solvent, *e.g.* pyridine and liquid ammonia. Water and alcohols have both acid–base properties and are called amphoteric solvents. The acidity and basicity of various solvents are determined based on the acidity and basicity of water.

Water, acetic acid, and alcohols can be classified as protonic solvents; benzene, chloroform, and hexane as aprotic solvents; acetic acid, 2,2,2-trifluoroethanol as acidic solvents; and pyridine and liquid ammonia as basic solvents.

7.7.7 Ionization Power

A solvent that has high polarity and can ionize a solute is called an ionizing solvent. In contrast, a solvent with no polarity is called a non-ionizing or inert solvent. In the case of water, it is an ionizing solvent that ionizes solutes. Water and *N*,*N*-dimethylformamide can be classified as ionizing solvents, and benzene and 1,2-dichloroethane can be classified as non-ionizing solvents (inert solvents).

7.7.8 Hydrophilicity–Hydrophobicity

This classification is based on the ease of mixing with water or the solubility in water. Methanol, ethanol, and acetic acid can be classified as hydrophilic solvents, whereas benzene and 1,2-dichloroethane can be classified as hydrophobic solvents.

7.8 Examples of Solvent Parameters

In Section 7.7, the entire solvent parameters proposed so far have been explained. Although various parameters have been presented, all of them are not necessarily used to understand complex formation reactions. Here, we will explain the solvent parameters that are most often used and considered important in explaining the solvent effects on the complex formation reaction. The values of the solvent parameters of typical solvents are shown in Appendix 1 (Table 7.2) at the end of the chapter. The specific usage is described in this book.

7.8.1 Relative Permittivity (Permittivity)

Permittivity is the most fundamental physical property for estimating the electrostatic interaction involving ions and dipoles. The vacuum permittivity is $\varepsilon_0 = 8.854 \times 10^{-12}$ F m^{-1}, and the ratio of the permittivity ε of the solvent to this is the relative permittivity $\varepsilon_r = \varepsilon/\varepsilon_0$. Note that relative permittivity is sometimes described as permittivity. The ratio of the capacitance of the capacitor filled with the solvent to the capacitance of the capacitor filled with a vacuum (approximately air) between the electrodes corresponds to the relative permittivity.

The force F acting between two ions (the electric charges z_+ and z_-) separated by a distance r in a solvent such as relative permittivity ε_r is expressed using the following equation:

$$F = \frac{z_+ z_- e^2}{4\pi\varepsilon_0\varepsilon_r r^2} \tag{7.1}$$

This electrostatic attraction is proportional to the electric charge and inversely proportional to the square of the distance and the relative permittivity. In this way, it can be seen that the electrostatic attraction between the ions decreases in proportion to $1/\varepsilon_r$ in a solvent of the relative permittivity ε_r. On the other hand, when an ion with an electric charge z_i and a radius r is transferred from a vacuum into a solvent with a relative permittivity ε_r, the Gibbs free energy change $\Delta_{solv}G^\circ$ (stabilization) per mole is estimated using the following Born equation:

$$\Delta_{solv}G^\circ = -\frac{N_A z_i^2 e^2}{8\pi\varepsilon_0 r}\left(1 - \frac{1}{\varepsilon_r}\right) \tag{7.2}$$

Here, it is assumed that the solvent is a uniform continuum, the ions are rigid spheres, and only purely electrostatic interactions are working. The solvation-free energy per mole of monovalent ions calculated using this formula is shown in Table 7.1.

Table 7.1 Effects of ionic radius r and relative permittivity ε_r on the solvation free energy $\Delta_{solv}G^\circ$ of monovalent ions (25.0 °C).

$\Delta_{solv}G^\circ$ (kJ mol^{-1})						
r/Å	ε_r	2	5	10	20	40	60
1		348	556	626	660	678	686
2		174	278	313	330	339	343
5		70	111	125	132	136	137

As the table shows, the smaller the ion and the larger the number of charges, and the higher the solvation's relative permittivity of the solvent, the more stable the solvation. The values estimated using the Born equation $(-\Delta_{solv}G°)$ tend to be slightly larger than the measured values when r is used as the crystal ionic radius (see Chapter 9).

Although not mentioned in detail here, the permittivity is greatly related to the activity coefficient of the electrolyte and the ion-pair formation (ion association). In particular, since the relative permittivity of water is as large as 78 (25 °C), the effect of water on lowering the activity coefficient and ion-pair formation is small. For example, according to the Debye–Hückel theory (see Chapter 9), for (1 : 1) electrolytes, the activity coefficient of ions is 0.9 to 1 for aqueous solutions with a concentration of 0.01 mol dm^{-3} or lower. Measurements that do not require rigorous treatments do not require activity coefficient correction. In this case, ion-pair formation is negligible. However, the ion-pair formation cannot be ignored in a solvent with a low dielectric constant; according to the ion-association theory, almost 50% of ions form the ion-pair in 10^{-5} mol dm^{-3} (1 : 1) electrolyte solution in a solvent with a relative permittivity of about 10.

7.8.2 E_T Value

Various solvent parameters have been proposed as values representing the effect of polarity on different organic reactions, but the E_T value presented by Dimrosth–Reichardt is the most commonly used.[16] It is a solvent parameter, $E_T = 28\,590$ kcal mol^{-1} nm/λ_{max} (nm) estimated from the maximum absorption wavelength λ_{max} of the visible absorption spectrum of pyridinium-*N*-phenol betaine in various solvents. The E_T values obtained using the dye shown in Figure 7.3 are most often used and are sometimes indicated as E_T (30). This molecule has a resonance structure as shown in Figure 7.3, and in a high acceptor solvent the electrons move to the right

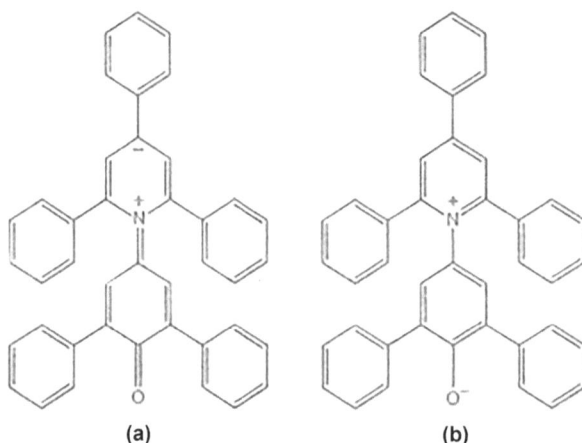

(a)   (b)

Figure 7.3   Resonant structures (a) and (b) of betain dye for E_T measurements.

(Figure 7.3(b)) (localization), and the absorption spectrum blue-shifts. The E_T value uses this maximum absorption energy as an index of the solvent's acceptor property. As can be seen from the structure in Figure 7.3(b), the solvent molecule mainly interacts with the polarized betaine molecule at the negatively charged O^-. The stronger the solvent, the more polarized the betaine molecule will be, and the higher the E_T value will be.

7.8.3 Dissolution Parameter δ

There is a regular solution theory as a theory that quantitatively deals with a system in which the dispersion force (London force) between molecules is the main interaction. In this theory, the aggregation energy (corresponding to the intermolecular force) of the solvent–solute interaction is assumed to be the geometric mean of the respective energies between the solvent–solvent and the solute–solute energies. However, it is not necessary to distinguish between the solvent and the solute, instead denoted as solvent (1) and solvent (2). For example, the square root of the aggregation energy (ΔE_T) of solvent (1) corrected for the effect of the molar volume (V_1) is the solubility parameter δ.

$$\delta_1 = \left(\frac{\Delta E_1}{V_1}\right)^{\frac{1}{2}} \tag{7.3}$$

Here, ΔE_1 has a relationship of enthalpy of vaporization ΔH_1 as $\Delta E_1 = \Delta H_1 - RT$ and is conventionally expressed in cal mol^{-1}, and the unit of δ_1 is $(\text{cal cm}^{-3})^{1/2}$. This dissolution parameter can quantitatively explain the thermodynamics of the two-correlation distribution. Even if other interactions are involved in the reaction, the regular solution theory can be applied to the system as long as the change in the interaction is negligible before and after the reaction.

Problem 1

The O–H stretching frequency of phenol shifts to the low frequency side in the order of nitromethane (NM) < nitrobenzene (NB) < acetonitrile (AN) < 1,4-dioxane < ethyl acetate < diethyl ether < dimethyl formamide (DMF) < pyridine. Explain the order of the O–H frequency shift from the donor number of the solvents.

Problem 2

The pK_s $(= -\log K_s$, K_s being the solubility product) values of NaCl in nitromethane (NM), acetonitrile (AN), propylene carbonate (PC), *N,N*-dimethylformamide (DMF), and dimethyl sulfoxide (DMSO) are 10.0, 8.2, 7.8, 4.7, and 2.7, respectively. Explain the order of the solubility of NaCl in the solvents from the donor and accepter numbers.

Problem 3

A thiocyanate ion (SCN^-) has two coordination sites of S and N. Dimethyl sulfoxide (DMSO) also has two coordination sites of S and O. d^{10} ions, Zn^{2+}, Cd^{2+}, and Hg^{2+} form tetrahedral complexes with four SCN^- in water and DMSO.

   Predict which atoms of SCN^- coordinate to Zn^{2+}, Cd^{2+}, and Hg^{2+} in water and DMSO.

Answer to Problem 1

In Appendix 1, the donor number is 2.7 (NM), 4.4 (NB), 14.1 (AN), 14.8 (1,4-dioxane), 17.1 (ethyl acetate), 19.2 (diethyl ether), 26.6 (DMF), and 33.1 (pyridine). The solvent with a high donor number strongly interacts with the phenol OH group, resulting in the low stretching frequency of the O–H group.

Answer to Problem 2

In solvents with similar accepter numbers, the solubility of NaCl increases with increasing donor number. DMSO > PC > AN > NM. On the other hand, in solvents with similar donor numbers, the solubility of NaCl increases with increasing accepter number. DMSO > DMF.

Answer to Problem 3

Zn^{2+} is classified as a hard acid. Thus, the N atoms (hard base) of 4 SCN^- are bonded to Zn^{2+} in water and DMSO. Hg^{2+}, a soft acid, prefers the soft base and thus the S atoms of 4 SCN^- coordinate Hg^{2+}. Cd^{2+} is in between Zn^{2+} and Hg^{2+}. In water, Cd^{2+} is coordinated with two N and two S atoms of 4 SCN^-. In DMSO, Cd^{2+} is bonded with 3 N atoms and 1 S atom of 4 SCN^-. This difference in the coordinating atoms is caused by the donor and accepter properties of the cation and anion, *i.e.* ion solvation.

Appendix 1

Table 7.2 Typical solvation parameters (specific permittivity ε, donor number D_N, acceptor number A_N, E_T/kcal mol^{-1} values, solubility parameters δ/(cal cm^{-3})$^{1/2}$.

Solvent	ε (25 °C)	D_N	A_N	E_T	δ
Acetonitrile (AN)	36.0	14.1	19.3	45.6	11.9
Acetone	20.7	17.0	12.5	42.2	9.62
Aniline	6.9^a	35		44.3	10.3
Isopropylether	3.9	19		34.1	7.06
Ethanol	24.5	20	37.1	51.9	12.78
Ethylene glycol	37.7	20	43.4	56.3	14.6

Table 7.2 (*Continued*)

Solvent	ε (25 °C)	D_N	A_N	E_T	δ
1-Octanol	10.3^a	32	30.4	48.1	10.3
n-Octane	1.95^a	0		31.1	7.55
Chlorobenzene	5.62	3.3	11.9	36.8	9.5
Chloroform	4.8^a	4	23.1	39.1	9.3
Acetic acid	6.15^a	20	52.9	55.2	13
Ethyl acetate	6.0	17.1	9.3	38.1	8.91
Methyl acetate	6.7	16.3	10.7	38.9	9.46
Benzyl cyanide	18.4	15.1	17.7	42.7	
Diethyl ether	4.3^a	19.2	3.9	34.5	7.53
Carbon tetrachloride	2.23	0	8.6	32.4	8.6
1,4-Dioxane	2.21	14.8	10.8	36.0	9.73
Cyclohexane	2.02	0	1.6	30.9	8.2
1,2-Dichloroethane (DCE)	10.36	0	16.7	41.3	9.8
Dichloromethane	8.9	1	20.4	40.7	
N,N-Dimethylacetamide (DMA)	37.8	27.8	13.6	42.9	10.8
Dimethyl sulfoxide (DMSO)	46.7	29.8	19.3	45.1	13
N,N-Dimethylformamide (DMF)	36.7	26.6	16	43.2	12
N-Methyl formamide (NMF)	182.4	27	32.1	54.1	
Ethylene carbonate	89.6^c	16.4		48.6	
Propylene carbonate (PC)	64.4 (66.1^a)	15.1	28.3	46.0	
Tetrahydrofuran (THF)	7.39	20.0	8.0	37.4	9.32
Tetramethylurea	23.1^a	29.6	9.2	40.9	
2,2,2-Trifluoroethanol (TFE)	27.68^a	~0	53.3	59.8	
Toluene	2.38	0.1	6.8	33.9	8.9
Nitrobenzene (NB)	34.8	4.4	14.8	41.2	10
Nitromethane (NM)	37.3^a	2.7	20.5	46.3	12.6
Carbon disulphide	2.64^a	2	7.5	32.8	10
Pyridine	12.3	33.1	14.2	40.5	10.8
1-Butanol	17.51	29	32.2	49.7	11.6
2-Butanol	16.56		30.5	47.1	
Isobutanol	17.93	37	35.5	48.6	
t-Butanol	12.47	38	27.1	43.3^b	10.6
n-Butyronitrile	20.3^a	16.6		42.5	10.5
1-Propanol	20.3	30	33.7	50.7	11.9
2-Propanol	19.9	36	33.5	48.4	11.5
Propionitrile	27.2^a	16.1	19.7	43.6	10.7
Hexamethylphosphoramide (HMPa)	30^a	38.8	9.8		10.9
n-Hexane	1.88	0	0	31.0	7.3
Benzene	2.28	0.1	8.2	34.3	9.15
Benzonitrile	25.2	11.9	15.5	41.5	
1-Pentanol	13.9	25	31	49.1	10.9
Formamide (FA)	111.0^a	24	39.8	56.8	17.8
Water	78.3	18.0 (32^d)	54.8	63.12	23.4
Acetic anhydride	20.7^a	10.5	18.5	43.9	
Methanol	33.7	19	41.3	55.4	14.5
Tributyl phosphate	6.8 (8.3^a)	23.7	9.9	38.9	
Trimethyl phosphate	20.6^a	23.0	16.3	43.6	

^{a} 20 °C.
^{b} 30 °C.
^{c} 40 °C.
^{d} Empirically obtained value.

References

1. B. B. Hansen, S. Spittle, B. Chen, D. Poe, Y. Zhang, J. M. Klein, A. Horton, L. Adhikari, T. Zelovich, B. W. Doherty, B. Gurkan, E. J. Maginn, A. Ragauskas, M. Dadmun, T. A. Zawodzinski, G. A. Baker, M. E. Tuckerman, R. F. Savinell and J. R. Sangoro, *Chem. Rev.*, 2021, **121**, 1232.
2. J. A. Riddick, W. B. Bunger and T. K. Sakano, *Organic Solvents*, Wiley, New York, 4th edn, 1986.
3. K. Burger, H. Ohtaki and S. Yamada, *Japanese translation, Hisuiyoueki no Kagaku (Chemistry of non-aqueous solutions)*, Gakkai Shuppan Center, Tokyo, 1988.
4. H. Ohtaki, *Youeki Kagaku (Solution Chemistry)*, Shokabou, Tokyo, 1985.
5. *Directory of solvents*, ed. B. P. Whim and P. G. Johnson, Blackie Academic & Professional, London, 1996.
6. Y. Marcus, *The properties of solvents*, Wiley, New York, 1998.
7. S. Ahrland, J. Chatt and N. R. Davies, *Q. Rev., Chem. Soc.*, 1958, **12**, 265–276.
8. R. G. Pearson, *J. Am. Chem. Soc.*, 1963, **85**, 3533–3539.
9. R. G. Pearson, *J. Chem. Educ.*, 1968, **45**, 643–648.
10. R. G. Pearson, *Inorg. Chem.*, 1988, **27**, 734–740.
11. U. Mayer, V. Gutmann and W. Gerger, *Monatsh. Chem.*, 1975, **106**, 1235–1257.
12. V. Gutmann, *Coord. Chem. Rev.*, 1976, **18**, 225–255.
13. I. Lindqvist and M. Zackrisson, *Acta Chem. Scand.*, 1960, **14**, 453–456.
14. M. Sandström, I. Persson and P. Persson, *Acta Chem. Scand.*, 1990, **44**, 653–675.
15. M. Katayama, M. Shinoda, K. Ozutsumi, S. Funahashi and Y. Inada, *Anal. Sci.*, 1995, **28**, 103–106.
16. C. Reichardt, *Solvents and solvent effects in organic chemistry*, Wiley-VCH, London, 3rd Updated and Enlarged Edition, 2003.

CHAPTER 8

Solvent Structures

TOSHIO YAMAGUCHI[a,b]

[a] Key Laboratory of Salt Lake Resources Chemistry of Qinghai Province, Qinghai Institute of Salt Lakes, Chinese Academy of Sciences, Xining, Qinghai 810008, China; [b] Professor Emeritus, Fukuoka University, 8-19-1 Nanakuma, Jonan, Fukuoka 814-0180, Japan
Email: yamaguch@fukuoka-u.ac.jp

8.1 Introduction

The physicochemical properties of a solvent originate from the microscopic structure. The Gibbs free energy change, enthalpy change, and entropy change of complex formation reactions that proceed in solution are governed by the combined effect of solvent–solvent, solvent–solute, and solute–solute interactions. Therefore, to understand the complex formation reaction and control it according to the purpose, it is necessary to understand the microscopic structure of the solution. It is also important when we consider various interactions of stable dissolved complexes. To understand the physicochemical properties of some common solvents, the structures and important physical parameters are presented.

8.2 Water

The most common solvent used in coordination chemistry is water. Water becomes hexagonal ice (I_h) at 0.1013 MPa (1 atm) and 0 °C. In ice I_h, the central water molecule is hydrogen-bonded to the surrounding four water molecules, and the tetrahedral structural units are connected three-dimensionally. The heat of fusion required to change ice at 0 °C to liquid water at the same temperature is 6.01 kJ mol^{-1}. Since the energy to break one

Coordination Chemistry Fundamentals Series No. 2
Metal Ions and Complexes in Solution
Edited by Toshio Yamaguchi and Ingmar Persson
© Japan Society of Coordination Chemistry 2024
Published by the Royal Society of Chemistry, www.rsc.org

hydrogen bond is about 17.5 kJ mol^{-1}, only 17% of the hydrogen bonds in ice are broken when the ice turns into liquid water of 0 °C. Even considering the energy change of a water molecule itself due to melting, about 70% of hydrogen bonds remain in the water at 0 °C. Thus, within about 10 Å (1 Å $= 10^{-10}$ m $= 100$ pm) from a given water molecule, the hydrogen-bonded network structure similar to ice remains as a time average structure. Figure 8.1 (left) shows the three-dimensional structure of water at 25 °C and 0.1 MPa obtained from X-ray scattering experiments and empirical potential structure refinement (EPSR) calculations.

Adjacent water molecules are hydrogen-bonded to a central water molecule in a tetrahedral fashion in the first coordination shell. Water molecules hydrogen-bonded to these water molecules constitute the second coordination shell. When the temperature rises from 0 °C, some water molecules with broken hydrogen bonds penetrate the interstitial sites of the ice-like structure, and the density of water increases to reach the maximum at around 4 °C. Therefore, the average coordination number of liquid water is about 4.4, slightly higher than that of ice (4). A water molecule has a structure with an H–O bond distance of 0.96 Å (96 pm) and an H–O–H bond angle of 104.5°, and its volume is approximated as that of a sphere with a radius of 1.4 Å (about 11.5 Å^3). On the other hand, since the average volume

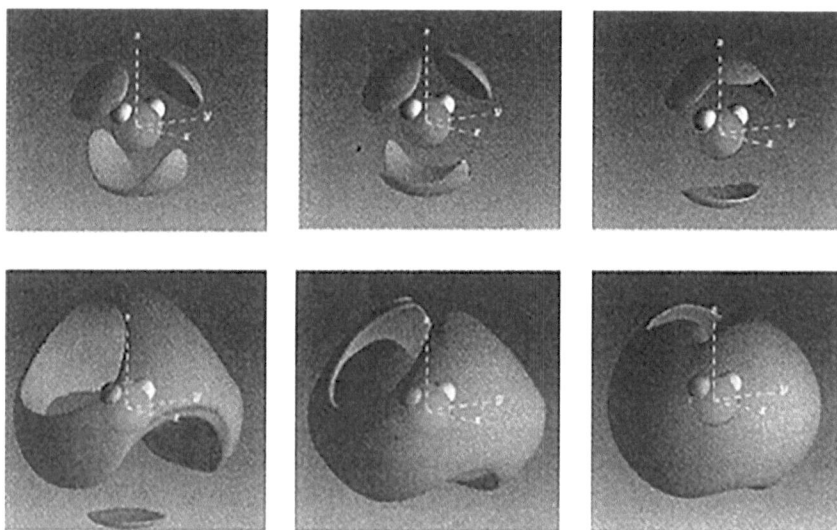

Figure 8.1   Three-dimensional structure of liquid water: (left) 25 °C, 0.1 MPa, (middle) 200 °C, 30 MPa, and (right) 300 °C, 30 MPa. The grey and white balls of a central water molecule represent the oxygen and hydrogen atoms, respectively. The surrounding grey lobs denote the nearest neighbour water oxygen atoms. The upper frames show the first coordination shell at 1.0–3.3 Å, and the lower frames show the second coordination shell at 3.3–4.9 Å. Reproduced from ref. 1 with permission from Elsevier, Copyright 2012.

of one isolated water molecule is 30 $Å^3$, about 62% of the inside of liquid water is an intermolecular void, which is slightly less than in ice (about 65%). These facts are consistent with the results of the above-mentioned structural analysis. Although a water molecule is small, its dipole moment is as large as 1.85 D (1 D (Debye) = 3.3356×10^{-30} C m), and it has a large relative permittivity ($\varepsilon_r = 78.3$) due to a polarization effect and directionality of the hydrogen-bonded network structure. As can be seen from eqn (7.1) in Chapter 7, the interionic force in crystals is weakened to about 1/80 in water so that a charged metal complex or a highly polar but uncharged metal complex is strongly hydrated and thereby dissolved in water. As the temperature and pressure increase, hydrogen bonds are gradually broken and deformed, so the relative permittivity (or others) and ion product (at 25 °C and 1 atm $[H^+][OH^-] = 10^{-14}$ $mol^2 dm^{-6}$) changes greatly. When the temperature and pressure are increased along the gas–liquid equilibrium line, two phases become one phase above the critical point of water (374.1 °C, 22.1 MPa), and it becomes a supercritical fluid that is neither a gas nor a liquid (in the case of water, the fluid is called supercritical water). For example, the relative permittivity (ε_r) is about 6 at 400 °C and 30 MPa, which is about the same as that of an organic solvent. Therefore, the solubility of salts is low in supercritical water, but water becomes completely miscible with non-polar gases, such as oxygen and methane. Since the ionic product of water rises to 10^{-11} $mol^2 dm^{-6}$ before the critical point (subcritical) at 250 °C and 30 MPa, acid-catalyzed reactions are promoted in subcritical water. Figure 8.3 (middle) and (right) show the three-dimensional structure of water at 200 °C, 30 MPa, and 300 °C, 30 MPa, respectively, as revealed by energy-dispersive X-ray diffraction and EPSR calculations.[1] From this figure, it can be seen that the structure of the second coordination shell is disrupted at higher temperatures. In other words, the characteristic of subcritical water is that the hydrogen bond is bent or broken, and the amount of small oligomeric water increases. In the supercritical state, it is possible to change from a high-temperature liquid-like state to a high-temperature gas-like state by lowering the pressure and reducing the density. Ionic reactions proceed in high-temperature and high-pressure water, and radical reactions proceed in high-temperature and low-pressure water. In recent years, subcritical and supercritical water has attracted attention as a reaction medium in nanotechnology applications such as oxidative decomposition of persistent chlorine compounds such as freon (chlorofluorocarbons) and PCB (poly chloro biphenyl), decomposition/recovery of waste plastics, generation of metal oxide nanoparticles, and hydrothermal synthesis.[2] See Section 5.1 for a general explanation of supercritical fluids. On the other hand, when the temperature of the water is lowered to a supercooled state, a clathrate-like structure based on a regular pentagonal hydrogen bond, as seen in the crystal structure of methane hydrate, is formed as reported from X-ray diffraction experiments on supercooled water up to −15 °C.[3] A similar hydrogen-bonded network structure of water has also been found in the crystal structure of metal complexes with nanotube spaces.[4]

8.3   Alcohol and Alcohol–Water Mixed Solvents

Alcohols are the second most important protic solvents after water. Since the alkyl group of alcohol cannot form a hydrogen bond, the intermolecular interactions are weaker than those of water. Therefore, the boiling point and melting point of low molecular weight alcohols are lower than those of water. Primary alcohols have a chain-like association structure in which the hydroxyl groups form hydrogen bonds. Alcohol is an amphipathic solvent with a hydrophilic hydroxyl group and a lipophilic alkyl group. Therefore, by changing the chain length of the alkyl group or adding water and changing the mixing ratio, the balance between the hydrophilicity and lipophilicity of the solution can be continuously adjusted so that the solubility of complexes can be altered due to the polarity and hydrophobicity of the complex. For methanol, from neutron isotope substitution experiments and EPSR calculation studies,[5] the average number of hydrogen bonds and the average number of chain-associated molecules of 1.95 ± 0.07 and (6.3 ± 0.7) molecules at -80 °C and 1.77 ± 0.07 and (5.5 ± 1.0) molecules at $+25$ °C have been reported. Its chain association structure is not straight but zigzag. The critical point of methanol is 239.4 °C and 8.1 MPa. Under the subcritical (202 °C and 74.7 MPa) and supercritical conditions (254 °C, 117.7 MPa) at the same density (0.700 g cm^{-3}) as under ambient conditions, the average number of hydrogen bonds decreases to 1.6 ± 0.1 and the average number of chain-associated molecules decreases to (3.1 ± 0.4) molecules. Furthermore, under supercritical (253 °C, 14.3 MPa) conditions at a low density (0.453 g cm^{-3}), the average number of hydrogen bonds is 1.0 ± 0.1 and the average number of chain-associated molecules is (1.8 ± 0.2) molecules (a dimer formed).

Ethanol has a chain association structure that is almost the same as that for methanol. From the X-ray scattering study,[6] a structure model in which the ethyl groups are arranged on the same side concerning the hydrogen bond chain reproduces the experimental values well.

As for 1-propanol, from the study of X-ray diffraction,[7] the 1-propyl group has a gauche structure in the molecule, and the chain association model with the 1-propyl groups arranged in a *cis* configuration reproduces the experimental values well compared to that with a *trans* structure (Figure 8.2).

Since a *cis* structural model is also suggested in the case of 2-propanol, the interaction between the propyl groups is considered to stabilize the chain association structure. Linear alcohols up to propanol are miscible with water at room temperature in all proportions. 1-Butanol has a compositional region that is immiscible with water, but tertiary butanol (TBA) does not. The structures of various alcohol–water mixed solvents have been studied from X-ray and neutron diffraction experiments. As a result, it has been clarified that alcohol and water appear to be mixed uniformly from a macroscopic view but have a microscopic phase-separated structure consisting of a domain in which alcohol molecules are aggregated and a domain in which water molecules are aggregated. As the mole fraction of alcohols

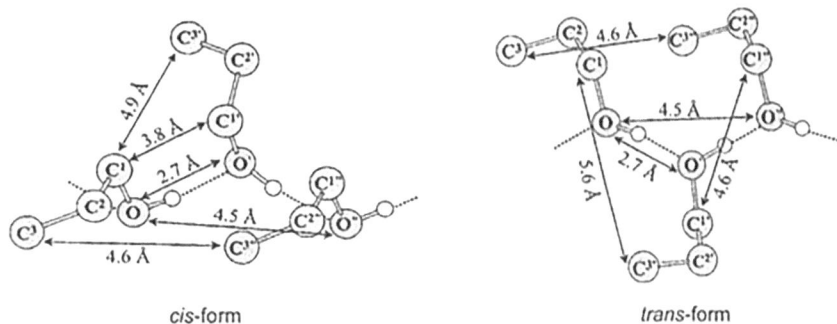

cis-form   trans-form

Figure 8.2   Structure model of liquid 1-propanol. The *cis*-conformation reproduced X-ray scattering data well. Reproduced from ref. 7 with permission from Springer Nature, Copyright 1969.

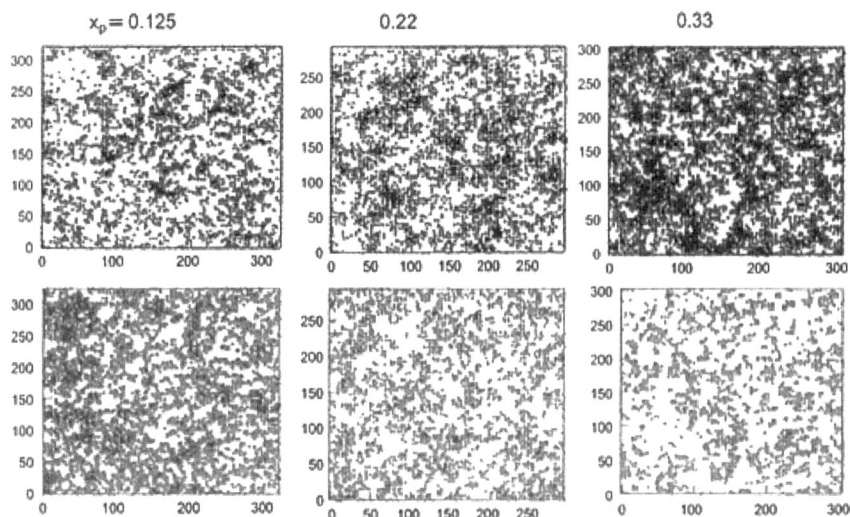

Figure 8.3   Mesoscopic structure of 1-propanol–water mixtures (a cell length of about 300 Å). The upper frames show the distribution of the hydrophobic groups of 1-propanol. The bottom frames show the distribution of water. The thick parts represent the dense packing of each molecule. The units of horizontal and vertical axes are in Å. x_p is the mole fraction of 1-propanol. Reproduced from ref. 8 with permission from AIP publishing, Copyright 2004.

increases, a tetrahedral hydrogen-bonded network structure of water is transformed to a chain-assembled alcohol cluster-based structure at a characteristic mole fraction. In a mixed solvent of methanol, ethanol, 1-propanol, or 2-propanol and water, this structural change occurs when the mole fraction of alcohol is about 0.3, 0.2, 0.1, and 0.1, respectively. Figure 8.3 shows the structure of 1-propanol–water mixed solvent obtained from small-angle neutron scattering experiments and reverse Monte Carlo (RMC) calculations.[8]

At a specific mole fraction of structure transformation, extreme values are seen in various thermodynamic and physicochemical quantities, such as mixed enthalpy, partial molar volume, and NMR chemical shift. Life science studies have also revealed that adding alcohol to proteins in aqueous solution causes α-helix structural transitions in this solution composition.

8.4   Amide Solvents

Typical amide-based solvents include formamide (FA) and *N*-methyl formamide (NMF), in which one H atom of an amino group (*cis* position concerning the O atom of an aldehyde group) is replaced with a methyl group, and *N,N*-dimethylformamide (DMF) in which another H atom is also substituted with a methyl group (Figure 8.4).

In formamide, an NH···O hydrogen bond is formed between the NH of the amino group and the O atom of the aldehyde group of the molecules. The distance between N–O atoms is about 3 Å. Because of this hydrogen bond, an ordered structure exists near 10 Å. On the other hand, although this hydrogen bond exists in NMF since there is only one NH hydrogen atom, the structural correlation over a long distance is smaller than that in FA. DMF, which is not able to form hydrogen bonds, has almost no long-range structure. A liquid structure model has been proposed to explain the radial distribution function by X-ray diffraction for the three amide liquids.[9-11] For formamide, as shown in Figure 8.5 (top), it is proposed that a cyclic dimer structure formed by N–H···O hydrogen bonds and a chain association structure are three-dimensionally combined.

Due to the presence of such a developed hydrogen-bond network, formamide has a higher boiling point (210.5 °C), melting point (2.55 °C), and viscosity (3.30 mPa s, 25 °C) than other amide liquids, despite its small molecular weight. It also has a high relative permittivity ($\varepsilon = 111.0$, 20 °C). On the other hand, NMF has a molecular structure, as shown in Figure 8.5 (middle). Since it is difficult to form a cyclic dimer, a one-dimensional chain association structure is mainly formed (Figure 8.5 (middle)). NMF has a larger relative permittivity than FA (182.4, 25 °C) because it does not form a cyclic dimer (the dipole moment is not offset). The melting point (-3.8 °C) and viscosity (1.65 mPa s, 25 °C) are lower than those of formamide because

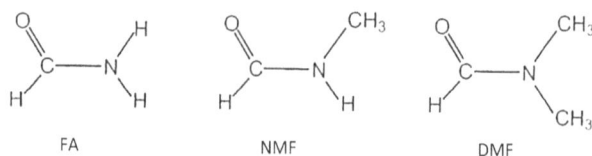

Figure 8.4 Structure of formamide (FA), *N*-methyformamide (NMF), and *N,N*-dimethylformamide (DMF).

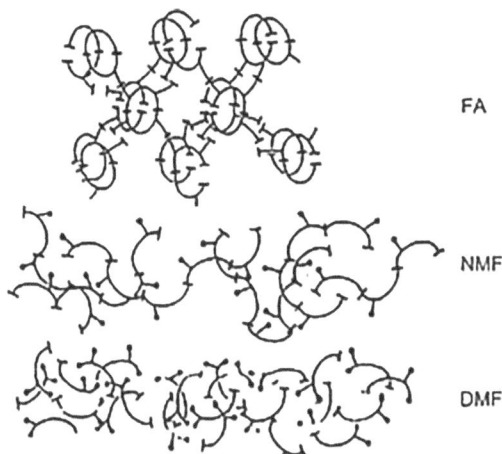

Figure 8.5 Structure model of liquid formamide (FA), *N*-methylformamide (NMF), and *N,N*-dimethylformamide (DMF). Symbol (●–) denotes the methyl group, (···) hydrogen bond (–NH···O=CH–). Reproduced from ref. 10 with permission from The Chemical Society of Japan.

of their relatively low association. Between molecules in DMF, which is difficult to hydrogen-bond, the structure like FA and NMF is hardly seen (Figure 8.5 (bottom)). This is the reason why the dipole moment (3.82 D) of DMF is about the same as those of FA (3.73 D) and NMF (3.83 D), but its relative permittivity ($\varepsilon_r = 36.7$, 25 °C) is low. Therefore, although DMF has the largest molecular weight among the three amide liquids, it has the lowest boiling point. It has a melting point (−61 °C) and a viscosity (0.796 mPa, 25 °C). From these facts, it can be said that FA and NMF are suitable for dissolving a metal complex having a charged or hydrogen bonding ability, and DMF is suitable for a complex having a low polarity.

8.5 Protic Polar Solvents

Aprotic solvents cannot form hydrogen bonds between solvent molecules, and dipole–dipole interactions dominate the liquid structure.

8.5.1 Dimethyl Sulfoxide (DMSO; $(CH_3)_2SO$)

DMSO is a solvent widely used in coordination chemistry and in the fields of industry, biology, organic chemistry, pharmacy, and life sciences. A DMSO molecule has a triangular pyramidal structure (G-symmetrical) with the S atom as the apex, two methyl groups, and one O atom as the other apex. The S atom has a lone pair of electrons. Many DMSO molecules are usually coordinated with O atoms to metal ions but may be coordinated with S atoms to metal ions classified as soft Lewis acids. DMSO has a large dipole moment (3.96 D) and a relatively high relative permittivity ($\varepsilon_r = 46.6$, 25 °C); its boiling point (189.0 °C) and melting point (18.55 °C) are high. From neutron diffraction experiments and EPSR calculation studies,[12] the most likely model is proposed such as the dipoles of DMSO arranged in an antiparallel fashion. The dipoles are oriented in the same direction at a long distance. DMSO is mixed with water in any composition, and the freezing point drops to −70 °C at DMSO : water = 1 : 2 (molar ratio), so the DMSO–water mixed solvent is used as a refrigerant.

In addition, with this mixed composition, the mixed enthalpy takes the minimum value, and the dielectric relaxation time increases. At DMSO : water = 1 : 20 (molar ratio), the tetrahedral network structure of water is not much different from that of pure water.[13] On the other hand, in the 1 : 2 mixed composition, the first coordination shell maintains a tetrahedral arrangement. Still, the second coordination shell of water is slightly broken, and the number of hydrogen bonds between water molecules decreases by about 30% compared with pure water.

8.5.2 Acetonitrile (AN; CH_3CN)

Acetonitrile (AN) has a large dipole moment (3.92 D) close to DMF. In the liquid, the acetonitrile molecules are arranged antiparallel to each other as in the crystal. Acetonitrile, similar to DMSO, has a long-range structure due to dipole–dipole interactions. Acetonitrile has a relatively high relative permittivity (36.0, 25 °C), dissolves many inorganic and organic compounds well, and has high electrochemical stability. Therefore, it is often used as a solvent for electrochemical measurements in coordination chemistry. In acetonitrile, the ion association constants of salts are considerably larger than that in DMF with a similar relative permittivity and that in ethanol with a low relative permittivity ($\varepsilon_r = 24.5$, 25 °C) due to the low basicity and inferior proton accepting capacity of acetonitrile.

Acetonitrile is mixed with water in any composition at room temperature, similar to organic solvents such as DMSO and acetone. From X-ray diffraction and IR measurements, acetonitrile and water molecules are not uniformly mixed in acetonitrile–water mixtures. Still, they are segregated into clusters consisting of water molecules and clusters consisting of acetonitrile molecules at a microscopic level.[14] A mixed solvent having a molar fraction of acetonitrile of 0.38 separates into two liquid phases when cooled to 0 °C. This occurs because the water molecules at the cluster's interface are incorporated into the water molecule cluster grown at low temperatures.

8.5.3 Acetone ($(CH_3)_2CO$)

The chemical formula of acetone is similar to that of DMSO, but the chemical properties and molecular structure are different because the central atom is different. An acetone molecule has a planar triangular (C_{2v})

structure, the dipole moment is relatively large at 2.88 D, and the oxygen atom of the carbonyl can accept a hydrogen bond with a protic solvent such as water. A study of neutron diffraction experiments[12] has proposed a short-range structure and a long-range structure similar to DMSO. As in the case of DMSO, a neutron scattering experiment[13] with a molar ratio of 1:2 was carried out for the acetone–water mixed solvent, and it was clarified that microscopic phase separation occurs.

8.5.4 Chloroform ($CHCl_3$)

The liquid structure of chloroform has been proposed in a study[15] that combines X-ray/neutron diffraction and RMC (reverse Monte Carlo) calculations. The mutual orientation of two adjacent tetrahedral molecules can be divided into six types (Figure 8.6(a)).[16] When the distance between molecules (between central carbon atoms) is 4 Å or more, the edge-to-edge (2:2) orientation is taken, and, among them, (H, Cl)–(Cl, Cl) (the H–Cl side is facing the Cl–Cl side) is the most frequent orientation (Figure 8.6 (b, left)). On the other hand, when the intermolecular distance is 4 Å or less, the edge-to-face (2:3) orientation is taken and the (H, Cl)–(H, C1, C1) arrangement is particularly frequent.

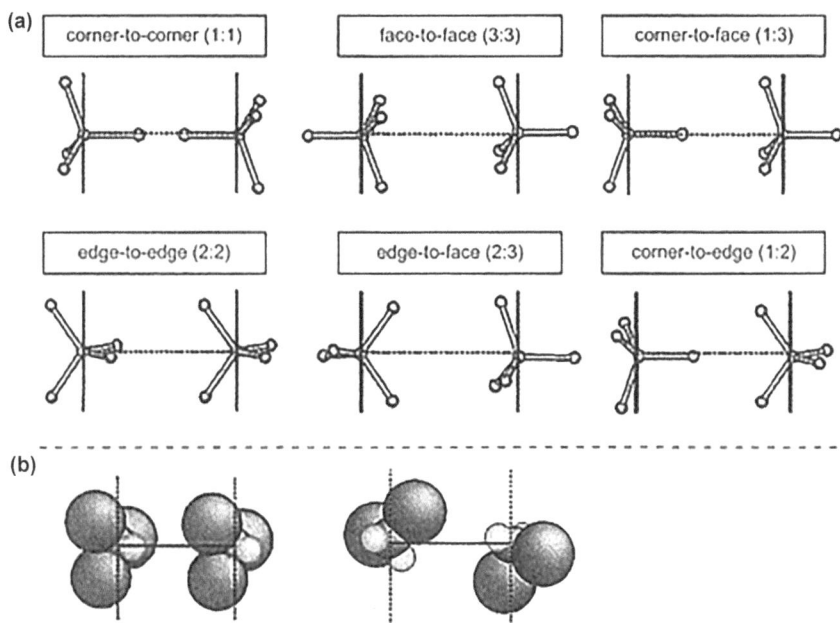

Figure 8.6 (a) Six kinds of configuration between the neighboring tetrahedral molecules. In liquid CCl_4, edge-to-edge (Cl, Cl)–(Cl, Cl) is dominant. (b) In liquid $CHCl_3$, edge-to-edge (H, Cl)–(Cl, Cl) (left), and in CH_2Cl_2, edge-to-edge (H, Cl)–(H, Cl) (right). Reproduced from ref. 16 with permission from AIP publishing, Copyright 2007.

8.5.5   Dichloromethane (CH_2Cl_2)

Dichloromethane is a solvent often used for synthesis in organic chemistry and coordination chemistry. Its liquid structure has been proposed in a study[15] that combines X-ray/neutron diffraction and RMC calculations. The intermolecular orientation in liquids is edge-to-edge (2:2), with (H, Cl)–(H, Cl) being the most frequent (Figure 8.6 (b, right)).

8.5.6   Cyclic Ether: Dioxane (($CH_2)_4O_2$) and Tetrahydrofuran (THF; ($CH_2)_4O$)

From X-ray diffraction, small-angle neutron scattering, and NMR relaxation measurements, the structures of 1,4-dioxane–water, 1,3-dioxane–water, and tetrahydrofuran–water mixed solvents have been determined.[17] When the mole fraction of the cyclic ether is 0.1 or less, the tetrahedral network structure of water is dominant, and when the mole fraction is 0.3 or more, the ordered structure of the cyclic ether itself is dominant. At a mole fraction of 0.1–0.2, the water structure and the ordered structure of the cyclic ether are broken. The concentration fluctuations of the water aggregate and cyclic ether aggregate domains are maximized near the mole fraction of 0.3. This concentration fluctuation decreases in the order of (THF–water)>(1,3-dioxane–water)>(1,4-dioxane–water). In the measurements of NMR relaxation time, the rotational motion of water molecules is most restricted at a mole fraction around 0.3. This is due to the strongest hydrogen bond between water molecules in this composition.

8.6   Low Polarity Solvents

8.6.1   Benzene (C_6H_6) and Toluene ($C_6H_5CH_3$)

Benzene and toluene contain aromatic rings and are the simplest liquids that involve intermolecular π–π interactions. There are four possible types of

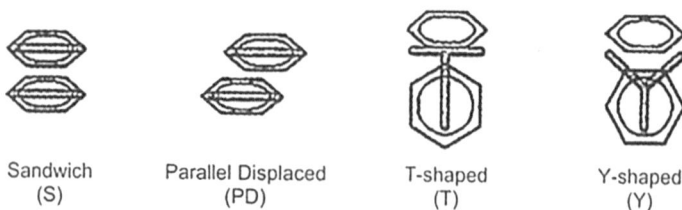

Sandwich (S)	Parallel Displaced (PD)	T-shaped (T)	Y-shaped (Y)

Figure 8.7   Four possible configurations for the benzene dimer. In liquid benzene, the PD configuration is formed within 5 Å of intermolecular distance, whereas the Y configuration is favorable beyond 5 Å apart. Reproduced from ref. 18 with permission from American Chemical Society, Copyright 2010.

mutual orientation of the dimer formed by these interactions, as shown in Figure 8.7. From the study of neutron diffraction and EPSR modelling,[18] the average distance between the centers of aromatic rings is 5.75 Å. There are about 12 adjacent molecules around the central molecule. The mutual orientation between the molecules is similar to the graphite structure, with the structure in which the center of the aromatic ring deviates when the intermolecular distance is within 5 Å. On the other hand, when the intermolecular distance is 5 Å or more, it has a Y-oriented structure, in which two hydrogen atoms are oriented toward the π-orbital of the acceptor. Conventionally, the T-oriented structure estimated from the molecular orbital calculation exists as a saddle point. The above structure is higher in benzene than in toluene.

8.6.2   Carbon Tetrachloride (CCl_4)

Neutron diffraction experiments over the temperature range from the melting point to 160 °C along the vapor–liquid equilibrium line have been reported.[19] The mutual orientation between adjacent molecules is dominated by the edge-to-edge (2 : 2) structure in Figure 8.6(a) as a result of analysis by recent X-ray/neutron diffraction experiments and RMC calculations.[20] There are also a few previously proposed corner-to-face (1 : 3) and corner-to-edge (1 : 2) structures. In this way, there is a specific orientation correlation between non-polar molecules. Still, the liquid structure explains the experimental values with a model close to the cubic close-packed structure, which regards CCl_4 molecules as rigid spheres with an effective diameter of about 5 Å. The molecular weight of a CCl_4 molecule is nearly four times that of acetonitrile, but the boiling point and melting point do not change much. This is because only a weak interaction like van der Waals forces acts between the CCl_4 molecules.

8.6.3   Ionic Liquids (ILs)

Many studies use X-ray and neutron diffraction, molecular dynamics simulations, and Raman scattering for examining the structures of ionic liquids.[21] The structural features of ionic liquids are (1) the short-range order in which the Coulomb interaction between cations and anions acts, (2) the long-range interaction in which the interaction involving polar and non-polar groups acts, (3) the presence of significant structural inhomogeneity due to the association of cations with alkyl groups or of anions. The 0.2–0.5 Å^{-1} pre-peak in small-angle X-ray scattering suggests this structural heterogeneity. From the study of X-ray scattering and molecular dynamics (MD) simulations,[22] the correlation length reflecting structural heterogeneity is 14 Å for [bmim]PF_6 (bmim = 1-butyl-3-methylimidazolium ion) and is 26 Å for [C_{10}mim]BF_4 (C_{10}mim = 1-decyl-3-methylimidazolium ion), showing microscopic phase separation behavior (Figure 8.8).

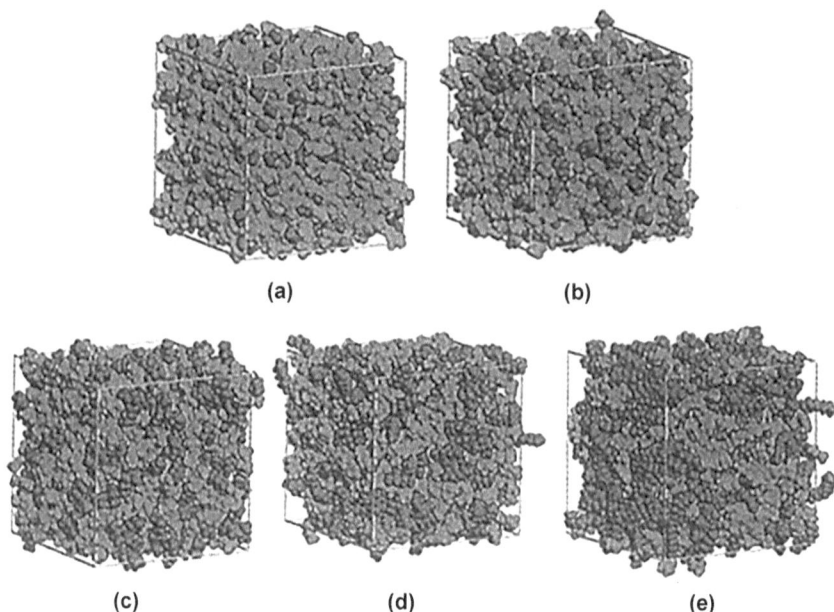

(a)   (b)

(c)   (d)   (e)

Figure 8.8   Snapshots of a MD simulation box containing $[C_n\text{min}][PF_6]$ ionic liquids: (a) $[C_2\text{mim}][PF_6]$, (b) $[C_4\text{min}][PF_6]$, (c) $[C_6\text{min}][PF_6]$, (d) $[C_8\text{min}][PF_6]$, and (e) $[C_{12}\text{min}][PF_6]$. Dark grey represents polar groups and light grey, non-polar groups. When a non-polar group becomes longer from C_2 to C_{12}, the polar and non-polar groups are more segregated. Reproduced from ref. 22 with permission from American Chemical Society, Copyright 2006.

8.6.4   Deep Eutectic Solvents (DESs)

A DES made of alkyl amide $+ \text{Li}^+/\text{ClO}_4^-$ was investigated by molecular dynamics simulations.[23] The simulated X-ray and neutron scattering structure functions of the DES display a pre-peak at $0.1 < q/\text{Å}^{-1} < 0.4$, a signature of nanoscale structural organization/heterogeneity, as shown in Figure 8.9. The liquid structure of a DES (1 : 2 molar mixture of choline chloride and urea) has been determined by neutron scattering with H/D isotopic substitution combined with EPSR modelling.[24] Figure 8.9 shows the spatial density functions showing the 3D structure of the DES. Yellow surfaces depict choline cations, the purple surface represents urea molecules, and the green surfaces show chloride anions. Choline interacts very strongly with chloride by hydrogen bonding. This structure is stabilized by the complementary hydrogen bond formation of choline and urea with chloride. This set of interactions causes the formation of a radially layered sandwich structure. This sandwich structure is also visualized as a charge-delocalized, locally stoichiometric cage centered on choline. The delicate balance of strong forces between all species is sufficient to prevent the crystallization of the mixture at room temperature (Figure 8.10).

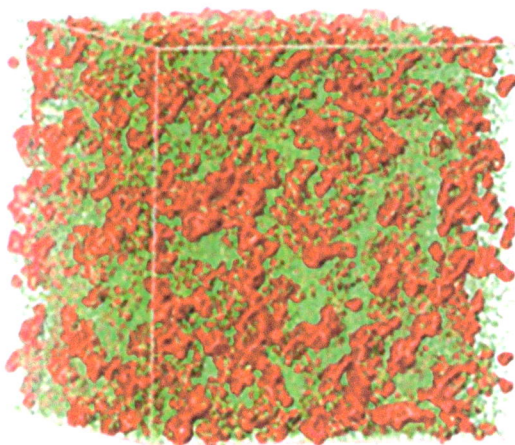

Figure 8.9 Structure of the acetamide $+ Li^+/ClO_4^-$ DES. The Li^+/ClO_4^- groups are shown as red isofurfaces and the acetamide molecules are shown in green. Reproduced from ref. 23 with permission from American Chemical Society, Copyright 2016.

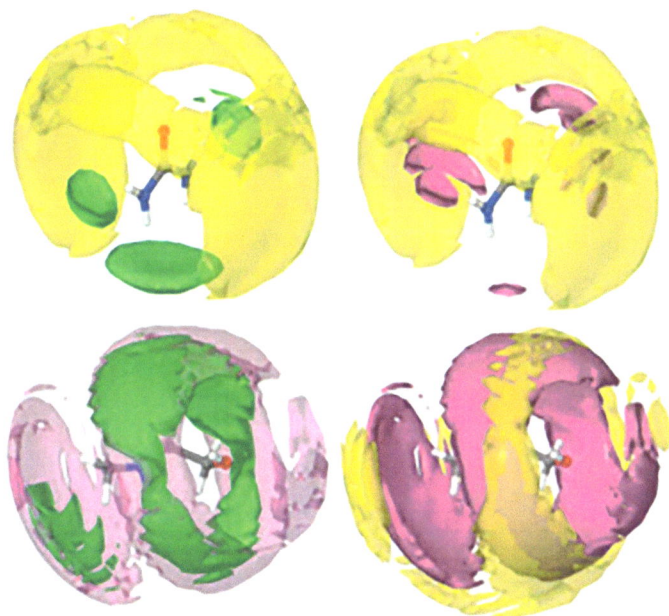

Figure 8.10 Spatial density functions showing the 3D structure of the DES (1 : 2 molar mixture of choline chloride and urea). Yellow surfaces represent choline cations, purple surfaces represent urea molecules and green ones show chloride anions. Reproduced from ref. 24 with permission from the Royal Society of Chemistry.

Problem 1

Explain the characteristics of the structure of liquids and solutions, compared with the crystalline structure.

Problem 2

Explain the typical structure of hydrogen-bonded solvents (water and methanol) and non hydrogen-bonded solvents (carbon tetrachloride).

Answer to Problem 1

X-ray and neutron scattering techniques of liquids and solutions give a halo pattern. Since ions and molecules are moving around in liquids and solutions, the liquid structure seen by X-ray and neutron scattering techniques is a time- and space-average structure, 1D structure. On the other hand, the crystalline structure shown by X-ray and neutron scattering is a periodic 3D structure.

Answer to Problem 2

Liquid water has a tetrahedral hydrogen-bonded open structure, similar to ice. Methanol has a zig-zag chain structure connected by hydrogen bonds. Carbon tetrachloride is an aprotic solvent and has no dipole moment. Thus, carbon tetrachloride molecules are arranged in a closed packed structure with some preferred orientation of the molecules.

References

1. T. Yamaguchi, K. Fujimura, K. Uchi, K. Yoshida and Y. Katayama, *J. Mol. Liq.*, 2012, **176**, 44.
2. T. Sako, *Supercritical Fluids*, Agune Shofu Sha, Tokyo, 2001.
3. H. Yokoyama, M. Kannami and H. Kanno, *Chem. Phys. Lett.*, 2008, **463**, 99.
4. M. Tadokoro, C. Iida, T. Saitoh and Y. Miyazato, *Chem. Lett.*, 2010, **39**, 186.
5. (a) T. Yamaguchi, K. Hidaka and A. K. Soper, *Mol. Phys.*, 1999, **96**, 1159; Erratum, *Mol. Phys.*, 1999, **97**, 603; (b) T. Yamaguchi, C. J. Benmore and A. K. Soper, *J. Chem. Phys.*, 2000, **112**, 8976.
6. T. Yamaguchi, T. Takamuku and A. K. Soper, *J. Neutron Res.*, 2005, **13**, 129.
7. T. Takamuku, H. Maruyama, K. Watanabe and T. Yamaguchi, *J. Solution Chem.*, 2004, **33**, 641.
8. M. Misawa, I. Dairoku, A. Honma, Y. Yamada, T. Sato, K. Maruyama, K. Mori, S. Suzuki and T. Otomo, *J. Chem. Phys.*, 2004, **121**, 4716.
9. H. Ohtaki, A. Funaki, B. M. Rode and G. J. Reibnegger, *Bull. Chem. Soc. Jpn.*, 1983, **56**, 2116.

10. H. Ohtaki, S. Itoh and B. M. Rode, *Bull. Chem. Soc. Jpn.*, 1986, **59**, 271.

11. H. Ohtaki, S. Itoh, T. Yamaguchi, S. Ishiguro and B. M. Rode, *Bull. Chem. Soc. Jpn.*, 1983, **56**, 3406.

12. S. E. Maclain, A. K. Soper and A. Luzar, *J. Chem. Phys.*, 2006, **124**, 074502.

13. S. E. Maclain, A. K. Soper and A. Luzar, *J. Chem. Phys.*, 2007, **127**, 174515.

14. T. Takamuku, M. Tabata, A. Yamaguchi, J. Nishimoto, M. Kumamoto, H. Wakita and T. Yamaguchi, *J. Phys. Chem. B*, 1998, **102**, 8880.

15. S. Pothoczki, L. Temleitner and L. Puszati, *J. Chem. Phys.*, 2010, **132**, 164511.

16. R. Ray, *J. Chem. Phys.*, 2007, **126**, 164506.

17. T. Takamuku, A. Nakamizo, M. Tabata, K. Yoshida, T. Yamaguchi and T. Otomo, *J. Mol. Liq.*, 2003, **103–104**, 143.

18. T. F. Headen, C. A. Howard, N. T. Skipper, M. A. Wilkinson, D. T. Bowron and A. K. Soper, *J. Am. Chem. Soc.*, 2010, **132**, 5735.

19. (a) M. Misawa, *J. Chem. Phys.*, 1989, **91**, 5648; (b) M. Misawa, *J. Chem. Phys.*, 1990, **93**, 6774.

20. S. Pothoczki, L. Temleitner, P. Jóvári, S. Kohara and L. Puszatai, *J. Chem. Phys.*, 2009, **130**, 064503.

21. E. W. Castner Jr and J. F. Wishart, *J. Chem. Phys.*, 2010, **132**, 120901.

22. J. N. A. C. Lopes and A. A. H. Pàdua, *J. Phys. Chem. B*, 2006, **110**, 3330.

23. S. Kaur, A. Gupta and H. K. Kashyap, *J. Phys. Chem. B*, 2016, **120**, 6712.

24. O. S. Hammond, D. T. Bowlon and K. J. Edler, *Green Chem.*, 2016, **18**, 2736.

CHAPTER 9

Hydration and Solvation of Metal Ions

INGMAR PERSSON

Department of Molecular Sciences, Swedish University of Agricultural Sciences, P.O. Box 7015, SE-750 07 Uppsala, Sweden
Email: ingmar.persson@slu.se

9.1 Introduction

Knowledge about the structures of hydrated and solvated metal ions and complexes in solution is of the utmost importance in order to understand their chemical behavior and physico-chemical constants in the reaction medium. This chapter gives an overview of the experimentally determined structures of hydrated and solvated metal ions in the solid state and solution reported to date, *i.e.* early 2022. A short presentation of the available methods for accurate structure determination and physico-chemical constants in solution is given in Chapter 1. The solvents selected in this review are the oxygen donor solvents water (aq), dimethylsulfoxide (dmso), *N,N*-dimethylformamide (dmf) and *N,N'*-dimethylpropyleneurea (dmpu), and the nitrogen donor solvents acetonitrile (AN) and liquid ammonia (NH_3). These solvents have been selected because a relatively large number of solvate structures, in both the solid state and solution, have been reported, and their bonding characteristics are different.[1] Common oxygen donor solvents, as, *e.g.* alcohols, have similar bonding characteristics to water, and, from a structural point of view, their structures are in most cases very similar to those of the corresponding hydrates. *N,N'*-Dimethylpropyleneurea has been included in this survey due to its space-demanding properties upon co-ordination causing a different coordination chemistry of metal ion solvates

Coordination Chemistry Fundamentals Series No. 2
Metal Ions and Complexes in Solution
Edited by Toshio Yamaguchi and Ingmar Persson
© Japan Society of Coordination Chemistry 2024
Published by the Royal Society of Chemistry, www.rsc.org

than most other oxygen donor solvents. The impact of lower coordination numbers with low symmetry on the physico-chemical behavior is discussed. The nitrogen donor solvents acetonitrile and liquid ammonia have significantly different bonding characteristics to metal ions being weak and strong electron-pair donors, respectively.[1] In order to give as much information as possible, this overview will mainly be in the form of two tables. Only striking differences in the coordination number, geometry or bond distances will be discussed in more detail.

9.2 Structures of Hydrated and Solvated Metal Ions in the Solid State and Solution

9.2.1 Oxygen Donor Solvents

The structures of hydrated and solvated metal ions in the selected oxygen donor solvents in the solid state and solution are summarized in Table 9.1. The structures of hydrated, and dmso and dmf solvated metal ions are in most cases very similar with the same coordination number and figure, and similar M–O bond distances, while the solvates of the space-demanding solvent dmpu have in many cases lower coordination numbers with lower symmetry (Table 9.1).

Lithium, sodium and silver(ı) ions are the only monovalent metal ions, which are sufficiently strongly hydrated to crystallize with a full hydration shell, while potassium, rubidium, cesium and thallium(ı) ions crystallize in most cases without a water shell, or the reported structures are not in agreement with the expected M–O bond distances from established ionic radii,[88] and the size of the oxygen atom in coordinated water molecules, 1.34 Å.[89] This radius of oxygen can also be used for most other oxygen donor solvents as seen in a comparison of M–O bond distances in metal ion solvates (Table 9.1). This means that the hydrate and solvate structures of these metal ions must be determined in solution.

Lithium, sodium, potassium, calcium, strontium, barium, cobalt(ıı), nickel and cadmium ions are reported to form hydrated complexes of dimers or higher oligomers.[2,3] However, such complexes have not, for any of these metal ions, been observed in aqueous solution.

The hydrated silver(ı) ion is two-coordinated in a linear fashion with a short mean Ag–O bond distance, 2.13 Å, in the solid state.[2,3] The hydrated silver(ı) ion is also two-coordinated in aqueous solution, but the mean Ag–O bond distance is *ca.* 0.2 Å longer than that in the solid state (Table 9.1). The reason for the longer Ag–O bond distance in aqueous solution is that additional water molecules are present at a distance between the first and second hydration spheres, but it has not been possible to describe how they are arranged or how many they are.[59,60]

For a number of metal ions, several coordination numbers and geometries are reported in the solid state. However, in aqueous solution the highest coordination number is the dominating one, *e.g.* the hydrated calcium ion is

Table 9.1 Summary of mean M–O bond distances in hydrated and dimethylsulfoxide, N,N-dimethylformamide and N,N'-dimethylpropyleneurea solvated metal ions in the solid state and solution. The following abbreviations are used in this table: coordination number and configuration: two – linear, $D_{\infty h}$ and bent, C_{2v}; three – triangular, $D_{\infty h}(3)$, and trigonal pyramidal, C_{3v}; four – tetrahedral, T_d, and square-planar, $D_{4h}(4)$; five – square pyramidal, C_{4v}, and trigonal bipyramidal, $D_{3h}(5)$; six – octahedral, O_h, and tetragonally elongated octahedral (Jahn–Teller distorted octahedral), C_{4v}; seven – pentagonal bipyramidal, D_{5h}; eight – square antiprism, $D_{4h}(8)$; nine – tricapped prismatic, $D_{3h}(9)$, and for compounds without a well-defined coordination geometry, the coordination number is given by a figure.

Metal ion	Water		Dimethylsulfoxide		N,N-Dimethylformamide		N,N'-Dimethylpropyleneurea	
	(s)[2,3]	(aq)[4,5]	(s)[2,3]	(dmso)	(s)[2,3]	(dmf)	(s)[2,3]	(dmpu)
Li^+	1.941 – T_d	1.95 – T_d[6,7]	1.918 – T_d		1.926 – T_d; 1.847 – $D_{3h}(3)$; 1.906 – C_{3v}			
Na^+	2.417 – O_h; 2.341 – D_{3h}	2.43 – O_h[7,8]	2.460 – O_h; 2.333 – C_{4v}	2.43 – O_h[7]				
K^+		2.81 – 7,8		2.79 – 7 – 7[7]				
Rb^+		2.98 – $D_{4h}(8)$[7-9]		2.99 – $D_{4h}(8)$[9]				
Cs^+		3.10 – $D_{4h}(8)$[7,8]		3.06 – $D_{4h}(8)$[7]				
Be^{2+}	1.613 – T_d	1.62 – T_d[10]	1.619 – T_d		1.618 – T_d			
Mg^{2+}	2.066 – O_h	2.07 – O_h	2.070 – O_h	2.062 – O_h[71]	2.057 – O_h			
Ca^{2+}	2.476 – $D_{4h}(8)$; 2.401 – $C_{3v}(7)$	2.48 – $D_{4h}(8)$[11,12]						
Sr^{2+}	2.324 – O_h; 2.613 – D_{4h}	2.61 – $D_{4h}(8)$[13-15]	2.305 – O_h; 2.589 – 8; 2.531 – 7; 2.445 – O_h	2.303 – O_h[71]; 2.54 – 7[13]	2.290 – O_h; 2.507 – $C_{3v}(7)$	2.555 – 7[15]	2.331 – O_h	
Ba^{2+}	2.80 – D_{4h}	2.81 – $D_{4h}(8)$[13,16,17]	2.757 – $D_{4h}(8)$	2.76 – $D_{4h}(8)$[13]	2.801 – $D_{4h}(8)$		2.497 – O_h[a]	
Sc^{3+}	2.170 + 2.459 – $D_{3h}(9)$	2.17 + 2.33 – $D_{3h}(9)$[18]		2.09 – O_h[72]	2.072 – O_h		2.633 – O_h[a]	
Y^{3+}	2.353 – $D_{4h}(8)$	2.367 – $D_{4h}(8)$[19]	2.073 – O_h; 2.348 – $D_{4h}(8)$; 2.291 – 7	2.364 – $D_{4h}(8)$[73]	2.338 – $D_{4h}(8)$	2.364 – $D_{4h}(8)$[73]	2.074 – O_h; 2.219 – O_h	2.24 – O_h[75]

Ion						
La^{3+}	2.53 + 2.61 – $D_{3h}(9)$ 2.556 – 9 2.495 – $D_{4h}(8)$	2.53 + 2.60 – $D_{3h}(9)$[20]	2.545 – 9 2.481 – $D_{4h}(8)$	2.500 – $D_{4h}(8)$[74]	2.558 – 9 2.469 – $D_{4h}(8)$	2.45 – 7[85]
Ce^{3+}	2.49 + 2.60 – $D_{3h}(9)$ 2.535 – 9 2.485 – $D_{4h}(8)$	2.49 + 2.62 – $D_{3h}(9)$[20]	2.535 – 9 2.471 – $D_{4h}(8)$		2.507 – 9 2.464 – $D_{4h}(8)$ 2.368 – O_h	2.44 – 7[85]
Pr^{3+}	2.48 + 2.57 – $D_{3h}(9)$ 2.506 – 9 2.476 – $D_{4h}(8)$	2.48 + 2.58 – $D_{3h}(9)$[20]	2.442 – $D_{4h}(8)$	2.450 – $D_{4h}(8)$[74]	2.500 – 9 2.452 – $D_{4h}(8)$ 2.426 – 7	2.42 – 7[85]
Nd^{3+}	2.46 + 2.57 – $D_{3h}(9)$ 2.505 – 9 2.439 – $D_{4h}(8)$	2.47 + 2.60 – $D_{3h}(9)$[20]	2.427 – $D_{4h}(8)$	2.430 – $D_{4h}(8)$[74]	2.428 – $D_{4h}(8)$ 2.343 – O_h	2.41 – 7[85]
Sm^{2+}					2.485 – O_h	
Sm^{3+}	2.44 + 2.55 – $D_{3h}(9)$ 2.465 – 9 2.430 – $D_{4h}(8)$	2.43 + 2.53 – $D_{3h}(9)$[20]	2.389 – $D_{4h}(8)$	2.41 – $D_{4h}(8)$[74]	2.397 – $D_{4h}(8)$ 2.338 – 7 2.319 – O_h	
Eu^{2+}	2.548 – 7[15]					
Eu^{3+}	2.41 + 2.54 – $D_{3h}(9)$ 2.463 – 9 2.416 – $D_{4h}(8)$	2.42 + 2.57 – $D_{3h}(9)$[20]	2.396 – $D_{4h}(8)$	2.40 – $D_{4h}(8)$[74]	2.389 – $D_{4h}(8)$	2.37 – 7[85]
Gd^{3+}	2.40 + 2.54 – $D_{3h}(9)$ 2.444 – 9 2.403 – $D_{4h}(8)$	2.41 + 2.53 – $D_{3h}(9)$[20]	2.370 – $D_{4h}(8)$	2.39 – $D_{4h}(8)$[74]	2.381 – $D_{4h}(8)$ 2.263 – O_h	2.34 – 7[85]

Table 9.1 (*Continued*)

Metal ion	Water		Dimethylsulfoxide		N,N-Dimethylformamide		N,N'-Dimethylpropyleneurea	
	(s)[2,3]	(aq)[4,5]	(s)[2,3]	(dmso)	(s)[2,3]	(dmf)	(s)[2,3]	(dmpu)
Tb^{3+}	$2.39 + 2.51 - D_{3h}(9)$ $2.428 - 9$ $2.381 - D_{4h}(8)$	$2.40 + 2.53 - D_{3h}(9)$[20]	$2.368 - D_{4h}(8)$	$2.37 - D_{4h}(8)$[74]	$2.365 - D_{4h}(8)$		$2.247 - O_h$	$2.33 - 7$[85]
Dy^{3+}	$2.37 + 2.52 - D_{3h}(9)$ $2.422 - 9$ $2.371 - D_{4h}(8)$	$2.38 + 2.50 - D_{3h}(9)$[20]	$2.360 - D_{4h}(8)$	$2.36 - D_{4h}(8)$[74]	$2.355 - D_{4h}(8)$			$2.32 - 7$[85]
Ho^{3+}	$2.38 + 2.49 - D_{3h}(9)$ $2.356 - D_{4h}(8)$	$2.37 + 2.50 - D_{3h}(9)$[20]		$2.34 - D_{4h}(8)$[74]	$2.365 - D_{4h}(8)$			$29 - 7$[85]
Er^{3+}	$2.35 + 2.52 - D_{3h}(9)$ $2.345 - D_{4h}(8)$	$2.36 + 2.48 - D_{3h}(9)$[20]	$2.345 - D_{4h}(8)$ $2.279 - D_{5h}$	$2.33 - D_{4h}(8)$[74]	$2.332 - D_{4h}(8)$			$2.300 - 7$[85]
Tm^{3+}	$2.33 + 2.51 - D_{3h}(9)$	$2.35 + 2.47 - D_{3h}(9)$[20]	$2.332 - D_{4h}(8)$ $2.265 - D_{5h}$		$2.351 - O_h$		$2.216 - O_h$	
Yb^{2+}								
Yb^{3+}	$2.94 + 2.501 - D_{3h}(9)$ $2.326 - D_{4h}(8)$	$2.32 + 2.43 - D_{3h}(9)$[20]			$2.306 - D_{4h}(8)$			$2.270 - 7$[85]
Lu^{3+}	$2.30 + 2.51 - D_{3h}(9)$ $2.318 - D_{4h}(8)$	$2.28 + 2.37 - D_{3h}(9)$[20]	$2.302 - D_{4h}(8)$	$2.29 - D_{4h}(8)$[74]	$2.198 - O_h$ $2.305 - D_{4h}(8)$		$2.181 - O_h$	
Ac^{3+}	$2.498 - 10$	$2.691 - 11$[21]						$2.18 - O_h$[85]
Th^{4+}	$2.448 - 9$	$2.45 - 9$[22-25]	$2.457 - 9$	$2.447 - 9$[75]				$2.404 - D_4$[75]

Ion								
U^{3+}	2.51 + 2.60 – D_{3h}(9)[26]	2.50 + 2.58 – D_{3h}(9)[26]				2.52 – 7		
U^{4+}	2.41 – 9[24,27]							
UO_2^{2+}	2.395 – D_{4h}(8)	1.77 + 2.42 – D_{5h}[27–30]	2.406 – D_{4h}(8)	2.352 – D_{4h}(8)	1.762 + 2.376 – D_{5h}	2.39 – D_{4h}	1.769 + 2.377 – D_{5h}	1.78 + 2.36 – D_{5h}
Np^{3+}	1.760 + 2.444 – D_{5h}	2.48 + 2.56 – D_{3h}[26]						
Np^{4+}	2.491 + 2.571 – D_{3h}	2.40 – 9[28]						
NpO_2^+	2.335 – D_{4h}(8)		2.340 – D_{4h}(8)					
NpO_2^{2+}	1.744 + 2.416 – D_{5h}	1.83 + 2.50 – D_{5h}[28,31,32]			1.741 + 2.368 – D_{5h}			
Pu^{3+}	2.475 + 2.571 – D_{3h}	1.75 + 2.42 – D_{5h}[32]						
Pu^{4+}	2.39 – 9[33]	2.45 + 2.55 – D_{3h}(9)[28,33,34]						
PuO_2^+	2.395 – D_{4h}(8)	2.39 – 9[33]	2.323 – D_{4h}(8)					
PuO_2^{2+}	1.732 + 2.409 – D_{3h}	1.81 + 2.47 – D_{5h}[33]			1.741 + 2.368 – D_{5h}			
Am^{3+}	2.466 + 2.578 – D_{3h}(9)	1.75 + 2.41 – D_{5h}[33,34]			1.744 + 2.379 – D_{5h}			
Cm^{3+}	2.453 + 2.555 – D_{3h}(9)	2.43 + 2.54 – D_{3h}(9)[26,33,34]				2.047 – O_h		
Bk^{3+}		2.40 + 2.53 – D_{3h}(9)[26,34]						
Cf^{3+}	2.423 + 2.550 – D_{3h}(9)	2.38 + 2.50 – D_{3h}(9)[26]	2.37 + 2.49 – D_{3h}(9)[26]					
Ti^{3+}	2.02 – O_h	2.03 – O_h[35]						
TiO^{2+}				1.644 + 2.034 + 2.239	1.646 + 2.035 + 2.24[76]			
Zr^{4+}	2.131 – O_h	2.19 – D_{4h}(8)[36]		2.196 – D_{4h}(8)				
Hf^{4+}	1.994 – O_h	2.16 – D_{4h}(8)[36]		2.179 – D_{4h}(8)				
V^{2+}		2.13 – O_h[34]				2.047 – O_h		
V^{3+}	1.580 + 2.026 + 2.187	1.99 – O_h[37]				2.047 – O_h		
VO^{2+}		1.60 + 2.03 + 2.20[38]		1.602 + 2.028 + 2.180	1.595 + 2.019 + 2.156	1.605 + 2.028 + 2.24[38]	1.577 + 1.970	
VO_2^+	2.083 + 2.333 – C_{4v}	1.62 + 1.99 + 2.22[38]		1.628 + 2.104 + 2.25[38]				
Cr^{2+}	1.965 – O_h	2.08 + (2.3) – C_{4v}[37]		1.969 – O_h				
Cr^{3+}	2.089 – O_h	1.97 – O_h[39]						
Mo^{3+}	2.174 – O_h	2.09 – O_h[40]						
Mn^{2+}		2.17 – O_h[41]		2.174 – O_h	2.161 – O_h	2.16 – O_h[41]	2.087 – 5[41]	
Mn^{3+}	2.088 – O_h							
Fe^{2+}	2.120 – O_h	2.12 – O_h[42,43]		2.125 – O_h	2.113 – O_h	2.101 – O_h[42]	2.048 – 5[42]	

Table 9.1 (*Continued*)

Metal ion	Water		Dimethylsulfoxide		N,N-Dimethylformamide		N,N'-Dimethylpropyleneurea	
	$(s)^{2,3}$	$(aq)^{4,5}$	$(s)^{2,3}$	(dmso)	$(s)^{2,3}$	(dmf)	$(s)^{2,3}$	(dmpu)
Fe^{3+}	$1.995 - O_h$	$2.00 - O_h$[42]	$2.000 - O_h$	$2.004 - O_h$[42]	$1.996 - O_h$		$1.98 - T_d$[a]	$1.99 - 5$[42]
Ru^{2+}	$2.111 - O_h$	$2.11 - O_h$	$2.124 + 2.252 - O_h$		$2.089 - O_h$			
Ru^{3+}	$2.021 - O_h$	$2.03 - O_h$			$2.016 - O_h$			
Co^{2+}	$2.087 - O_h$	$2.09 - O_h$[44–48]	$2.096 - O_h$	$2.096 - O_h$[77]	$2.083 - O_h$	$2.09 - O_h$[83,84]		$1.996 - 5$[86]
Co^{3+}	$1.873 - O_h$		$2.005 - O_h$					
Rh^+			$2.106 + 2.164 - cis\text{-}O_h$					
Rh^{3+}	$2.014 - O_h$	$2.04 - O_h$[49–51]	$2.043 + 2.222 - O_h$		$1.996 - O_h$			
Ir^{3+}	$2.04 - O_h$							
Ni^{2+}	$2.055 - O_h$	$2.055 - O_h$[52]	$2.066 - O_h$	$2.064 - O_h$[77]	$2.048 - O_h$			$2.000 - C_{4v}$[52]
Pd^{2+}	$2.029 - D_{4h}(4)$	$2.01 + 2.8 - D_{4h}(6)$[53,54]	$2.059 + 2.240 - cis\text{-}D_{4h}(4)$	$2.04 + 2.23 - D_{2h}$[78]				
Pt^{2+}	$2.012 - D_{4h}(4)$	$2.02 + 2.8\ D_{4h}(6)$[55,56]	$2.046 + 2.206 - cis\text{-}D_{4h}(4)$	$2.07 + 2.21 - D_{2h}$[78]				
Cu^+		$2.14 - T_d$[a,57]						
Cu^{2+}	$1.980 + 2.322 - C_{4v}(6)$ $1.944 + 2.248 - C_{4v}(5)$	$1.96 + 2.32 - D_{4h}(6)$[58]	$1.979 + 2.276 - C_{3v}(4)$ $1.929 - D_{4h}(4)$	$1.95 + 2.33 - D_{4h}(6)$[58]	$1.975 + 2.2327 - D_{4h}(6)$ $1.96 + 2.28 - C_{4v}(5)$ $1.947 - D_{4h}(4)$			
Ag^+	$2.129 - D_{\infty h}$	$2.32 + 2.54 - 2 + 2$[59] $2.32 - D_{\infty h}$[60]	$2.321 - C_{2v}$	$29 + 2.48 - 2 + 2$[59]	$2.267 - T_d$		$1.939 - D_{4h}$[58]	$29 + 2.54 - 2 + 2$[59]
Zn^{2+}	$2.088 - O_h$ $2.017 - D_{3h}(5)$	$2.09 - O_h$	$2.097 - O_h$ $1.925 - T_d$	$2.100 - O_h$[79]	$2.089 - O_h$			$1.99 - 5$[61]

Ion						
Cd^{2+}	2.266 – O_h	2.27 – O_h[61]	2.273 – O_h	2.29 – O_h[61]	2.269 – O_h	2.26 – O_h[61]
Hg_2^{2+}	2.33 – T_d[62]	2.26 – T_d[62]		2.275 – T_d[62]		
Hg^{2+}	2.32 – 7[63], 2.342 – O_h	2.32 – 7[63], 2.34 – O_h[64]	2.343 – O_h	2.28 – 7[63], 2.339 – O_h[80]		
Al^{3+}	1.883 – O_h	1.89 – O_h	1.892 – O_h		1.878 – O_h	
Ga^{3+}	1.946 – O_h	1.96 – O_h[39]	1.967 – O_h	1.955 – O_h[81]	1.960 – O_h	1.925 – 5[87]
In^{3+}	2.124 – O_h	2.14 – O_h[39]	2.142 – O_h	2.135 – O_h[81]		2.146 – O_h[87]
Tl^{+}	2.73 + 98 – 2 + 2[65]	2.73 + 98 – 2 + 2[65]		2.66 + 98 – 2 + 2[65]		2.72 + 3.27 – 2 + 2[65]
Tl^{3+}	2.23 – O_h[66,67]	2.23 – O_h[66,67]	2.224 – O_h	2.22 – O_h[82]		
Sn^{2+}	2.204 – C_{3v}	2.21 – C_{3v}[68]	2.168 + 2.343 – C_{4v}			
Pb^{2+}	2.53 – 6[69]	2.53 – 6[69]	2.612 – 7, 2.536 – 6	2.52 – 6[69]	2.520 – O_h, 2.56 – 6[69]	2.49 – O_h[69]
Bi^{3+}	2.440 + 2.582 – $D_{3h}(9)$, 2.41 – $D_{4h}(8)$[70]	2.41 – $D_{4h}(8)$[70]	2.437 – $D_{4h}(8)$	2.411 – $D_{4h}(8)$[70], 2.438 – $D_{4h}(8)$	297 – O_h	2.32 – O_h[70]

[a] Denotes an extrapolated value.

eight-coordinated in aqueous solution, even though several structures of six-, seven- and eight-coordinated hydrated calcium ions are reported in the solid state (Table 9.1).[2,3] The only exception from this trend is the hydrated bismuth(III) ion which is eight-coordinated in aqueous solution,[70] while it is nine-coordinated in the only reported crystal structure of the hydrated bismuth(III) ion with a full hydration shell, $[Bi(H_2O)_9](CF_3SO_3)_3$.[90,91]

The coordination number and configuration are in most cases predictable from the electronic structure and the ionic radius of the metal ion. There are however some cases where the structure of the hydrated metal ion is hard to predict. The scandium(III) and yttrium(III) ions in group 3 of the periodic table are both eight-coordinated, but they have different coordination geometries. The hydrated scandium(III) ion has a bicapped trigonal prismatic configuration with the third capping position being empty,[18] while the hydrated yttrium(III) ion has a square antiprismatic configuration,[19] in both aqueous solution and the solid state. Why the smaller scandium(III) ion adopts a coordination figure typical for larger ions has not been explained and is an area for further research. The hydrated lighter lanthanoid(III) ions, through dysprosium(III), and the hydrated actinoid(III) ions are nine-co-ordinated in a tricapped trigonal prismatic configuration in aqueous solution.[20] This basic configuration is maintained for the hydrated heavier lanthanoid(III) ions, starting at holmium(III), but with an increasing water deficit in the capping positions.[20] One of the capping water molecules when water deficit is present is more strongly bound than the other capping water molecules causing the metal ion to be out of the center of gravity.[20] This is also the case for capping water molecules in the hydrated scandium(III) ion.[18] In the solid state, the hydrated lanthanoid(III) ions are either eight- or nine-coordinated. Nine-coordination is dominating for the hydrated lighter lanthanoid(III) and the hydrated actinoid(III) ions in the solid state, while eight-coordination is dominating for the heavier lanthanoid(III) ions.[2,3] This flexibility in the coordination number and geometry of the lanthanoid(III) ions and the importance of symmetry at crystallization are clearly shown by the dmso and dmpu solvates in the solid state and solution. The dmso solvated lanthanoid(III) ions are all eight-coordinated in a square antiprismatic fashion in both dmso solution and the solid state.[74,92] The dmpu solvated lanthanoid(III) ions are all seven-coordinated in dmpu solution with exception for the lutetium(III) ion, which is six-coordinated in an octahedral fashion.[85] On the other hand, the dmpu solvated lanthanoid(III) ions are all six-coordinated with a regular octahedral configuration in the solid state.[85]

The hydrated palladium(II) and platinum(II) ions are four-coordinated in a square-planar fashion in the solid state.[93–96] Combined EXAFS and theoretical simulations have shown that in aqueous solution an additional one or two water molecules are weakly bound in the axial positions forming elongated tetragonally elongated square pyramidal or octahedral complexes.[53–56]

Octahedral complexes of metal ions with d^4 and d^9 electron configurations are expected to display Jahn–Teller distortion with four M–ligand bond distances in a square-planar configuration with significantly shorter M–O bond distances than the remaining two axial ones.[97,98] This is well-known for copper(II) complexes, as seen in numerous cases in both solution and the solid state, and it is expected to also be visible in octahedral chromium(II) and manganese(III) complexes. Recent studies have shown that the hydrated and solvated copper(II) ion is non-centrosymmetric with significantly different Cu–O bond distances to the ligands in the axial positions in solution.[58,99] A survey of crystallographic studies of copper(II) complexes showed that copper(II) complexes crystallizing in non-centrosymmetric space groups or copper not placed in the center of symmetry, and $[Cu(H_2O)_5(O–ligand)]$ complexes all showed different Cu–O bond distances in the axial positions.[58] Whether this is general for Jahn–Teller distorted complexes, as for chromium(II) and manganese(III) complexes, needs further studies. EXAFS studies of complexes with identical Cu–O bond distances to the axial ligands as determined crystallographically have shown that they indeed are different.[58] The reason is that the axial positions are randomly oriented in the solid state, and crystallography will give a mean structure, which has a higher symmetry compared to the individual complexes. On the other hand, EXAFS is a lattice-independent method, which can reveal the actual bond distances.

The mercury(II) ion binds six[63] or seven water molecules[64] in a wide and highly asymmetric bond distance distribution in aqueous solution with mean Hg–O bond distances of 2.33 and 2.32 Å, respectively. $[Hg(OS(CH_3)_2)_6](ClO_4)_2$ crystallizes in a low-symmetry space group ($P–1$) and shows that four of the dmso molecules have a significantly shorter Hg–O bond distance than the remaining two, 2.319 and 2.377 Å, respectively.[100] EXAFS data of six-coordinated homoleptic mercury(II) complexes all show a significant asymmetric bond distance distribution.[101] This strongly indicates a splitting of the Hg–O bond distances as in solid $[Hg(OS(CH_3)_2)_6](ClO_4)_2$, and that all six-coordinated mercury(II) complexes indeed have this bond distance distribution. However, it is only observed in solid compounds crystallizing in low-symmetry space groups as discussed for the hydrated copper(II) ion, *vide ultra*. This bond distance distribution can be described as a so-called pseudo-Jahn–Teller effect.[102] The pseudo-Jahn–Teller effect is a destabilization of, *e.g.* six-coordinated mercury(II) complexes, including small monodentate ligands, attributed to a contribution of the mercury(II) $5d_{z^2}$ atomic orbital to the bonding molecular orbitals by vibronic coupling.[97,98,102] The significant asymmetry of the Hg–O bond distance distribution in both the hydrated mercury(II) ion in aqueous solution and solid $[Hg(H_2O)_6](ClO_4)_2$ as well as in dmso solution and solid $[Hg(Opy)_6](ClO_4)_2$ strongly supports that these mercury(II) complexes have a pseudo-Jahn–Teller distorted octahedral configuration.[63]

The hydrated and dmso solvated tin(II) and lead(II) complexes are strongly affected by electrons in an anti-bonding orbital causing a significant gap in their coordination spheres. The hydrated tin(II) ion is three-coordinated in both aqueous solution and in solid $[Sn(H_2O)_3](ClO_4)_2$, with a mean

Sn–O bond distance of 2.20 Å, and a huge gap in the coordination sphere with all O–Sn–O bond angles being *ca.* 77°.[68] This means that less than $\frac{1}{4}$ of the coordination sphere is occupied by binding water molecules. The dmso solvated tin(II) ion is five-coordinated in a square-pyramidal fashion with Sn–O bond distances of 2.20 and 2.35, and 2.14 and 2.45 Å to the axially and equatorially bound dmso molecules, in dmso solution and the solid state, respectively.[68] The dmso solvated lead(II) ions have diffuse coordination of *ca.* six water and dmso molecules at a mean Pb–O bond distance of *ca.* 2.53 Å in aqueous and dmso solutions, most likely with a significant gap in the coordination spheres.[69]

Dmpu is space-demanding upon coordination to metal ions causing a lower coordination number and lower symmetry than in the corresponding metal ion hydrates, and dmso and dmf solvates (Table 9.1). This may cause a significantly higher reactivity, *e.g.* a much stronger ability to form metal complexes with, *e.g.* halide ions.[41,87,103,104]

9.2.2 Nitrogen Donor Solvents

The structures of hydrated and solvated metal ions in the nitrogen donor solvents acetonitrile and liquid ammonia in the solid state and solution are summarized in Table 9.2. The structures of hydrated and acetonitrile solvated metal ions are in most cases similar with the same coordination number and figure, while ammonia solvated metal ions often have lower coordination numbers (Table 9.2). The reason for the lower coordination number in the ammine solvates is the bonding characteristic as the ammonia molecules have no spatial demands except the size of the nitrogen atom. Acetonitrile is a weak electron-pair donor, slightly weaker than the oxygen donor solvents in this survey, while ammonia is a very strong electron-pair donor.[1] It is therefore expected that ammonia will solvate the most soft metal ions much more strongly than acetonitrile and the oxygen donor solvents. Strong covalent bonding seems to favor a lower coordination number as seen for the copper(I) and silver(I) ions, which are three-coordinated in a triangular fashion in liquid ammonia, but four-coordinated in a tetrahedral fashion in acetonitrile, and gold(I) has a linear configuration in liquid ammonia, but it has a tetrahedral structure in acetonitrile (Table 9.2). The ammine solvated mercury(II) ion is four-coordinated in a tetrahedral fashion in liquid ammonia and the solid state (Table 9.2).

The differences in M–N and M–O bond distances for ammonia and acetonitrile solvated and hydrated metal ions are included in Table 9.2. It is clearly seen that for metal ions forming mainly electrostatic interactions the difference in M–N and M–O bond distances is *ca.* 0.07 and 0.12 Å for acetonitrile and ammonia. However, for typical soft metal ions as the platinum group and coinage metal ions, there is almost no difference in M–N and M–O bond distances (Table 9.2). This is due to strong covalent bonding which may include pi-back-bonding.

Table 9.2 Summary of mean M–N bond distances in acetonitrile and ammonia solvated metal ions in the solid state and solution. The same abbreviations as in Table 9.1 are used. The information of the hydrated metal ions is given for comparison. The difference in M–N and M–O bond distances is given to show the ability of metal ions to bind strongly to nitrogen and oxygen donor solvents. For compounds without a well-defined coordination geometry, the coordination number is given by a figure.

Metal ion	Water (s)	Water (aq)	Acetonitrile (s)[2,3]	Acetonitrile (AN)	Liquid ammonia (s)[2,3]	Liquid ammonia (NH$_3$ (l))	Differences NH$_3$(s) – aq(s)	Differences NH$_3$(s) – AN
Li$^+$	1.941 – T_d	1.95 – T_d	2.024 – T_d		2.078 – T_d		0.137	0.054
Na$^+$	2.417 – O_h 2.366 – 5	2.43 – O_h	2.452 – O_h		2.497 – 5		0.131	
K$^+$					2.758 – O_h			
Rb$^+$	2.98 – $D_{4h}(8)$[7-9]			3.06[105]	3.179 – D_4			
Mg^{2+}	2.066 – O_h	2.07 – O_h	2.152 – O_h		2.177 – O_h		0.111	0.025
Ca^{2+}	2.476 – D_{4h} 2.401 – 7 2.324 – O_h	2.48 – D_{4h}	2.514 – D_{4h}		2.619 – D_{4h} 2.553 – 7 2.505 – O_h		0.143 0.152 0.181	0.105
Sr^{2+}	2.613 – D_{4h}	2.61 – D_{4h}	2.682 – D_{4h}	2.665 – D_{4h}[105,106]	2.753 – D_{4h}		0.140	0.071
Ba^{2+}	2.80 – D_{4h}	2.81 – D_{4h}			2.962 – 9		0.130	
Y^{3+}	2.353 – D_{4h}	2.367 – D_{4h}			2.487 – 7 2.474 – O_h			
La^{3+}	2.53 + 2.61 – $D_{3h}(9)$ 2.495 – D_{4h}	2.53 + 2.60 – $D_{3h}(9)$	2.636 – 9	2.58 – 9[107]	2.706 – 9		0.150	0.070
Ce^{3+}	2.49 + 2.60 – $D_{3h}(9)$ 2.485 – D_{4h}	2.49 + 2.62 – $D_{3h}(9)$	2.611 – 9		2.678 – O_h			
Pr^{3+}	2.48 + 2.57 – $D_{3h}(9)$	2.48 + 2.58 – $D_{3h}(9)$		2.581 – 9				
Nd^{3+}	2.46 + 2.57 – $D_{3h}(9)$	2.47 + 2.60 – $D_{3h}(9)$	2.573 – 9		2.654 – O_h			
Sm^{3+}	2.44 + 2.55 – $D_{3h}(9)$ 2.430 – D_{4h}	2.43 + 2.53 – $D_{3h}(9)$	2.534 – 9					
Eu^{2+}	2.548 – D_4			2.640 – 7[15]	2.752 – D_4			

Table 9.2 (Continued)

Metal ion	Water (s)	Water (aq)	Acetonitrile (s)[2,3]	Acetonitrile (AN)	Liquid ammonia (s)[2,3] (NH₃ (l))	NH₃(s) − aq(s)	NH₃(s) − AN
Tb^{3+}	$2.39 + 2.51 - D_{3h}(9)$	$2.40 + 2.53 - D_{3h}(9)$	$2.503 - 9$				
	$2.381 - D_{4h}$						
Dy^{3+}	$2.37 + 2.52 - D_{3h}(9)$	$2.38 + 2.50 - D_{3h}(9)$	$2.422 - 9$	$2.42 - 9$[107]			
Tm^{3+}	$2.339 - D_{4h}$	$2.35 + 2.47 - D_{3h}(9)$	$2.411 - D_{4h}$				
Yb^{2+}			$2.648 - D_{4h}$				
U^{3+}	$2.509 + 2.596 - D_{3h}$	$2.50 + 2.58 - D_{3h}$	$2.614 - 9$				
Pu^{3+}	$2.475 + 2.571 - D_{3h}$	$2.45 + 2.55 - D_{3h}$					
Zr^{4+}		$2.19 - D_4$	$2.572 - 9$				
V^{2+}	$2.131 - O_h$	$2.13 - O_h$	$2.407 - D_4$				
Cr^{2+}	$2.083 + 2.333$	$2.08 + (2.3)$	$2.113 - O_h$	$2.224 - O_h$	$2.225 - O_h$	0.094	0.112
			$2.079-2.423 - O_h$				
			$2.066 - D_{4h}$				
Cr^{3+}	$1.965 - O_h$	$1.97 - O_h$	$2.000 - O_h$		$2.073 - O_h$	0.108	0.073
Mn^{2+}	$2.174 - O_h$	$2.17 - O_h$	$2.228 - O_h$	$2.193 - O_h$[41]	$2.285 - O_h$	0.111	0.057
Tc^{2+}			$2.062 - O_h$				
Re^{2+}			$2.055 - 7$				
Fe^{2+}	$2.120 - O_h$	$2.12 - O_h$	$2.155 - O_h$		$2.218 - O_h$	0.098	0.063
Fe^{3+}	$1.995 - O_h$	$2.000 - O_h$			$2.173 - O_h$	0.178	
Ru^{2+}	$2.111 - O_h$	$2.11 - O_h$	$2.025 - O_h$		$2.14 - O_h$	0.029	0.115
Ru^{3+}	$2.021 - O_h$	$2.03 - O_h$			$2.099 - O_h$	0.078	
Os^{2+}					$2.111 - O_h$		
Co^{2+}	$2.087 - O_h$	$2.09 - O_h$	$2.112 - O_h$		$2.176 - O_h$	0.092	0.067
Rh^{3+}	$2.014 - O_h$	$2.04 - O_h$	$1.987 - O_h$		$2.064 - O_h$	0.050	0.077
Ir^{3+}	$2.04 - O_h$				$2.087 - O_h$	0.045	

Ion								
Ni²⁺	$2.055 - O_h$	$2.055 - O_h$	$2.067 - O_h$	$2.062 - O_h$[52]	$2.135 - O_h$		0.080	0.068
Pd²⁺	$2.029 - D_{4h}$		$1.967 - D_{4h}$		$2.038 - D_{4h}$		0.009	0.071
Pt²⁺	$2.012 - D_{4h}$		$1.972 - D_{4h}$	$2.000 - D_{4h}$[78]	$2.047 - D_{4h}$		0.035	0.075
Cu⁺	$2.14 - T_d^{a}$		$1.995 - T_d$; $1.938 - D_{3h}$; $1.887 - D_{\infty h}$	$1.99 - T_d$[57]		$2.004 - D_{3h}(3)$[111]		
Cu²⁺	$1.980 + 2.322 - C_{4v}(6)$; $1.944 + 2.248 - C_{4v}$	$1.96 + 2.32 - D_{4h}(6)$	$2.00 + 2.28 - D_{4h}(6)$; $1.997 + 2.378 - C_{4v}$	$1.903 + 2.382 - D_{4h}(6)$[108]		$2.065 + 2.29 - D_{4h}(6)$[112]; $2.00 + 2.19 - C_{4v}$[113]		
Ag⁺	2.32		$2.275 - T_d$; $2.227 - D_{3h}$; $2.117 - D_{\infty h}$	$2.25 - T_d$[109]	$2.348 - T_d$; $2.28 - D_{3h}$; $2.125 - D_{\infty h}$	$2.26 - D_{3h}$[114]	-0.004	0.073; 0.038; 0.008
Au⁺	2.129		$1.960 - D_{\infty h}$	$2.19 - T_d$[110]	$2.038 - D_{\infty h}$	$2.025 - D_{\infty h}$[114]	0.078	
Au³⁺	$2.097 - D_{4h}$				$2.024 - D_{4h}$		0.111	0.066
Zn²⁺	$2.088 - O_h$	$2.09 - O_h$	$2.133 - O_h$	$2.12 - O_h$[108]	$2.199 - O_h$	$2.177 - 5$[112]		0.038
Cd²⁺	$2.266 - O_h$	$2.27 - O_h$	$1.988 - T_d$		$2.026 - T_d$; $2.367 - O_h$	$2.350 - O_h$[112]	0.101	
Hg²⁺	$2.342 - O_h$	$2.34 - O_h$						
Al³⁺	$1.883 - O_h$	$1.89 - O_h$	$1.946 - O_h$		$2.246 - T_d$; $2.048 - O_h$	$2.225 - T_d$[115]	0.165	0.102
Ga³⁺	$1.946 - O_h$	$1.96 - O_h$			$2.081 - O_h$		0.135	

[a] Denotes an extrapolated value.

Problem 1

Which hydrated metal ion(s) has (have) the lowest and highest coordination numbers, respectively.

Problem 2

Which hydrated metal ions are expected to have Jahn–Teller distorted coordination?

Problem 3

Which solvated metal ions have a lower coordination number in solution than possible from the steric point of view?

Answer to Problem 1

Tin(II) with a coordination number of three and actinium(III) with a coordination number of eleven.

Answer to Problem 2

Copper(II), chromium(II) and manganese(III).

Answer to Problem 3

Ammonia solvated copper(I) – three, silver(I) – three, gold(I) – two, zinc(II) – four and mercury(II) – four.

Acknowledgements

The support from the Swedish Research Council to my projects concerning coordination chemistry of hydrated and solvated metal ions and complexes is gratefully acknowledged.

References

1. M. Sandström, I. Persson and P. Persson, *Acta Chem. Scand.*, 1990, **44**, 653.
2. F. H. Allen, *Acta Crystallogr., Sect. B: Struct. Sci.*, 2002, **58**, 380; Cambridge Structure Database (Conquest 2022.1.0).
3. *Inorganic Crystal Structure Database* 1.4.6, release 2021-2, FIZ Karlsruhe, Eggenstein-Leopoldshafen, Germany, 2022.
4. H. Ohtaki and T. Radnai, *Chem. Rev.*, 1993, **99**, 157, and references therein.
5. G. Johansson, *Adv. Inorg. Chem.*, 1992, **39**, 159, and references therein.
6. P. R. Smirnov and V. N. Trostin, *Russ. J. Gen. Chem.*, 2006, **76**, 175, and references therein.

7. J. Mähler and I. Persson, *Inorg. Chem.*, 2012, **51**, 425, and references therein.
8. P. R. Smirnov, *Russ. J. Inorg. Chem.*, 2020, **90**, 1693, and references therein.
9. P. D'Angelo and I. Persson, *Inorg. Chem.*, 2004, **43**, 3543.
10. T. Yamaguchi, H. Ohtaki, E. Spohr, G. Pálinkás, K. Heinzinger and M. M. Probst, *Z. Naturforsch., A: Phys. Sci.*, 1986, **41**, 1175.
11. F. Jalilehvand, D. Spångberg, P. Lindqvist-Reis, K. Hermansson, I. Persson and M. Sandström, *J. Am. Chem. Soc.*, 2001, **123**, 431.
12. P. D'Angelo, P.-E. Petit and N. V. Pavel, *J. Phys. Chem. B*, 2004, **108**, 11857.
13. I. Persson, M. Sandström, H. Yokoyama and M. Chaudhry, *Z. Naturforsch., A: Phys. Sci.*, 1995, **50**, 21.
14. P. D'Angelo, V. Migliorati, F. Sessa, G. Mancini and I. Persson, *J. Phys. Chem. B*, 2016, **123**, 4114.
15. G. Moreau, R. Scopelliti, L. Helm, J. Purans and A. E. Merbach, *J. Phys. Chem. A*, 2002, **106**, 9612.
16. V. Migliorati, A. Caruso and P. D'Angelo, *Inorg. Chem.*, 2019, **58**, 14551.
17. P. D'Angelo, N. V. Pavel, D. Roccatano and H.-F. Nolting, *Phys. Rev. B: Condens. Matter Mater. Phys.*, 1996, **54**, 12129.
18. P. Lindqvist-Reis, I. Persson and M. Sandström, *Dalton Trans.*, 2006, 3868.
19. P. Lindqvist-Reis, K. Lamble, S. Pattanaik, M. Sandström and I. Persson, *J. Phys. Chem. B*, 2000, **104**, 402.
20. I. Persson, P. D'Angelo, S. De Panfilis, M. Sandström and L. Eriksson, *Chem. – Eur. J.*, 2008, **14**, 3056.
21. M. G. Ferrier, B. W. Stein, E. R. Batista, J. M. Berg, E. R. Birnbaum, J. W. Engle, K. D. John, S. A. Kozimor, J. S. Lezama Pacheco and L. N. Redman, *ACS Cent. Sci.*, 2017, **3**, 176.
22. N. Torapava, I. Persson, L. Eriksson and D. Lundberg, *Inorg. Chem.*, 2009, **48**, 11712.
23. R. Spezia, C. Beuchat, R. Vuilleumier, P. D'Angelo and L. Gagliardi, *J. Phys. Chem. B*, 2012, **116**, 6465.
24. H. Moll, M. A. Denecke, F. Jalilehvand, M. Sandström and I. Grenthe, *Inorg. Chem.*, 1999, **38**, 1795.
25. C. Hennig, K. Schmeide, V. Brendler, H. Moll, S. Tsushima and A. C. Scheinost, *Inorg. Chem.*, 2007, **46**, 5882.
26. P. D'Angelo, F. Martelli, R. Spezia, A. Filipponi and M. A. Denecke, *Inorg. Chem.*, 2013, **52**, 10318.
27. A. Ikeda-Ohno, C. Hennig, S. Tsushima, A. C. Scheinost, G. Bernhard and T. Yaita, *Inorg. Chem.*, 2009, **48**, 7201.
28. P. G. Allen, J. J. Bucher, D. K. Shuh, N. M. Edelstein and T. Reich, *Inorg. Chem.*, 1997, **36**, 4676.
29. H. A. Thompson, G. E. Brown Jr. and G. A. Parks, *Am. Mineral.*, 1997, **82**, 483.

30. J. Neuefeind, L. Soderholm and S. Skanthakumar, *J. Phys. Chem. A*, 2004, **108**, 2733.

31. J. M. Combes, C. J. Chisholm-Brause, G. E. Brown Jr., G. A. Parks, S. D. Conradson, I. R. Triay, P. E. Hobart and A. Meijer, *Environ. Sci. Technol.*, 1992, **36**, 376.

32. S. Skanthakumar, M. R. Antonio and L. A. Soderholm, *Inorg. Chem.*, 2008, **47**, 4591.

33. S. D. Conradson, K. D. Abney, B. D. Begg, E. D. Brady, D. L. Clark, C. den Auwer, M. Ding, P. K. Dorhout, F. J. Espinosa-Faller, P. L. Gordon, R. G. Haire, N. J. Hess, R. F. Hess, D. W. Keogh, G. H Lander, A. J. Lupinetti, L. A. Morales, M. P. Neu, P. D. Palmer, P. Paviet-Hartmann, S. D. Reilly, W. H. Runde, C. D. Tait, D. K. Veirs and F. Wastin, *Inorg. Chem.*, 2004, **43**, 116.

34. P. Lindqvist-Reis, C. Apostolidis, J. Rebizant, A. Morgenstern, R. Klenze, O. Walter, T. Fanghänel and R. G. Haire, *Angew. Chem., Int. Ed.*, 2007, **46**, 919.

35. T. Miyanaga, I. Watanabe and S. Ikeda, *Bull. Chem. Soc. Jpn.*, 1990, **63**, 3282.

36. C. Hagfeldt, V. Kessler and I. Persson, *Dalton Trans.*, 2004, 2142.

37. T. Miyanaga, I. Watanabe and S. Ikeda, *Chem. Lett.*, 1988, **17**, 1073.

38. J. Krakowiak, D. Lundberg and I. Persson, *Inorg. Chem.*, 2012, **51**, 9598.

39. P. Lindqvist-Reis, S. Díaz-Moreno, A. Munoz-Páez, S. Pattanaik, I. Persson and M. Sandström, *Inorg. Chem.*, 1998, **37**, 6675.

40. S. P. Cramer, P. K. Eidem, M. T. Pafett, J. R. Winkler, Z. Dori and H. B. Gray, *J. Am. Chem. Soc.*, 1983, **105**, 799.

41. H. Konieczna, D. Lundberg and I. Persson, *Polyhedron*, 2021, **195**, 114961.

42. D. Lundberg, A.-S. Ullström, P. D'Angelo and I. Persson, *Inorg. Chim. Acta*, 2007, **360**, 1809.

43. E. Kálmán, T. Radnai, G. Pálinkás, F. Hajdus and A. Vertes, *Electrochim. Acta*, 1988, **39**, 1223.

44. P. D'Angelo, V. Barone, G. Chillemi, N. Sanna, W. Meyer-Klauche and N. V. Pavel, *J. Am. Chem. Soc.*, 2002, **124**, 1958.

45. R. Spezia, M. Duvail, P. Vitorge, T. Cartailler, J. Tortajada, G. Chillemi, P. D'Angelo and M.-P. Gaigeot, *J. Phys. Chem. A*, 2006, **110**, 13081.

46. C. Chen and K. Hayes, *Geochim. Cosmochim. Acta*, 1999, **63**, 3205.

47. H. Ohtaki, T. Yamaguchi and M. Maeda, *Bull. Chem. Soc. Jpn.*, 1976, **49**, 701.

48. D. Z. Caralampio, B. Reeves, M. R. Beccia, J. M. Martínez, R. R. Pappalardo, C. den Auwer and E. Sánchez Marcos, *Mol. Phys.*, 2019, **117**, 3320.

49. R. Caminiti, D. Atzei, P. Cucca, A. Anedda and G. Bongiovanni, *J. Phys. Chem.*, 1986, **90**, 238.

50. R. Caminiti and P. Cucca, *Phys. Chem. Lett.*, 1984, **108**, 51.

51. M. C. Read and M. Sandström, *Acta Chem. Scand.*, 1992, **46**, 1177.

52. O. Kristiansson, I. Persson, D. Bobicz and D. Xu, *Inorg. Chim. Acta*, 2003, **344**, 15.

53. J. Purans, B. Fourest, C. Cannes, V. Sladkov, F. David, L. Venault and M. Lecomte, *J. Chem. Phys. B*, 2005, **109**, 11074.
54. T. S. Hofer, B. R. Randolf, A. Ali Shah, B. M. Rode and I. Persson, *Chem. Phys. Lett.*, 2007, **445**, 193.
55. F. Jalilehvand and L. J. Laffin, *Inorg. Chem.*, 2008, **47**, 3248.
56. T. S. Hofer, B. R. Randolf, B. M. Rode and I. Persson, *Dalton Trans.*, 2009, 1512.
57. I. Persson, J. E. Penner-Hahn and K. O. Hodgson, *Inorg. Chem.*, 1993, **32**, 2497, and references therein.
58. I. Persson, D. Lundberg, É. G. Bajnoczi, K. Klementiev, J. Just and K. G. V. Sigfridsson Clauss, *Inorg. Chem.*, 2020, **59**, 9538.
59. I. Persson and K. B. Nilsson, *Inorg. Chem.*, 2006, **45**, 7428.
60. M. Busato, A. Melchior, V. Migliorati, A. Colella, I. Persson, G. Mancini, D. Veclaniz and P. D'Angelo, *Inorg. Chem.*, 2020, **59**, 10291.
61. P. D'Angelo, G. Chillemi, V. Barone, G. Mancini, N. Sanna and I. Persson, *J. Phys. Chem. B*, 2005, **109**, 9178.
62. J. Rosdahl, I. Persson, L. Kloo and K. Ståhl, *Inorg. Chim. Acta*, 2004, **357**, 2624.
63. G. Chillemi, G. Mancini, N. Sanna, V. Barone, S. Della Longa, M. Benfatto, N. V. Pavel and P. D'Angelo, *J. Am. Chem. Soc.*, 2007, **129**, 5430.
64. I. Persson, L. Eriksson, P. Lindqvist-Reis, P. Persson and M. Sandström, *Chem. – Eur. J.*, 2008, **14**, 6687.
65. I. Persson, F. Jalilehvand and M. Sandström, *Inorg. Chem.*, 2002, **41**, 192.
66. J. Glaser and G. Johansson, *Acta Chem. Scand., Ser. A*, 1982, **36**, 125.
67. J. Blixt, J. Glaser, J. Mink, I. Persson, P. Persson and M. Sandström, *J. Am. Chem. Soc.*, 1995, **117**, 5089.
68. I. Persson, P. D'Angelo and D. Lundberg, *Chem. – Eur. J.*, 2016, **22**, 18583.
69. I. Persson, K. Lyczko, D. Lundberg, L. Eriksson and A. Płaczek, *Inorg. Chem.*, 2011, **50**, 1058.
70. J. Näslund, I. Persson and M. Sandström, *Inorg. Chem.*, 2000, **39**, 4012.
71. A.-S. Ullström, D. Warminska and I. Persson, *J. Coord. Chem.* 2005, **58**, 611.
72. M. Y. Skripkin, P. Lindqvist-Reis, A. Abbasi, J. Mink, I. Persson and M. Sandström, *Dalton Trans.*, 2004, 4038.
73. P. Lindqvist-Reis, J. Näslund, I. Persson and M. Sandström, *J. Chem. Soc., Dalton Trans.*, 2000, 2703.
74. I. Persson, E. Damian-Risberg, P. D'Angelo, S. De Panfilis, M. Sandström and A. Abbasi, *Inorg. Chem.*, 2007, **46**, 7742.
75. N Torapava, D. Lundberg and I. Persson, *Eur. J. Inorg. Chem.*, 2011, 5273.
76. I. Persson and D. Lundberg, *J. Solution Chem.*, 2017, **46**, 476.
77. D. Lundberg, D. Warmińska, A. Fuchs and I. Persson, *Phys. Chem. Chem. Phys.*, 2018, **20**, 14525, and references therein.

78. B. Hellquist, L. Bengtsson, B. Holmberg, B. Hedman, I. Persson and L. I. Elding, *Acta Chem. Scand.*, 1991, **45**, 449.
79. I. Persson, *Acta Chem. Scand., Ser. A*, 1982, **36**, 7.
80. M. Sandström, I. Persson and S. Ahrland, *Acta Chem. Scand., Ser. A*, 1978, **32**, 607.
81. A. Molla-Abbassi, M. Skripkin, M. Kritikos, I. Persson, J. Mink and M. Sandström, *Dalton Trans.*, 2003, 1746.
82. G. Ma, A. Molla-Abbassi, M. Kritikos, A. Ilyukhin, F. Jalilehvand, V. G. Kessler, M. Skripkin, M. Sandström, J. Glaser, J. Näslund and I. Persson, *Inorg. Chem.*, 2001, **40**, 6432.
83. H. Yokoyama, S. Suzuki, M. Goto, K. Shinozaki, Y. Abe and S.-I. Ishiguro, *Z. Naturforsch., A: Phys. Sci.*, 1995, **50**, 301.
84. O. Kazuhiko, T. Kazuyuki, Y. Udagawa and S.-I. Ishiguro, *Bull. Chem. Soc. Jpn.*, 1991, **64**, 1528.
85. D. Lundberg, I. Persson, L. Eriksson, P. D'Angelo and S. De Panfilis, *Inorg. Chem.*, 2010, **49**, 4420.
86. D. Lundberg, P. Lindqvist-Reis, K. Lyczko, L. Eriksson and I. Persson, manuscript in preparation.
87. Ö. Topel, I. Persson, D. Lundberg and A.-S. Ullström, *Inorg. Chim. Acta*, 2010, **363**, 988.
88. R. D. Shannon, *Acta Crystallogr., Sect. A: Cryst. Phys., Diffr., Theor. Gen. Crystallogr.*, 1976, **32**, 751.
89. J. K. Beattie, S. P. Best, B. W. Skelton and A. H. White, *J. Chem. Soc., Dalton Trans.*, 1981, 2105.
90. W. Frank, G. J. Reiss and J. Schneider, *Angew. Chem., Int. Ed.*, 1995, **34**, 2416.
91. J. Näslund, I. Persson and M. Sandström, *Inorg. Chem.*, 2000, **39**, 4012.
92. A. Abbasi, E. Damian-Risberg, L. Eriksson, J. Mink, I. Persson, M. Sandström, Y. V. Sidorov, M. Y. Skripkin and A.-S. Ullström, *Inorg. Chem.*, 2007, **46**, 7731.
93. Z. Mazej and E. Goreshnik, *Z. Kristallogr. – Cryst. Mater.*, 2017, **232**, 339.
94. M. N. Sokolov, I. V. Kalinina, E. V. Peresypkina, N. K. Moroz, D. Y. Naumov and V. P. Fedin, *Eur. J. Inorg. Chem.*, 2013, 1772.
95. F. R. Fortea-Perez, D. Armentano, M. Julve, G. De Munno and S.-E. Stiriba, *J. Coord. Chem.*, 2014, **67**, 4003.
96. K. Seppelt, *Z. Anorg. Allg. Chem.*, 2010, **636**, 2391.
97. I. B. Bersuker, *The Jahn–Teller Effect and Vibronic Interactions in Modern Chemistry*, Plenum Press, New York, 1984, ch. 1–5, ISBN: 978-1461296546.
98. I. B. Bersuker and V. Z. Polinger, *Vibronic Interactions in Molecules and Crystals*, Springer, Berlin, 1989, ch. 3 and 4, ISBN: 978-3-642-83479-0.
99. P. Frank, M. Benfatto and M. Qayyum, *J. Chem. Phys.*, 2018, **148**, 204302.
100. M. Sandström and I. Persson, *Acta Chem. Scand., Ser. A*, 1978, **32**, 95.
101. I. Persson, L. Eriksson, P. Lindqvist-Reis, P. Persson and M. Sandström, *Chem. – Eur. J.*, 2008, **14**, 6687.

102. D. Strömberg, M. Sandström and U. Wahlgren, *Chem. Phys. Lett.*, 1990, **172**, 49.

103. D. Bobicz, O. Kristiansson and I. Persson, *J. Chem. Soc., Dalton Trans.*, 2002, 4201.

104. D. Lundberg, A.-S. Ullström, P. D'Angelo, D. Warminska and I. Persson, *Inorg. Chim. Acta*, 2007, **360**, 2744.

105. P. D'Angelo and N. V. Pavel, *J. Chem. Phys.*, 1999, **111**, 5107.

106. P. D'Angelo and N. V. Pavel, *J. Synchrotron Radiat.*, 1999, **6**, 281.

107. V. Migliorati, A. Filipponi, F. Sessa, A. Lapi, A. Serva and P. D'Angelo, *Phys. Chem. Chem. Phys.*, 2019, **12**, 13058.

108. V. Migliorati and P. D'Angelo, *J. Phys. Chem. B*, 2015, **119**, 4061.

109. K. B. Nilsson and I. Persson, *Acta Chem. Scand., Ser. A*, 1987, **41**, 139.

110. S. Ahrland, K. Nilsson, I. Persson, A. Yuchi and J. E. Penner-Hahn, *Inorg. Chem.*, 1989, **28**, 1833.

111. K. B. Nilsson and I. Persson, *Dalton Trans.*, 2004, 1312.

112. K. B. Nilsson, L. Eriksson, V. G. Kessler and I. Persson, *J. Mol. Liq.*, 2007, **131–132**, 113.

113. M. Valli, S. Matsuo, H. Wakita, T. Yamaguchi and M. Nomura, *Inorg. Chem.*, 1996, **35**, 5642.

114. K. B. Nilsson, V. G. Kessler and I. Persson, *Inorg. Chem.*, 2006, **45**, 6912.

115. K. B. Nilsson, M. Maliarik, I. Persson, A. Fischer, A.-S. Ullström, L. Eriksson and M. Sandström, *Inorg. Chem.*, 2008, **47**, 1953.

CHAPTER 10

Ion Solvation in Non-aqueous and Mixed Solvents

YASUHIRO UMEBAYASHI*[a] AND INGMAR PERSSON[b]

[a] Graduate School of Science and Technology, Niigata University,
8050 Ikarashi, 2-no-cho, Nishi-ku, Niigata City, 950-2181, Japan;
[b] Department of Molecular Sciences, Swedish University of Agricultural
Sciences, P.O. Box 7015, SE-750 07 Uppsala, Sweden
*Email: yumescc@chem.sc.niigata-u.ac.jp

10.1 Introduction

In non-aqueous solvents, which generally have a larger molecular volume than water, the bulkiness of the solvent molecules, or the steric factor, is a major factor controlling the solvation structure and reactivity of metal ions, in addition to the coordination ability of the solvent as described in previous sections. Knowledge of the structure of the solvated metal ion, *i.e.*, structural parameters such as bond distances and coordination numbers, is vital in the understanding of its thermodynamic properties and reactivity. This is based on several empirical rules. In this section, the bonding characteristics and the importance of the steric factors of solvent molecules for the solvation structure and reactivity of metal ions, empirical rules and the structure of metal ion solvates in some solvents space-demanding upon coordination will be presented.

10.2 Bond Length Variation Rule

Gutmann summarized the first three bond length variation rules, albeit qualitatively, based on his discussion of donor and acceptor properties.[1] The

Coordination Chemistry Fundamentals Series No. 2
Metal Ions and Complexes in Solution
Edited by Toshio Yamaguchi and Ingmar Persson
© Japan Society of Coordination Chemistry 2024
Published by the Royal Society of Chemistry, www.rsc.org

first bond length variation rule is related to donor–acceptor interactions between molecules to conformational changes in the molecules caused by such interactions. This can be exemplified with the variation in the S=O bond distance in free, oxygen and sulfur bound dimethylsulfoxide, $(CH_3)_2SO$. The S=O bond distance in uncoordinated dmso is *ca.* 1.50 Å.[2–6] In complexes with platina group metal ions, dmso binds through its sulfur atom and the S=O bond becomes shorter, *ca.* 1.48 Å,[6] due to partial electron transfer from the free electron-pairs to the S=O bond. On the other hand, in metal complexes with dmso binding through its oxygen atom, an electron transfer to the M–O bond takes place causing a lengthening of the S=O bond to *ca.* 1.53 Å.[6]

The bond length change rule, which is derived by extending the donor–acceptor concept, is based on changes in the electronic structure and includes issues related to geometric configuration. The mean M–ligand bond distance is strongly related to the coordination number and figure as shown in the compiled ionic radii of metal ions by Shannon.[7] This pattern can also be seen Tables 9.1 and 9.2. In order to estimate a metal–solvent bond distance, a good starting point is to take the ionic radius of the metal ion and add the atomic radius of the donor atom of the solvent. The atomic radius of oxygen in water has been estimated to be 1.34 Å.[8] As the bonding character of oxygen in different oxygen donor solvents is reasonably similar, the atomic radius of water oxygen can also be applied for most other oxygen donor solvents, even though small deviations of ±0.01 Å are observed. Amide oxygens might be slightly smaller and sulfoxide oxygens are slightly larger than water oxygen (Table 9.1). For nitrogen donor solvents, it is more difficult to make accurate estimations as the atomic radii of, *e.g.*, acetonitrile, ammonia/amines and pyridines differ significantly, and their bonding characteristics are different. The acetonitrile nitrogen is a weak electron-pair donor and has a low electron density on nitrogen due to the C≡N triple bond, while ammonia, amine and pyridine nitrogens are strong electron-pair donors.[9] Solvents with a strong electron-pair donor ability, as phosphorus and sulfur donor solvents and liquid ammonia and amines, bind strongly to metal ions able to form strong covalent interactions (soft Lewis acids). Due to the significant electron overlap in such metal ion solvates, the ionic and atomic radii of the metal ion and donor atom cannot be applied, and shorter metal ion–solvate bond distances are observed (see Table 9.2). For coordination of acetonitrile and ammonia to metal ions forming mainly electrostatic interactions, the atomic radii of acetonitrile and ammonia nitrogen atoms are estimated to be, *ca.* 1.41 and 1.47 Å, respectively, based on the values in Table 9.2. Metal ion solvates with a large degree of covalency display often lower coordination numbers than sterically possible, *e.g.* copper(II) is three-coordinate and mercury(II) is four-coordinate in liquid ammonia (Table 9.2).

Estimation of the coordination number from the mean M–O bond distance is of special value in LAXS and EXAFS studies of solvated metal ions in solution as M–O bond distances can be much more accurately determined than the number of distances. For metal ion solvates with nitrogen, sulfur and phosphorus donor solvents, comparisons with solid-state structures are recommended for estimation of the coordination number from studies in solution.

10.3   Ligand Cone Angle

Tolman introduced the concept of ligand cone angle with respect to steric factors.[10] He separated the steric factor from the electronic factor for phosphine ligands and defined the ligand cone angle as the angle at the top of the cone that is circumscribed to the substituent attached to the phosphorus atom (donor atom) with the metal at the top. The ligand cone angle is well known for tertiary phosphine ligands, but can be determined for other ligand systems as well. Tolman's concept of the ligand cone angle introduces a structural parameter, the cone angle, which is useful in understanding solvation from a structural point of view.

However, for N,N'-dimethylpropyleneurea (dmpu) it has been observed that the M–O–C bond angle is flexible. In solvates without any steric restrictions, the M–O–C bond angle is *ca.* 125°, while in the hexakis(dmpu)yttrium(III) solvate the Y–O–C bond angle is *ca.* 169° in order to accommodate as many solvate molecules as possible.[11] For dmpu solvated metal ions and complexes, a strict correlation between the M–O–C bond angle and crowdedness of the dmpu ligands in the complex is observed.

10.4   Hammett Rule and Steric Factors

According to the Bell–Evans–Polanyi principle,[12,13] there is a linear relationship between the activation energy, E_a, and the heat of reaction, ΔH, for many similar systems:

$$E_a = A + B\Delta H \tag{10.1}$$

where A and B are constants specific to the reaction system. As a specific example, there is a linear relationship between the logarithm of the equilibrium constant for the acid dissociation reaction of many substituted benzoic acids and the logarithm of the rate constant for another reaction of molecules with the same substituent, which has been proposed as the Hammett rule.[14,15] If the rate constant and equilibrium constant are k_H and K_H, respectively, with reference to the case when the substituent is hydrogen, and the rate constant and equilibrium constant are k_X and K_X, respectively, when the substituent is X, then

$$\log(k_X/k_H) = \log(K_X/K_H) = \rho \cdot \sigma_X \tag{10.2}$$

where ρ and σ_X represent the reaction and substituent constants, respectively. On the other hand, Taft introduced the stereo substituent constant E_S

when he extended the Hammett rule to aliphatic compounds,[16,17] and Charton found that E_S has a linear relationship with the steric parameter ν, which is estimated by the van der Waals radius used in molecular mechanics.[18,19]

The ν values for various substituents are summarized elsewhere.[20] Steric repulsion is an exchange repulsion based on Pauli's exclusion principle and is not limited to organic chemical reactions, but applicable when solvent molecules coordinate metal ions. Molecular mechanics and molecular simulations are often used to study the solvation structure of metal ions, and the fact that E_S has a linear relationship with ν suggests that destabilization and increased reactivity due to steric repulsion between solvent molecules in the first solvation shell of the metal ion can be quantified.

10.5 Metal Ion Solvates with Solvents Space-demanding Upon Coordination

A series of amide and urea solvents with substituents are good model solvents for considering steric factors in solvent effects. To facilitate comparison of bulkiness, Figure 10.1 shows the solvents used here on the same scale, with the ligand cone angle increasing from DMF to HMPA.

10.5.1 First Row Transition Metal(II) Ions

Table 10.1 summarises the structural parameters of the hydrated and solvated divalent transition metal ions in water, DMF, DMA, DMP, TMU, DMPU and HMPA in solution and the solid state. DMF, DMA and DMP seem not to have any steric demands upon coordination and have the coordination numbers and figures similar to hydrates. On the other hand, TMU, DMPU and HMPA have larger steric demands and lower coordination numbers are observed (Table 10.1). Manganese(II), iron(II), cobalt(II) and nickel(II) are five-coordinate in TMU and DMPU solution and they are very difficult to crystallise, and no crystal structures have been reported so far. Copper(II) is four-coordinate in a square-planar fashion in TMU and DMPU forming characteristic intense green solutions. Due to the bulkiness of these solvent molecules, the axial positions are partly covered, making it impossible for the ligands to bind in these positions. Zinc(II) is five-coordinate in TMU and

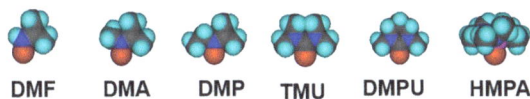

DMF   DMA   DMP   TMU   DMPU   HMPA

Figure 10.1   Structures of various amide solvents: *N,N*-dimethylformamide (DMF), *N,N*-dimethylacetamide (DMA), *N,N*-dimethylpropionamide (DMP), 1,1,3,3-tetramethylurea (TMU), *N,N*-dimethylpropyleneurea (DMPU) and hexamethylphosphoric triamide (HMPA).

Table 10.1   Bond distances (Å) and coordination numbers (within parentheses) for the first solvation shell of the first transition metal(II) ions in non-aqueous solvents with increasing steric hindrance upon coordination in solution and the solid state (values in italics) by EXAFS and crystallography. The three Cu–O bond distances reported refer to the equatorial and the two different axial positions, see Table 9.1.

	Mn	Fe	Co	Ni	Cu	Zn
Water[a]	2.17 (6)	2.12 (6)	2.09 (6)	2.055 (6)	1.957, 2.14, 2.32 (6)	2.09 (6)
	2.174 (4)	*2.120 (6)*	*2.087 (6)*	*2.055 (6)*	*1.963, 2.184, 2.343 (6)*	*2.088 (6)*
DMF[a]	2.16 (6)	2.10 (6)	2.08 (6)	2.01 (6)	1.968, 2.178, 2.33 (6)	2.08 (6)
	2.161 (4)	*2.113 (6)*	*2.083 (6)*	*2.048 (6)*	*1.975, 2.327 (6)*	*2.089 (6)*
DMA	2.16 (6)[b]		2.07 (6)[b]	2.04 (6)[b]	1.956, 2.14, 2.32 (6)[c]	1.99 (5)[b]
	2.161 (4)[d]		*2.088 (6)[d]*	*2.060 (6)[d]*	*1.975 (4)[d]*	*2.089 (6)[d]*
DMP			2.02 (5)[e]	2.04 (6)[e]	1.952, 2.18, 2.33 (6)[c]	1.99 (5)[e]
TMU	2.09 (5)[f]	2.05 (5)[f]	2.00 (5)[f]	2.00 (5)[f]	1.929 (4)[f]	1.99 (5)[f]
DMPU[a]	2.087 (5)	2.048 (5)	1.996 (5)	2.000 (5)	1.939 (4)	1.99 (5)
					1.901 (4)	*1.925 (4)*
HMPA	2.07 (5)[g]	1.98 (4)[g]	1.95 (4)[g]	1.97 (4)[g]	1.92 (4)[g]	1.93 (4)[g]
			1.947 (4)[d]			

[a] See Table 9.1.
[b] Ref. 21.
[c] Ref. 22.
[d] Ref. 23.
[e] Ref. 24.
[f] Ref. 25.
[g] Ref. 26.

DMPU solutions, while a tetrahedral $Zn(dmpu)_4^{2+}$ complex is formed in the solid state. HMPA is the most space-demanding solvent upon coordination, forming four-coordinate complexes with all divalent transition metal ions, except for the largest one, manganese(II), which is five-coordinate. Of these, only the HMPA solvated cobalt(II) ion has been characterized crystallographically.

10.5.1.1   Lanthanoid(III) Ions

The ionic radius of lanthanide ions (III) (Ln^{3+}) decreases with increasing atomic number due to the lanthanide contraction. This is caused by the poor shielding of the nuclear charge by the 4f electrons, and, thereby, the electrons in the electron shell outside the fourth shell are drawn towards the nucleus, resulting in a decreasing atomic/ionic radius along the lanthanoid series. An overview of the solid state structures containing hydrated lanthanide(III) ions shows that the lighter elements are nine-coordinate in a tricapped trigonal prismatic fashion in most compounds, while heavier lanthanides are eight-coordinate in a square antiprismatic fashion.

The heaviest lanthanoid(III) ions, Ho–Lu, with a tricapped trigonal prismatic configuration have increasing water deficit with increasing atomic number in the capping positions, $[Ho(H_2O)_{8.91}](CF_3SO_3)_3$, $[Er(H_2O)_{8.96}](CF_3SO_3)_3$, $[Tm(H_2O)_{8.8}](CF_3SO_3)_3$, $[Yb(H_2O)_{8.7}](CF_3SO_3)_3$ and $[Lu(H_2O)_{8.2}](CF_3SO_3)_3$.[27,28] The structures of dimethylsulfoxide solvated lanthanide(III) ions have all been reported to be eight-coordinate in both solution and the solid state.[29,30] The same pattern is observed for the dmf solvated lanthanide(III) ions in the solid state and solution with all being eight-coordinate in a square anti-prismatic fashion, except for some solid nine-coordinate lanthanum(III) dmf solvates.[6] The dmpu solvated lanthanide(III) ions are all seven-coordinate in dmpu solution, except for the heaviest and smallest one, lutetium(III), which is six-coordinate in a regular octahedral fashion. On the other hand, in the solid state all dmpu solvated lanthanide(III) ions are six-coordinate in a regular octahedral fashion. This shows that complexes with high symmetry are favored upon crystallization, while in solution the highest possible coordination from the steric point of view is obtained. This pattern is also observed for the divalent transition metal ions discussed above (Table 10.1).

10.6 Reactivity of Metal Ion Solvates with Low Symmetry

Complex formation is always a competition between solvation of the metal ion and ligand, and formation of new bonds between the metal and ligand. Strong solvation prevents or hampers complex formation, but the way in which low symmetry and steric crowding affect the metal ion affects its ability to form complexes. The complex formation of manganese(II), cobalt(II), nickel(II) and iron(III) with bromide has been studied in DMPU and HMPA solutions, and the results are compared with those of aqueous solution in Table 10.2. The stability constants for these systems are significantly higher in DMPU and HMPA than those in water in spite of the expected stronger solvation due to a higher electron-pair ability than hydration and DMSO solvation; no stability constants are reported for these systems in DMSO, most likely because no complexes are formed. This is supported by the fact that compounds crystallizing from concentrated solutions have the compositions $[Mn(dmso)_6]Br_2$[31] and $[Fe(dmso)_3]Br_2$,[32] and the D_N and D_S values, showing the electron-pair donor ability of DMSO, are between those of water and DMPU and HMPA.[9] The reported results indicate that the lower symmetry of the metal ions of DMPU and HMPA solvates are less stable compared to, *e.g.*, the octahedral hydrated and DMSO solvated metal ions, and the required desolvation for complex formation requires less energy and thereby more stable complexes are formed.

As mentioned above, the nearest-neighbor solvation number of lanthanide ions decreases with increasing atomic number in DMA, which greatly affects the complexation reaction. A series of bromo-complexation reactions have been investigated by precise calorimetry to determine the reaction

Table 10.2 Selected stability constants, $\log_{10} K_n$, for the stepwise formation of manganese(II), cobalt(II), nickel(II) and iron(III) bromide complexes in aqueous, DMPU and HMPA solutions at 298 K.

Complex	Water	DMPU	HMPA
D_N	18^a		38.8^a
D_S	17^a	34^b	34^a
$MnBr^+$	0.20^c	3.36^c	4.1^d
$MnBr_2$	-0.26^c	2.56^c	2.4^d
$CoBr^+$			5.55^e
$CoBr_2$			3.19^e
$CoBr_3^-$			0.79^e
$NiBr^+$	-0.11^f	3.30^g	
$NiBr_2$		2.56^g	
$FeBr^{2+}$	-0.12^h	2.76^i	
$FeBr_2^+$	-0.26^h	2.19^i	
$FeBr_3$		1.7^i	

[a] Ref. 33.
[b] Ref. 34.
[c] Ref. 35.
[d] Ref. 25 and 26.
[e] Ref. 36.
[f] Ref. 31.
[g] Ref. 37.
[h] Ref. 38.
[i] Ref. 39.

Gibbs free energy $\Delta G°$, reaction enthalpy $\Delta H°$, and reaction entropy $\Delta S°$. $\Delta H°$ and $\Delta S°$ are both positive and small regardless of the type of lanthanide ion, whereas in DMA they are both positive and large and strongly depend on the type of lanthanide ion.[40] This indicates that bromo complex formation in DMF produces exospheric complexes (outer sphere ion pairs or non-contact ion pairs) without desolvation, whereas in DMA inner-sphere complexes (contact ion pairs) are formed. In DMA, even though its dielectric constant, D_N and A_N are virtually equal to those of DMF, the steric hindrance between the solvated DMA molecules due to the β-position methyl group of a donor atom of the carbonyl oxygen causes a change from the outer-sphere type to the inner-sphere type of the complexes formed.

10.7 Solvation Structure of Metal Ions in Mixed Solvents

Mixtures of two or more solvents are used in all fields of chemistry to bring out properties not found in a single component. Mixtures of water and non-aqueous solvents have been of interest from various perspectives, such as hydrogen bonding properties, density fluctuations, and miscibility, and have been the subject of much research for many years. In Section 10.6, we discuss the solvation of metal ions in terms of not only metal ion–solvent molecular interactions, but also the importance of crowdedness of the metal ion solvates. In mixed solvents, selective solvation of metal ions is predicted, but it is extremely difficult to separate individual factors such as the donor–acceptor nature, steric factors, and bulk liquid structure. As mentioned above, water is not a suitable solvent for generalizing the steric factor involvement in this regard, as the steric factor is almost negligible due to its significantly smaller molecular volume.

Amide solvents are good model systems for selective solvation of metal ions as well as solvation structures of metal ions in non-aqueous solvents. The introduction of substituents can control the liquid structural properties due to steric factors and hydrogen bonding, and changes in the molecular structure affect the reference vibration in Raman/IR, *etc.*, giving characteristic vibrational bands for individual solvents. Therefore, Raman/IR vibrational spectroscopy is very useful for the study of selective solvation of metal ions. In the following, we will discuss selective solvation of metal ions in terms of steric factors and liquid structural properties.

10.7.1 Selective Solvation by the Steric Factor

Selective solvation of Mg^{2+}, Ca^{2+}, Sr^{2+} and Ba^{2+},[35] as well as Mn^{2+}, Ni^{2+} and Zn^{2+}, Nd^{3+}, Gd^{3+} and Tm^{3+} [41,42] in DMF–DMA mixed solvents has been investigated. Figure 10.2 plots the mole fraction of DMF in the first solvated shell of metal ions (x_{DMF}, M^{2+}_{inner}) against the mole fraction of bulk DMF (x_{DMF}). Mg^{2+} is six-coordinate in both DMF and DMA. On the other hand, Ca^{2+} with a larger ionic radius may be six-, seven- or eight-coordinate depending on the solvent, but is selectively solvated by DMA with a high donor nature. Furthermore, Sr^{2+} and Ba^{2+} with even larger ionic radii are eight-coordinate, and their selectivity due to the donor nature is reduced. The first transition metal(II) ions are all more selective for DMF than Mg^{2+}, except for Mn^{2+}, which has a relatively large ionic radius.[7] DMF is also selectively solvated by Nd^{3+}, Gd^{3+}, and Tm^{3+}, whose nearest neighbor solvation numbers change with solvent composition; in the formation of bromo

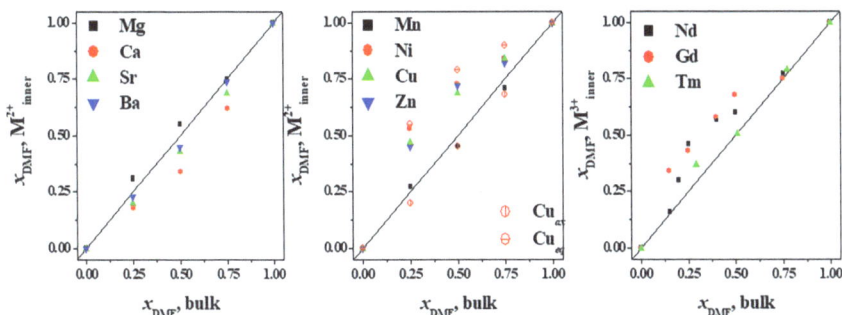

Figure 10.2 Selective solvation of various metal ions in DMF–DMA mixtures. The diagonal lines in the figure indicate that the bulk DMF mole fraction is equal to the DMF mole fraction of the metal ion solvation sphere on the vertical axis.

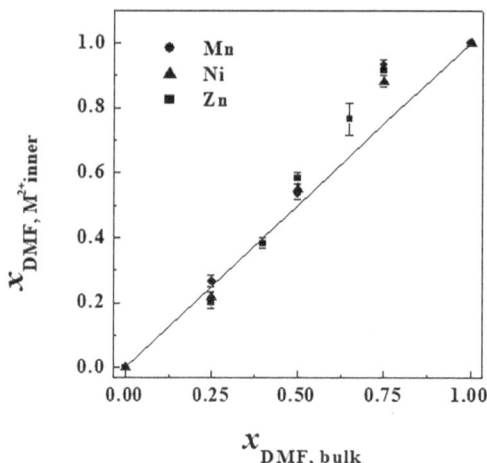

Figure 10.3 Selective solvation in DMF–NMF mixtures. Reproduced from ref. 44
with permission from American Chemical Society, Copyright 2006.

complexes in DMF–DMA mixtures, metal-ion independent, outer sphere
complex–inner sphere complex switching occurs at a DMF mole fraction of 0.6.
Careful examination of the solvation selectivity shows that switching occurs
when the number of DMA molecules in the first solvation shell reaches 3.[43]

10.7.2   Selective Solvation by Liquid Structural Properties

N-Methylformamide (NMF), in which the *N*-methyl group of DMF is replaced
by hydrogen, has a well-ordered liquid structure due to strong intermolecular hydrogen bonds giving a chain-like network structure.[44] In the
DMF–NMF mixed solvent system, in the region of high DMF mole fraction,
there are excess dipole–dipole interactions between DMF molecules, but
they are weaker than the hydrogen bonds between DMF–NMF and between
NMF–NMF. Such DMF molecules have an energetic advantage in the transition from the bulk to the first solvation shell of the metal ion, where DMF
selectively solvates the metal ion (Figure 10.3).

Problem 1

Why may complex formation be more strongly space-demanding upon coordination than in water, even though the metal–solvate bonds are stronger
than the corresponding metal–hydrate bonds?

Problem 2

Is it possible to extrapolate the structure of a solvated metal ion in solution
from the structure of the corresponding complex in the solid state. Give
some examples of solvated metal ions with different structures in solution
and the solid state by using Tables 9.1 and 9.2.

Answer to Problem 1

High symmetry stabilizes metal complexes, and complexes with low symmetry as five- and seven-coordinated ones are therefore more reactive. Metal ion solvates with low symmetry are therefore more reactive than the ones with higher symmetry, and formation of a complex with higher symmetry is therefore favorable; see Table 10.2 for some examples.

Answer to Problem 2

Structures can never be extrapolated from the solid state to solution or the opposite. In order to get information about structures in solution, they must be determined in solution and the actual solvent. The dmpu solvated lanthanoid(III) ions are seven-coordinate in dmpu solution and six-coordinate (octahedral) in the solid state, and the only reported bismuth(III) hydrate is nine-coordinate (tricapped trigonal prism) in the solid state, while it is eight-coordinate (square antiprism) in aqueous solution.

References

1. V. Gutmann, *The Donor-Acceptor Approach to Molecular Interactions*, Plenum Press, New York, 1978.
2. R. Thomas, C. Brink-Shoemaker and K. Eriks, *Acta Crystallogr.*, 1966, **21**, 12.
3. M. A. Viswamitra and K. K. Kannan, *Nature*, 1966, **219**, 1016.
4. R. Gajda and A. Katrusiak, *J. Phys. Chem. B*, 2009, **113**, 2436.
5. H. Reuter, *Acta Crystallogr., Sect. E: Crystallogr. Commun.*, 2017, **73**, 1405.
6. F. H. Allen, *Acta Crystallogr., Sect. B: Struct. Sci., Cryst. Eng. Mater.*, 2002, **58**, 380; Cambridge Structure Database (Conquest 2022.1.0).
7. R. D. Shannon, *Acta Crystallogr., Sect. A: Cryst. Phys., Diffr., Theor. Gen. Crystallogr.*, 1976, **32**, 751.
8. J. K. Beattie, S. P. Best, B. W. Skelton and A. H. White, *J. Chem. Soc., Dalton Trans.*, 1981, 2105.
9. M. Sandström, I. Persson and P. Persson, *Acta Chem. Scand.*, 1990, **44**, 653, and references therein.
10. C. A. Tolman, *Chem. Rev.*, 1977, **77**, 313.
11. P. Lindqvist-Reis, J. Näslund, I. Persson and M. Sandström, *J. Chem. Soc., Dalton Trans.*, 2000, 2703.
12. R. P. Bell, *Proc. R. Soc. London, Ser. A*, 1936, **154**, 414.
13. M. G. Evans and M. Polanyi, *Trans. Faraday Soc.*, 1938, **34**, 11.
14. L. P. Hammett, *J. Am. Chem. Soc.*, 1937, **59**, 96.
15. R. W. Taft, *J. Am. Chem. Soc.*, 1952, **74**, 2729.
16. R. W. Taft, *J. Am. Chem. Soc.*, 1952, **74**, 3120.
17. R. W. Taft, *J. Am. Chem. Soc.*, 1953, **75**, 4538.
18. M. Charton, *J. Am. Chem. Soc.*, 1969, **91**, 615.
19. M. Charton, *J. Am. Chem. Soc.*, 1975, **97**, 1552.

20. M. Charton, *J. Org. Chem.*, 1976, **41**, 2217.
21. K. Ozutsumi, M. Koide, H. Suzuki and S.-I. Ishiguro, *J. Phys. Chem.*, 1993, **97**, 500.
22. I. Persson, D. Lundberg, É. G. Bajnoczi, K. Klementiev, J. Just and K. G. V. Sigfridsson Clauss, *Inorg. Chem.*, 2020, **59**, 9538.
23. H. Yokoyama and K. Nakajima, *J. Solution Chem.*, 2004, **33**, 607.
24. Y. Inada, K. Sugimoto, K. Ozutsumi and S. Funahashi, *Inorg. Chem.*, 1994, **33**, 1875.
25. K. Ozutsumi, Y. Abe, R. Takahashi and S.-I. Ishiguro, *J. Phys. Chem.*, 1994, **98**, 9894.
26. K. Ozutsumi, K. Tohji, Y. Udagawa, Y. Abe and S.-I. Ishiguro, *Inorg. Chim. Acta*, 1992, **191**, 183.
27. A. Abbasi, P. Lindqvist-Reis, L. Eriksson, D. Sandström, S. Lidin, I. Persson and M. Sandström, *Chem. Eur. J.*, 2005, **11**, 4065.
28. I. Persson, P. D'Angelo, S. De Panfilis, M. Sandström and L. Eriksson, *Chem. - Eur. J.*, 2008, **14**, 3056.
29. A. Abbasi, E. Damian-Risberg, L. Eriksson, J. Mink, I. Persson, M. Sandström, Y. V. Sidorov, M. Y. Skripkin and A.-S. Ullström, *Inorg. Chem.*, 2007, **46**, 7731.
30. I. Persson, E. Damian-Risberg, P. D'Angelo, S. De Panfilis, M. Sandström and A. Abbasi, *Inorg. Chem.*, 2007, **46**, 7742.
31. D. Bobicz, O. Kristiansson and I. Persson, *J. Chem. Soc., Dalton Trans.*, 2002, 4201.
32. P. R. Martinez-Alanis, R. A. Toscano and I. Castillo, *Acta Crystallogr., Sect. E: Struct. Rep. Online*, 2005, **61**, m2179.
33. M. Sandström, I. Persson and P. Persson, *Acta Chem. Scand.*, 1990, **44**, 653–675 and references therein.
34. P. Smirnov, L. Weng and I. Persson, *Phys. Chem. Chem. Phys.*, 2001, **3**, 5248–5254.
35. H. Konieczna, D. Lundberg and I. Persson, *Polyhedron*, 2021, **195**, 114961 and references therein.
36. Y. Abe, K. Ozutsumi and S. Ishiguro, *J. Chem. Soc., Faraday Trans. 1*, 1989, **85**, 3747.
37. M. B. Lister and M. W. Kennedy, *Can. J. Chem.*, 1966, **47**, 1709.
38. D. F. C. Morris and A. R. Wilson, *J. Inorg. Nucl. Chem.*, 1969, **31**, 1532.
39. D. Lundberg, A.-S. Ullström, P. D'Angelo, D. Warminska and I. Persson, *Inorg. Chim. Acta*, 2007, **360**, 2744.
40. S. Ishiguro, Y. Umebayashi and M. Komiya, *Coord. Chem. Rev.*, 2002, **226**, 103.
41. M. Asada, T. Fujimori, K. Fujii, R. Kanzaki, Y. Umebayashi and S. Ishiguro, *J. Raman Spectrosc.*, 2007, **38**, 417.
42. Y. Umebayashi, K. Matsumoto, M. Watanabe and S. Ishiguro, *Phys. Chem. Chem. Phys.*, 2001, **3**, 5475.
43. Y. Umebayashi, K. Matsumoto, I. Mekata and S. Ishiguro, *Phys. Chem. Chem. Phys.*, 2002, **4**, 5599.
44. K. Fujii, T. Kumai, T. Takamuku, Y. Umebayashi and S. Ishiguro, *J. Phys. Chem. A*, 2006, **110**, 1798.

CHAPTER 11

Reactions of Metal Complexes in Solution

KOJI ISHIHARA*[a] AND MASAHIKO INAMO[b]

[a] Department of Chemistry and Biochemistry, School of Advanced Science and Engineering, Waseda University, Okubo, Shinjuku, Tokyo 169-8555, Japan; [b] Department of Chemistry, Aichi University of Education, Kariya, Aichi 448-8542, Japan
*Email: ishi3719@waseda.jp

11.1 Introduction

First, let us consider the phenomenon that occurs when an electrolyte containing metal ions is dissolved in a solvent, such as water. When the salt dissociates upon dissolution, the metal ions and counter-ions are solvated by the solvent molecules. Transition metal ions accept lone pairs of electrons (Lewis acids). In aqueous solution, water molecules act as a donor of lone pairs of electrons (Lewis bases) to hydrate transition metal ions. For instance, nickel(II) ions are hydrated by water to form $Ni(H_2O)_6{}^{2+}(aq)$, which is often abbreviated to $Ni^{2+}(aq)$. Species such as $Ni(H_2O)_6{}^{2+}$ are known as aqua-metal ions or aqua-metal complexes. A complex has a central metal ion coordinated by one or more Lewis bases, called ligands. In $Ni(H_2O)_6{}^{2+}$, the central nickel(II) ion is bound by six water ligands and the coordination number (number of coordinated atoms) is six.

In this chapter, the basic terminology of metal ions and metal complexes in aqueous solutions is explained. The types of reactions of metal ions in solution and the reaction mechanism of their metal complexes are introduced. Among these reactions, the solvent exchange reaction, complex formation reaction, acid dissociation reaction, and ligand substitution reaction

Coordination Chemistry Fundamentals Series No. 2
Metal Ions and Complexes in Solution
Edited by Toshio Yamaguchi and Ingmar Persson
© Japan Society of Coordination Chemistry 2024
Published by the Royal Society of Chemistry, www.rsc.org

will be described in more detail from a kinetic point of view to deepen the understanding of the reactions of metal ions in solution.

11.1.1 Solvation Model of Metal Ions

The environment surrounding solvated metal ions can be explained in more detail using the solvation model shown in Figure 11.1.[1] Region A is called the first hydration shell and contains a central metal ion and coordinated water molecules. Therefore, it is also referred to as the inner-coordination shell. Region B is called the second hydration shell, in which the electrostatic field of the ions and the dipole of the water molecules interact with one another. Region C is a disordered region and bulk water is ordinary water that is not affected by the charge on the ions. However, region B may not be clear for weakly hydrated metal ions. Hydrating water molecules in the first hydration shell are not necessarily dynamically stable and exchange with the outer water molecules at a rate inherent to the metal ion. In other words, a solvent exchange reaction constantly occurs for most metal ions. In addition to solvent exchange reactions and complex formation reactions, acid dissociation reactions, ligand substitution reactions, and redox reactions may occur for aqua-metal ions.

11.1.2 Formation Constant of a Metal Complex

When a stronger Lewis base (*e.g.*, NH_3) is added to an aqueous solution containing Ni^{2+}, the water molecules coordinated to the central Ni^{2+} ion may substitute with ammonia to form an ammine complex: $Ni(NH_3)_6^{2+}$.

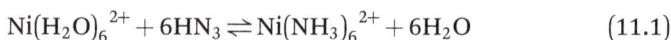

$$Ni(H_2O)_6^{2+} + 6HN_3 \rightleftharpoons Ni(NH_3)_6^{2+} + 6H_2O \qquad (11.1)$$

Ligands that donate only one electron pair to the central metal, for example, H_2O and NH_3, are called monodentate ligands. Some ligands that can donate two pairs of electrons to the metal, such as ethylenediamine (1,2-diaminoethane), oxalate ion, and bipyridine, are known as bidentate ligands. Ligands that can donate more than one pair of electrons are called

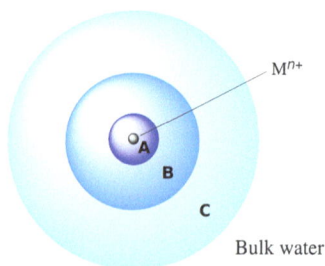

Figure 11.1   Model of aqua-metal ions in an aqueous solution: (A) first hydration shell (inner-sphere or inner-coordination sphere), (B) second hydration shell, and (C) disordered region.

polydentate ligands. A complex containing a polydentate ligand is called a chelate. The most popular polydentate ligand is the ethylenediamine-tetraacetate ion ($EDTA^{4-}$), which is known as a chelating agent and acts as a hexadentate ligand for most metal ions.

The formation of $Ni(NH_3)_6{}^{2+}$ (eqn (11.1)) occurs in a stepwise manner depending on the concentration of ammonia (uncoordinated), as shown by the following equations:

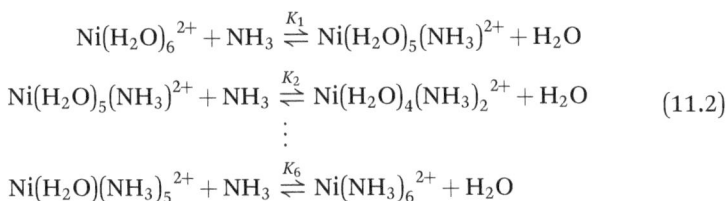

$$Ni(H_2O)_6{}^{2+} + NH_3 \overset{K_1}{\rightleftharpoons} Ni(H_2O)_5(NH_3)^{2+} + H_2O$$

$$Ni(H_2O)_5(NH_3)^{2+} + NH_3 \overset{K_2}{\rightleftharpoons} Ni(H_2O)_4(NH_3)_2{}^{2+} + H_2O \qquad (11.2)$$

$$\vdots$$

$$Ni(H_2O)(NH_3)_5{}^{2+} + NH_3 \overset{K_6}{\rightleftharpoons} Ni(NH_3)_6{}^{2+} + H_2O$$

The equilibrium constants (K_1, K_2..., K_6) for the reactions in eqn (11.2) (*i.e.*, the formation reaction of each complex ion) are called stepwise formation constants, which can be expressed as

$$K_n = [Ni(H_2O)_{6-n}(NH_3)_n{}^{2+}]/([Ni(H_2O)_{6-(n-1)}(NH_3)_{n-1}{}^{2+}] \cdot [NH_3]) \quad (n = 1\text{-}6).$$

while the overall equilibrium constant β_n for the reaction in eqn (11.3) is defined as

$$\beta_n = [Ni(H_2O)_{6-n}(NH_3)_n{}^{2+}]/([Ni(H_2O)_6{}^{2+}] \cdot [NH_3]^n) \quad (n = 1\text{-}6),$$

where $\beta_n = K_1 \cdot K_2 \ldots K_n$.

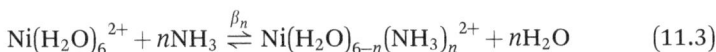

$$Ni(H_2O)_6{}^{2+} + nNH_3 \overset{\beta_n}{\rightleftharpoons} Ni(H_2O)_{6-n}(NH_3)_n{}^{2+} + nH_2O \qquad (11.3)$$

In general, the equilibrium constant K for the reaction (eqn (11.4)) in which the metal ion M^{n+} reacts with ligand L to form the complex ML^{n+} can be expressed by eqn (11.5) using molar concentrations. On the other hand, the equilibrium constant derived thermodynamically, *i.e.* the thermodynamic equilibrium constant, K^*, can be expressed by eqn (11.6) using the activity (a_i) of each chemical species (x_i). Substitution of this relationship (eqn (11.7)) between activity and molar concentration, where γ_i is the activity coefficient of x_i, into eqn (11.6) gives eqn (11.8). The thermodynamic equilibrium constant (K^*) is constant if temperature and pressure are constant, whereas the concentration equilibrium constant (K) is constant if the temperature, pressure, and activity coefficients or the ratio of the activity coefficients (Γ) are constant, as shown in eqn (11.4)–(11.8). The activity coefficient of each species is kept almost constant if the experiments are carried out under conditions in which the ionic strength is constant by adding a large excess of inert electrolyte (that does not participate or affect the chemical system under study), such as sodium perchlorate.

Therefore, the list of formation constants reported in the literature specifies the ionic strength in addition to the temperature and pressure (usually ambient pressure).

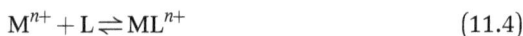

$$M^{n+} + L \rightleftharpoons ML^{n+} \tag{11.4}$$

$$K = \frac{[ML^{n+}]}{[M^{n+}][L]} \tag{11.5}$$

$$K^* = \frac{a_{ML}}{a_M a_L} \tag{11.6}$$

$$a_i = \gamma_i[x_i] \tag{11.7}$$

$$K^* = \frac{a_{ML}}{a_M a_L} = \frac{\gamma_{ML}[ML^{n+}]}{\gamma_M \gamma_L[M^{n+}][L]} = \Gamma \cdot K \tag{11.8}$$

11.1.3 Conditional Formation Constant of a Metal Complex

The formation constants tabulated in the literature are the values obtained for the reactions of aqua metal ions with fully deprotonated ligands such as Cl^-, the conjugate base of the strong acid, hydrochloric acid. On the other hand, ligands such as ammonia, NH_3, and acetate, CH_3COO^-, have acid–base properties and their concentration in aqueous solution depends on the pH value. As described below, the aqua metal ion is a Brønsted acid and the ligand is a Lewis base as well as a Brønsted base. Therefore, the values of the formation constants reported in the literature provide no information on the extent to which the complex is formed under the actual conditions of interest. Thus, it is convenient to use the concept of conditional or environmental formation constants, K_E, as proposed by Ringbom.[2]

Let us consider the general reaction of a divalent metal ion with a monodentate ligand (L^-), with acidic, basic or amphoteric properties, to form its corresponding metal complex (ML^+) (eqn (11.9)).

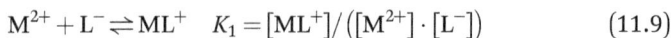

$$M^{2+} + L^- \rightleftharpoons ML^+ \quad K_1 = [ML^+]/([M^{2+}] \cdot [L^-]) \tag{11.9}$$

Depending on the pH, M^{2+} can hydrolyze to form $M(OH)^+$ and $M(OH)_2$ (eqn (11.10)), while L^- can be protonated to form HL and H_2L^+ (eqn (11.11)).

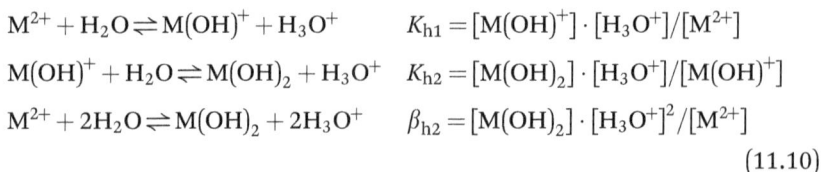

$$M^{2+} + H_2O \rightleftharpoons M(OH)^+ + H_3O^+ \qquad K_{h1} = [M(OH)^+] \cdot [H_3O^+]/[M^{2+}]$$

$$M(OH)^+ + H_2O \rightleftharpoons M(OH)_2 + H_3O^+ \qquad K_{h2} = [M(OH)_2] \cdot [H_3O^+]/[M(OH)^+]$$

$$M^{2+} + 2H_2O \rightleftharpoons M(OH)_2 + 2H_3O^+ \qquad \beta_{h2} = [M(OH)_2] \cdot [H_3O^+]^2/[M^{2+}]$$

$$\tag{11.10}$$

$$H_3O^+ + L^- \rightleftharpoons HL + H_2O \quad K_{L1} = 1/K_{a2} = [HL]/([H_3O^+] \cdot [L^-]) \tag{11.11}$$

$$H_3O^+ + HL \rightleftharpoons H_2L^+ + H_2O \quad K_{L2} = 1/K_{a1} = [H_2L]/([H_3O^+] \cdot [HL]) \tag{11.12}$$

$$C_M = [M'] + [ML^+];$$
$$[M'] = [M^{2+}] + [M(OH)^+] + [M(OH)_2] = [M^{2+}](1 + K_{h1}/[H^+] + \beta_{h2}/[H^+]^2);$$
$$1 + K_{h1}/[H^+] + \beta_{h2}/[H^+]^2 = \alpha_A \tag{11.13}$$

$$C_L = [L'] + [ML^+];$$
$$[L'] = [L^-] + [HL] + [H_2L^+] = [L^-](1 + K_{L1}[H_3O^+] + K_{L1} \cdot K_{L2} \cdot [H_3O^+]^2)$$
$$1 + K_{L1}[H_3O^+] + K_{L1} \cdot K_{L2} \cdot [H_3O^+]^2 = \alpha_H \tag{11.14}$$

$$K_E = [ML^+]/([M'] \cdot [L']) = [ML^+]/([M^{2+}] \cdot \alpha_A \cdot [L^-] \cdot \alpha_H) = K_1/(\alpha_A \cdot \alpha_H) \tag{11.15}$$

We will here calculate the conditional formation constant in an applied system: the titration of zinc(II) with EDTA in an ammonia/ammonium buffer with $[NH_3] = [NH_4^+] = 0.10$ M; for simplicity, the fully deprotonated form of EDTA, EDTA^{4-}, which is the form of the zinc(II)–EDTA complex, is abbreviated to Y^{4-}. Used chemical constants: pH = 9.00, the formation constants of the protonated forms of EDTA are the following: HY^{3-} – $10^{10.26}$ M^{-1}; H_2Y^{2-} – $10^{6.16}$ M^{-1}; H_3Y^- – $10^{2.67}$ M^{-1}; and H_4Y – $10^{2.00}$ M^{-1}; the formation constants used for the zinc–ammonia system are the following: $K_1 = 186$ M^{-1}, $K_2 = 219$ M^{-1}, $K_3 = 252$ M^{-1} and $K_4 = 113$ M^{-1} and for the ZnY^{2-} complex the used stability constant is $10^{16.52}$ M^{-1}.

$$Zn^{2+} + Y^{4-} \rightleftharpoons ZnY^{2-}; \quad K = [ZnY^{2-}]/([Zn^{2+}] \cdot [Y^{4-}])$$

$$H_3O^+ + Y^{4-} \rightleftharpoons HY^{3-} + H_2O; \quad K_{1Y} = [HY^{3-}]/([H_3O^+] \cdot [Y^{4-}]) = 10^{10.26} \text{ M}^{-1};$$

$$[HY^{3-}] = K_{1Y} \cdot [Y^{4-}] \cdot [H_3O^+]$$

$$H_3O^+ + HY^{3-} \rightleftharpoons H_2Y^{2-} + H_2O; \quad K_{2Y} = [H_2Y^{2-}]/([H_3O^+] \cdot [HY^{3-}]) = 10^{6.16} \text{ M}^{-1};$$

$$[H_2Y^{2-}] = K_{2Y} \cdot [HY^{3-}] \cdot [H_3O^+] = K_{1Y} \cdot K_{2Y} \cdot [Y^{4-}] \cdot [H_3O^+]^2$$

$$H_3O^+ + H_2Y^{2-} \rightleftharpoons H_3Y^- + H_2O; \quad K_{3Y} = [H_3Y^-]/([H_3O^+] \cdot [H_2Y^{2-}]) = 10^{2.67} \text{ M}^{-1};$$

$$[H_3Y^-] = K_{3Y} \cdot [H_2Y^{2-}] \cdot [H_3O^+] = K_{1Y} \cdot K_{2Y} \cdot K_{3Y} \cdot [Y^{4-}] \cdot [H_3O^+]^3$$

$$H_3O^+ + H_3Y^- \rightleftharpoons H_4Y + H_2O; \quad K_{4Y} = [H_4Y]/([H_3O^+] \cdot [H_3Y^-]) = 10^{2.00} \text{ M}^{-1};$$

$$[H_4Y] = K_{4Y} \cdot [H_3Y^-] \cdot [H_3O^+] = K_{1Y} \cdot K_{2Y} \cdot K_{3Y} \cdot K_{4Y} \cdot [Y^{4-}] \cdot [H_3O^+]^4$$

$$C_Y = [Y'] + [ZnY^{2-}];$$

$$[Y'] = [Y^{4-}] + [HY^{3-}] + [H_2Y^{2-}] + [H_3Y^-] + [H_4Y]$$

$$= [Y^{4-}](1 + K_{1Y} \cdot [H_3O^+] + K_{1Y} \cdot K_{2Y} \cdot [H_3O^+]^2$$

$$+ K_{1Y} \cdot K_{2Y} \cdot K_{3Y} \cdot [H_3O^+]^3 + K_{1Y} \cdot K_{2Y} \cdot K_{3Y} \cdot K_{4Y} \cdot [H_3O^+]^4)$$

and

$$a_H = 1 + K_{1Y} \cdot [H_3O^+] + K_{1Y} \cdot K_{2Y} \cdot [H_3O^+]^2 + K_{1Y} \cdot K_{2Y} \cdot K_{3Y} \cdot [H_3O^+]^3$$

$$+ K_{1Y} \cdot K_{2Y} \cdot K_{3Y} \cdot K_{4Y} \cdot [H_3O^+]^4$$

$$Zn^{2+} + NH_3 \rightleftharpoons Zn(NH_3)^{2+}; \quad K_1 = [Zn(NH_3)^{2-}]/([Zn^{2+}] \cdot [NH_3]) = 186 \ M^{-1}$$

$$Zn(NH_3)^{2+} + NH_3 \rightleftharpoons Zn(NH_3)_2^{2+};$$

$$K_2 = [Zn(NH_3)_2^{2+}]/([Zn(NH_3)^{2+}] \cdot [NH_3]) = 219 \ M^{-1}$$

$$Zn(NH_3)_2^{2+} + NH_3 \rightleftharpoons Zn(NH_3)_3^{2+};$$

$$K_3 = [Zn(NH_3)_3^{2+}]/([Zn(NH_3)_2^{2+}] \cdot [NH_3]) = 252 \ M^{-1}$$

$$Zn(NH_3)_3^{2+} + NH_3 \rightleftharpoons Zn(NH_3)_4^{2+};$$

$$K_3 = [Zn(NH_3)_4^{2+}]/([Zn(NH_3)_3^{2+}] \cdot [NH_3]) = 113 \ M^{-1}$$

$$C_{Zn} = [Zn'] + [ZnY^{2-}];$$

$$[Zn'] = [Zn^{2+}] + [Zn(NH_3)^{2+}] + [Zn(NH_3)_2^{2+}] + [Zn(NH_3)_3^{2+}] + [Zn(NH_3)_4^{2+}]$$

$$= [Zn^{2+}] \cdot (1 + K_1 \cdot [NH_3] + K_1 \cdot K_2 \cdot [NH_3]^2$$

$$+ K_1 \cdot K_2 \cdot K_3 \cdot [NH_3]^3 + K_1 \cdot K_2 \cdot K_3 \cdot K_4 \cdot [NH_3]^4)$$

and

$$\alpha_A = 1 + K_1 \cdot [NH_3] + K_1 \cdot K_2 \cdot [NH_3]^2 + K_1 \cdot K_2 \cdot K_3 \cdot [NH_3]^3$$

$$+ K_1 \cdot K_2 \cdot K_3 \cdot K_4 \cdot [NH_3]^4$$

$$K_E = [ZnY^{2-}]/([Zn'] \cdot [Y']) = [ZnY^{2-}]/([Zn^{2+}] \cdot \alpha_A \cdot [Y^{4-}] \cdot \alpha_H) = K_1/(\alpha_A \cdot \alpha_H)$$

Numerical calculations:

$$\alpha_H = 1 + 10^{10.26} \cdot (10^{-9.00}) + 10^{10.26 + 6.16} \cdot (10^{-9.00})^2$$

$$+ 10^{10.26 + 6.16 + 2.67} \cdot (10^{-9.00})^3 + 10^{10.26 + 6.16 + 2.67 + 2.00} \cdot (10^{-9.00})^4 = 19.2$$

$$\alpha_A = 1 + 186 \cdot (0.10) + 186 \cdot 219 \cdot (0.10)^2 + 186 \cdot 219 \cdot 252 \cdot (0.10)^3$$
$$+ 186 \cdot 219 \cdot 252 \cdot 113 \cdot (0.10)^4 = 1.27 \times 10^5$$
$$K_E = [ZnY^{2-}]/([Zn]' \cdot [Y]') = [ZnY^{2-}]/([Zn^{2+}] \cdot \alpha_A \cdot [Y^{4-}] \cdot \alpha_A)$$
$$= 10^{16.52}/(1.27 \times 10^5 \times 19.2) = 1.4 \times 10^{10} \ M^{-1}$$

This conditional formation constant is sufficiently large to allow an accurate determination of the zinc content in a sample using an EDTA titration at $pH = 9.00$ using an ammonia/ammonium buffer, as the analytical requirement for an accurate determination is 10^8.

11.1.4 The Types of Reactions of Metal Ions in Solution

The reactions of metal complexes, including aqua-metal ions in solution, can be classified as solvent exchange, complex formation, acid dissociation, ligand substitution, redox, and isomerization reactions (eqn (11.16)–(11.24), where M^{n+} is a six-coordinated metal ion with charge $n+$; L, L′, and L″ are uncharged monodentate ligands; and S is a monodentate solvent molecule). The solvent exchange reaction can be represented by eqn (11.16), in which the coordinated solvent molecule (*i.e.*, the leaving ligand) in the inner co-ordination sphere of the metal ion (M^{n+}) is replaced by a solvent molecule (*i.e.*, the entering ligand) in the outer coordination sphere. That is, the reactant and product are identical because the entering and leaving ligands are identical, and the Gibbs free energy change for the reaction is zero. When the entering ligand is a general ligand (L) other than a solvent molecule, the reaction is called a complex formation (complexation) reaction (eqn (11.17)). A water molecule coordinated in the inner sphere of the aqua-metal ion or aqua-metal complex tends to lose a proton to form a hydroxo-complex (eqn (11.18)), which is called a hydrolysis reaction (acid dissociation reaction) of the metal complex. The ligand substitution reaction is generally represented using eqn (11.19), in which the ligand (L′) coordinated to the central metal ion is replaced by a different ligand (L″). A special case of the ligand substitution reaction includes the solvent exchange reaction (eqn (11.16)), complex formation reaction (eqn (11.17)), aquation reaction (eqn (11.20)), anation reaction (eqn (11.21)), and ligand exchange reaction (eqn (11.22)). In eqn (11.19), it is called an aquation reaction (eqn (11.20)) when $L'' = H_2O$ and an anation reaction (eqn (11.21)) when $L'' = $ anion (X^-). The reduction–oxidation (redox) reaction (electron transfer reaction) can be expressed using eqn (11.23), for which inner-sphere and outer-sphere mechanisms have been proposed. When electron transfer occurs between two complexes, it occurs *via* one bridging ligand shared between two central metal ions in the inner-sphere mechanism, whereas electron transfer occurs without any change in the inner spheres of the two ions in the outer-sphere mechanism (see Examples). The isomerization reaction represented by eqn (11.24) includes two cases, with and without bond breaking (Examples 1 and 2, respectively) occurring during the isomerization process.

These reactions in solution are generally more likely to occur in combination with several reactions rather than occur alone.

Solvent exchange reaction:

$$[MS_6]^{n+} + {}^*S \rightarrow [MS_5{}^*S]^{n+} + S \tag{11.16}$$

Example:

$$[Ni(H_2O)_6]^{2+} + H_2O^* \rightarrow [Ni(H_2O)_5(H_2O^*)]^{2+} + H_2O$$

Complex formation reaction:

$$[MS_6]^{n+} + L \rightarrow [MS_5L]^{n+} + S \tag{11.17}$$

Example:

$$[Ni(H_2O)_6]^{2+} + Br^- \rightarrow [NiBr(H_2O)_5]^+ + H_2O$$

Acid dissociation reaction (hydrolysis reaction, proton transfer reaction):

$$[M(H_2O)_x(L)_y]^{n+} \rightarrow [M(OH)(H_2O)_{x-1}(L)_y]^{(n-1)+} + H^+ \tag{11.18}$$

Example:

$$[Ni(H_2O)_6]^{2+} \rightarrow [Ni(H_2O)_5(OH)]^+ + H^+$$

Ligand substitution reaction:

$$[ML_5L']^{n+} + L'' \rightarrow [ML_5L'']^{n+} + L' \tag{11.19}$$

Example:

$$[Ni(H_2O)_5(NH_3)]^{2+} + SCN^- \rightarrow [Ni(H_2O)_5(SCN)]^+ + NH_3$$

Aquation reaction:

$$[ML_5L']^{n+} + H_2O \rightarrow [ML_5(H_2O)]^{n+} + L' \tag{11.20}$$

Example:

$$[CoCl(NH_3)_5]^{2+} + H_2O \rightarrow [Co(H_2O)(NH_3)_5]^{3+} + Cl^-$$

Anation reaction:

$$[ML_5S]^{n+} + X^- \rightarrow [ML_5X]^{(n-1)+} + S \qquad (11.21)$$

Example:

$$[Co(H_2O)(NH_3)_5]^{3+} + Cl^- \rightarrow [CoCl(NH_3)_5]^{2+} + H_2O$$

Ligand exchange reaction:

$$[ML_6]^{n+} + {}^*L \rightarrow [ML_5{}^*L]^{n+} + L \qquad (11.22)$$

Example:

$$[Fe(CN)_6]^{4-} + {}^*CN^- \rightarrow [Fe(CN)_5({}^*CN)]^{4-} + CN^-$$

Redox reaction (electron transfer reaction):

$$[ML_6]^{n+} + [ML'_6]^{(n-1)+} \rightarrow [ML_6]^{(n-1)+} + [ML'_6]^{n+} \qquad (11.23)$$

Example:

$$[Fe(phen)_3]^{2+} + [Fe(phen)_3]^{3+} \rightarrow [Fe(phen)_3]^{3+} + [Fe(phen)_3]^{2+}$$

Isomerization:

Example 1:

$$cis\text{-}[CoCl_2(en)_2]^+ \rightarrow trans\text{-}[CoCl_2(en)_2]^+$$

Example 2:

$$[Co(ONO)(NH_3)_5]^{2+} \rightarrow [Co(NO_2)(NH_3)_5]^{2+} \qquad (11.24)$$

11.1.5 Mechanism of the Ligand Substitution Reaction

Langford and Gray proposed three possible types of stoichiometric mechanisms for the ligand substitution reaction of metal complexes: dissociative mechanism (D mechanism), associative mechanism (A mechanism), and interchange mechanism (I mechanism).[3]

D mechanism $[ML_n] \rightarrow [ML_{n-1}] + L$

                     $[ML_{n-1}] + X \rightarrow [ML_{n-1}X]$

A mechanism $[ML_n] + X \rightarrow [ML_nX]$

                     $[ML_nX] \rightarrow [ML_{n-1}X] + L$

I mechanism $[ML_n] + X \rightarrow [X\cdots ML_{n-1}\cdots L]^{\ddagger} \rightarrow [ML_{n-1}X] + L$

The D mechanism involves a reaction intermediate with a reduced coordination number resulting from the dissociation of the leaving ligand from the metal center, whereas the A mechanism involves a reaction intermediate with an increased coordination number resulting from the binding of the entering ligand to the metal center. A common feature of both reaction mechanisms is the presence of reaction intermediates. In contrast, in the I mechanism, the reaction proceeds concertedly, where the entering ligand moves from the outer to the inner coordination sphere, and the leaving ligand moves from the inner to the outer coordination sphere. The I mechanism may involve a variety of transition states and can be further subdivided into an associative interchange mechanism (I_a mechanism) and a dissociative interchange mechanism (I_d mechanism). The transition state of the I_a mechanism may display substantial bonding of both the entering and leaving ligands to the central metal, and the reaction rate may be more sensitive to changes in the entering ligand. In the I_d mechanism, both the entering and leaving ligands weakly bind to the central metal in the transition state, and the reaction rate may be sensitive to changes in the leaving ligand. The D and I mechanisms can be compared with the S_N1 and S_N2 mechanisms of the nucleophilic substitution reaction on the carbon atoms of organic compounds.

11.1.6   Solvent Exchange Reaction

Among the various types of ligand substitution reactions, the most fundamental reaction involving metal ions in solution is the solvent exchange reaction (eqn (11.16)). In the solvent exchange reaction, the solvent molecules in the first coordination sphere of the metal ion are exchanged for the bulk solvent molecules. This reaction is important for understanding the reactivity of metal ions, and the interaction between metal ions and solvents is closely related to the complex formation and general ligand substitution reactions.

There is no net chemical change in the solvent exchange reaction, and the standard free energy change is zero. Isotope labeling and NMR methods have been used to measure the rate of solvent exchange reactions. The isotope labeling method has been used for slow solvent exchange reactions, such as Cr(III) and Rh(III), while the NMR technique has been applied to metal ions with relatively fast exchange rates. In the NMR method, the signals of the nuclei of the solvent molecules are observed. In the case of diamagnetic metal ions, the signals of both the solvent molecules coordinated to the metal ion and bulk solvent molecules are observed and the rate of the solvent exchange reaction can be determined from the temperature change in the shape of the NMR signals. When the central metal ion is paramagnetic, the magnitude of the linewidth of the NMR signals is affected by the spin-relaxation rate of the bulk and coordinated solvent molecules, the difference in the chemical shift in these two environments, and the rate of the solvent exchange reaction. The transverse relaxation times and chemical shifts of the observed nuclei of the bulk solvent molecules can be analyzed using the Swift–Connick equation to determine the rate of the solvent exchange reaction.[4]

Figure 11.2 shows examples of the solvent exchange reaction rates reported to date. The rate constants for the solvent exchange reactions range from 10^{-10} to 10^{10} s^{-1}. The solvent exchange reaction rate has been measured not only in aqueous solutions but also in various non-aqueous solvents. However, the reactivity pattern observed in an aqueous solution is generally maintained in other solvents, and the rate is mainly governed by the type and charge of the central metal ion.

The relationship between the rate of the solvent exchange reaction and the electronic configuration of the central metal ion in an aqueous solution is shown in Figure 11.3. The rate is relatively slow for V^{2+} (t_{2g}^3) and Ni^{2+} ($t_{2g}^6 e_g^2$), and fast for Cr^{2+} ($t_{2g}^3 e_g^1$) and Cu^{2+} ($t_{2g}^6 e_g^3$). Basolo and Pearson considered the reactivity of the solvent exchange reaction based on crystal field theory;[6] the energy of the d orbitals was calculated for various geometries of the possible transition states for the ligand substitution reaction. The crystal field stabilization energy (CFSE) was calculated for the ground state and intermediate, and the difference between these parameters was defined as the crystal field activation energy (CFAE). Table 11.1 shows the values of the CFSE and CFAE when assuming a five-coordinate square pyramidal structure and a seven-coordinate octahedral wedge structure as the intermediate. Based on these energies, the order of the reactivity of the solvent exchange reaction can be

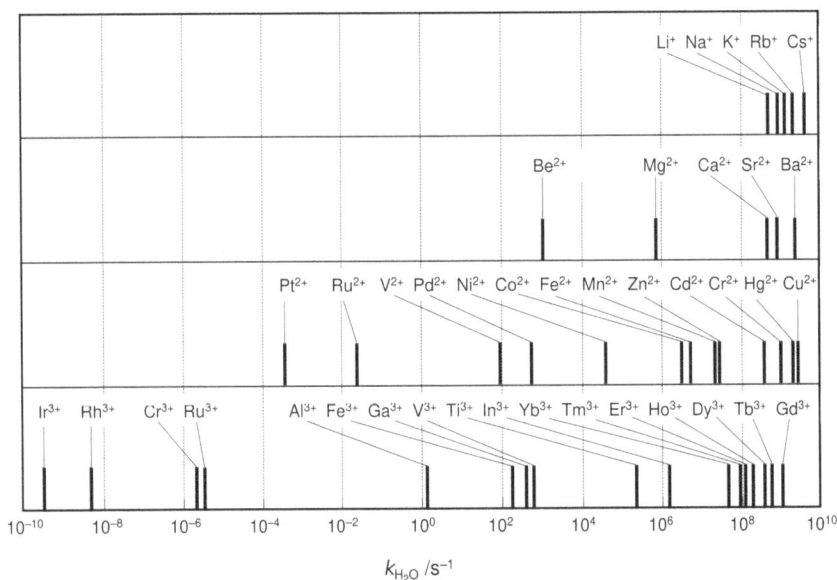

Figure 11.2 Rate constants for the water exchange reactions of metal ions (298 K). Reproduced from ref. 5 with permission from Swiss Chemical Society, Copyright 2019.

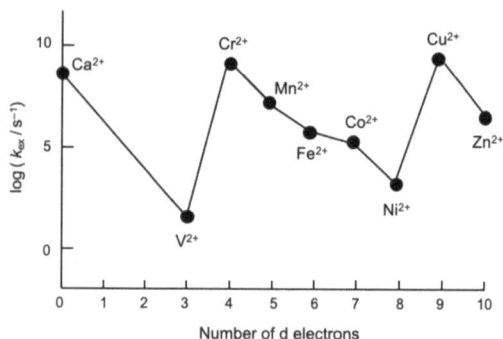

Figure 11.3 The relationship between the rate constants for the water exchange reaction of metal ions (298 K) and the number of d electrons.

Table 11.1 Crystal field activation energy for the two types of transition state with high-spin configurations.[a,b]

	CFSE			CFAE	
Number of d electrons	Octahedron	Square pyramid	Octahedral wedge	Square pyramid	Octahedral wedge
0, 10	0.00	0.00	0.00	0.00	0.00
1, 6	4.00	4.57	6.08	−0.57	−2.08
2, 7	8.00	9.14	8.68	−1.14	−0.68
3, 8	12.00	10.00	10.20	2.00	1.80
4, 9	6.00	9.14	8.97	−3.14	−2.79
5	0.00	0.00	0.00	0.00	0.00

[a] Units: D_q.
[b] Data are from ref. 6.

predicted. If the differences in D_q for different metal ions are disregarded, the following order of the reactivity for the transition state of the square pyramidal structure can be predicted: $d^4, d^9 > d^2, d^7 > d^1, d^6 > d^0, d^5, d^{10} > d^3, d^8$. It is also suggested that for d^1, d^3, d^6, and d^8 the transition state of the octahedral wedge structure is more energetically favorable than that of the square pyramidal structure. Although the D_q value varies with the metal ion and the contribution of the crystal field energy to the coordination bond is partial, such an argument is valid when considering the differences in the reactivity observed for similar complexes that differ only in the number of d electrons. The rate of the solvent exchange reaction of hydrated metal ions is shown in Figure 11.2, which shows that the solvent exchange reaction is slow for V^{2+} (d^3) and Ni^{2+} (d^8) and fast for Cr^{2+} (d^4) and Cu^{2+} (d^9), as expected from crystal field theory. In Cr^{2+} and Cu^{2+}, the hydrated metal ion has a distorted octahedral structure due to the Jahn–Teller effect, and the intramolecular conversion between the axial and equatorial positions as well as the solvent exchange reaction occurs in parallel. This effect is one of the reasons for the rapid solvent exchange reaction of these metal ions.

When comparing the rate of the solvent exchange reaction between divalent and trivalent metal ions with the same electron configuration, for example, V^{2+} *vs.* Cr^{3+} and Mn^{2+} *vs.* Fe^{3+}, the reaction rate of the trivalent metal ions is far slower. The relationship between the rate of the solvent exchange reaction and the ionic radius of the metal ion for the metal ions of groups 2 and 12, which have equal charges, shows that the rate of the solvent exchange reaction is faster for ions with larger ionic radii. This phenomenon may reflect the magnitude of the electrostatic interactions formed between the solvent molecules (water molecules) and metal ions.

11.2 Complex Formation Reaction

A reaction in which an aqua-metal ion reacts with a ligand in solution to form a metal complex is called a complex formation (complexation) reaction (eqn (11.17)). Because the metal ion is coordinated with solvent molecules in solution, the complex formation reaction is a type of ligand substitution reaction in which the coordinated solvent molecule is replaced by a ligand.

Consider the complexation reaction between a six-coordinate aqua-metal ion ($[M(H_2O)_6]^{n+}$) and ligand (L^{m-}) in an aqueous solution. The coordinating water molecules in $[M(H_2O)_6]^{n+}$ constantly exchange with the bulk water molecules (solvent exchange reaction), and the rate of this exchange depends on the nature of the metal ion and its valence (Figure 11.2). As $[M(H_2O)_6]^{n+}$ is a Brønsted acid whose strength changes depending on the nature of the metal ion and its charge, deprotonation occurs at the coordinated water molecules (acid dissociation reaction, hydrolysis, or proton transfer reaction) depending on the pH of the aqueous solution (Section 11.1.4). On the other hand, because the ligand is a Lewis base, the Lewis acid H^+ binds to the lone pair of electrons (protonation reaction) of the ligand, depending on the pH of the aqueous solution. Under the conditions that the hydrolysis reaction of a metal ion and the protonation reaction to a ligand do not occur, the complexation reaction of $[M(H_2O)_6]^{n+}$ with L^{m-} can be generally expressed using the following equation:

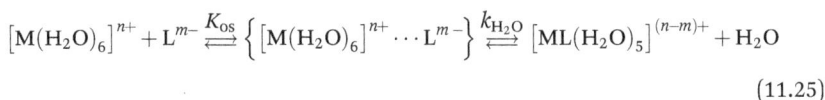

$$[M(H_2O)_6]^{n+} + L^{m-} \underset{K_{os}}{\rightleftharpoons} \left\{ [M(H_2O)_6]^{n+} \cdots L^{m-} \right\} \underset{k_{H_2O}}{\rightleftharpoons} [ML(H_2O)_5]^{(n-m)+} + H_2O$$

$$(11.25)$$

where $\{[M(H_2O)_6]^{n+} \cdots L^{m-}\}$ is an outer-sphere complex (also called an outer-sphere ion pair when ligand L has a negative charge), and its formation is a diffusion-controlled process. During the formation of the inner-sphere complex $[ML(H_2O)_5]^{(n-m)+}$ (corresponding to an ordinary complex) from the outer-sphere complex, the substitution reaction of the coordinated water molecule by ligand L takes place.

When the total concentration of the ligand (C_L) is in large excess ($C_M \ll C_L$) when compared to the total concentration of the metal ion (C_M) (pseudo-first-order conditions), higher-order complexes, such as $[ML_2(H_2O)_4]^{(n-2m)+}$

and $[ML_3(H_2O)_3]^{(n-3m)+}$, can be formed. However, the coordination rates of the second L^{m-}, third L^{m-},... are generally much faster than that of the first L^{m-}, so the formation of higher-order complexes is not rate-limiting. When L^{m-} is a multidentate ligand, chelate-ring closure is usually not rate-limiting. Therefore, when $[M(H_2O)_6]^{n+}$ and ligand L^{m-} react to form the inner-sphere complex, $[ML(H_2O)_5]^{(n-m)+}$, the observed pseudo-first-order rate constant (k_{obs}) can be expressed by eqn (11.26) using the formation constant (K_{os}) of the outer sphere complex and the rate constant (k_{H_2O}) of the rate-determining step.

$$k_{obs} = \frac{K_{os}k_{H_2O}}{1 + K_{os}C_L} C_L \tag{11.26}$$

Because the ligand concentration (C_L) is usually set to $<10^{-2}$ M (M $=$ mol L^{-1}), $1 \gg K_{os}C_L$ often holds, and $k_{obs} = k_f C_L$ $(C_M \ll C_L)$, the second-order rate constant (k_f) independent of the ligand concentration is given by eqn (11.27).

$$k_f = K_{os}k_{H_2O} \tag{11.27}$$

where the value of K_{os} is often estimated using the Fuoss equation.

Table 11.2 shows the k_f values observed for the complex formation reactions of Ni^{2+} ions. The k_f value changes depending on the charge of the ligand because the value of K_{os} changes depending on the charge of the ligand. Table 11.2 shows that the k_f values are almost constant for ligands with the same charge. Table 11.3 shows the k_{H_2O} values obtained using k_f and the K_{os} value estimated using the Fuoss equation. The values of k_{H_2O} were almost constant regardless of the type of ligand. This indicates that the dissociation process of the coordinated water molecules in $\{[M(H_2O)_6]^{n+}, L^{m-}\}$ is rate-limiting for the complexation of Ni^{2+} ions. Therefore, the complexation mechanism of Ni^{2+} ions corresponds to the I_d mechanism in the Langford and Gray classification (Section 11.1.5). The reaction mechanism, which is represented by eqn (11.25) and generally accepted, is called the Eigen mechanism or the Eigen–Wilkins mechanism.

Table 11.2 The formation rate constants (k_f) obtained for Ni(II) complexes (25 °C).

Charge of ligand	Ligand	$\log(k_f/$ M^{-1} s$^{-1})$	Charge of ligand	Ligand	$\log(k_f/$ M^{-1} s$^{-1})$
0	NH$_3$	3.6	1−	Glycine$^-$ ^{a}	4.3
	NH$_2$NH$_2$	3.4		Diglycine$^-$	4.6
	Imidazole	3.7		Triglycine$^-$	4.6
	Pyridine	3.7	2−	C$_2$O$_4^{2-}$	4.9
	Phen	3.6		CH$_2$(CO$_2$)$_2^{2-}$	4.8
	Bpy	3.2		ida^{2-}	4.9
	terpy	3.1	3−	HP$_2$O$_7^{3-}$	6.3
1−	SCN$^-$	3.7		Hedta^{3-}	5.3
	HC$_2$O$_4^-$	3.7	4−	HP$_3$O$_{10}^{4-}$	6.8
	HO$_2$CCH$_2$CO$_2^-$	3.5			

^{a} H$_2$NCH$_2$CO$_2^-$.

Table 11.3 The formation rate constants (k_{H_2O}) calculated for the inner-sphere complexes of Ni(II) (25 °C).[4,7]

Ligand	k_{H_2O} (25 °C)/10^4 s^{-1}
H_2O	2.7
Imidazole	1.6
SCN^-	0.6
$HC_2O_4^-$	0.9
Glycine$^-$	0.9
SO_4^{2-}	1.5
$C_2O_4^{2-}$	0.6
$HP_2O_7^{3-}$	1.2
$HP_3O_{10}^{4-}$	1.2

11.2.1 Complex Formation Reactions of Aqua-metal Ions (M^{n+}) and Hydroxo-metal Ions ($M(OH)^{(n-1)+}$)

According to the classification of Langford and Gray described in Section 11.1.5, the complexation of metal ions can be classified into two groups, group (i) and group (ii); group (i) involves metal ions such as Fe^{2+}, Co^{2+}, Ni^{2+}, Al^{3+}, and Ga^{3+}, in which the complexation rate constants are almost invariant irrespective of the type of ligand and the complex formation reaction appears to proceed *via* a dissociative mechanism; group (ii) involves metal ions such as Cr^{3+}, Mn^{2+}, and Fe^{3+}, in which the rate constants vary considerably depending on the type of ligand and the complex formation reaction appears to proceed *via* an associative mechanism. In contrast, the complexation rate constant of $M(OH)^{(n-1)+}$, which is the conjugate base of M^{n+}, is several orders of magnitude greater than the rate constant of M^{n+} and is almost constant regardless of the type of ligand. In other words, the complex formation reaction of hydrolyzed metal ions proceeds *via* a dissociative mechanism.

The fact that the rate constant of $M(OH)^{(n-1)+}$ is several orders of magnitude larger than that of M^{n+} means that the reaction of $M(OH)^{(n-1)+}$ cannot be ignored kinetically, even under sufficiently low pH conditions in which the amount of $M(OH)^{(n-1)+}$ is negligible when compared to M^{n+} (*i.e.*, the complex formation reactions of M^{n+} and $M(OH)^{(n-1)+}$ may occur concurrently).

As an example, we can consider the mechanism and reaction pathway of the complex formation reactions of Fe^{3+} and $Fe(OH)^{2+}$.[8] When the total concentration of Fe(III) ions (C_{Fe}) and pH are high, Fe^{3+}, which is a fairly strong Brønsted acid ($pK_a = 2.74$), tends to form polynuclear species and/or precipitates. Therefore, the complexation reaction of Fe^{3+} ions is carried out under considerably acidic conditions (for example, pH = 1), low C_{Fe}, and pseudo-first-order conditions, in which the ligand concentration C_L is larger than C_{Fe} ($C_{Fe} \ll C_L$). Under these conditions, the species to be considered is Fe^{3+}. However, the reaction of $FeOH^{2+}$ should be considered because

$Fe(OH)^{2+}$ is several orders of magnitude more reactive than Fe^{3+}. When the deprotonation of the neutral ligand (HL) occurs and both Fe^{3+} and $Fe(OH)^{2+}$ are potentially reactive under the experimental conditions, four complex formation pathways should be considered (eqn (11.28)). Higher-order complexes can be formed when $C_{Fe} \ll C_L$, but their formation processes are usually not rate-limiting. For simplicity, consider the case in which the contribution of the reverse reactions is negligible.

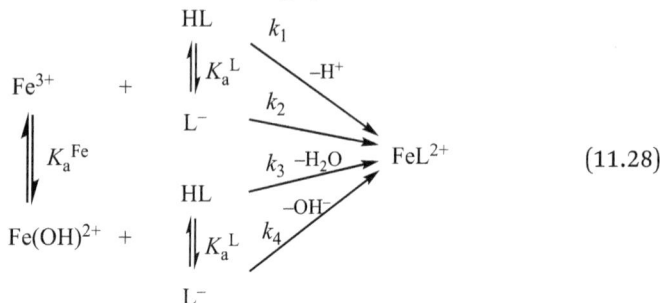

$$
\begin{array}{c}
\text{HL} \quad k_1 \\
Fe^{3+} \quad + \quad \Big\Vert K_a^L \quad \searrow \; -H^+ \\
\qquad\qquad\qquad k_2 \\
\Big\Vert K_a^{Fe} \qquad L^- \qquad k_3 \; -H_2O \longrightarrow FeL^{2+} \\
\qquad\qquad \text{HL} \qquad\qquad -OH^- \\
Fe(OH)^{2+} + \quad \Big\Vert K_a^L \; k_4 \\
\qquad\qquad\qquad L^-
\end{array}
\tag{11.28}
$$

In eqn (11.28), the formation rate of the product (FeL^{2+}) is given by eqn (11.29),

$$
\begin{aligned}
\frac{d[FeL^{2+}]}{dt} &= k_1[Fe^{3+}][HL] + k_2[Fe^{3+}][L^-] + k_3[Fe(OH)^{2+}][HL] \\
&\quad + k_4[Fe(OH)^{2+}][L^-] \\
&= (k_1[HL] + k_2[L^-])[Fe^{3+}] + (k_3[HL] + k_4[L^-])[Fe(OH)^{2+}]
\end{aligned}
\tag{11.29}
$$

where

$$
C_{Fe} = [Fe^{3+}] + [Fe(OH)^{2+}] + [FeL^{2+}]
\tag{11.30}
$$

$$
C_L = [HL] + [L^-] + [FeL^{2+}]
\tag{11.31}
$$

$$
K_a^{Fe} = \frac{[Fe(OH)^{2+}][H^+]}{[Fe^{3+}]}
\tag{11.32}
$$

$$
K_a^L = \frac{[L^-][H^+]}{[HL]}.
\tag{11.33}
$$

By substituting these equations into eqn (11.29) gives eqn (11.34).

$$
\begin{aligned}
\frac{d[FeL^{2+}]}{dt} &= \frac{k_1 + k_2 K_a^L[H^+]^{-1} + k_3 K_a^{Fe}[H^+]^{-1} + k_4 K_a^{Fe}[H^+]^{-2}}{(1 + K_a^L[H^+]^{-1})(1 + K_a^{Fe}[H^+]^{-1})} \\
&\quad \times (C_L - [FeL^{2+}])(C_{Fe} - [FeL^{2+}])
\end{aligned}
\tag{11.34}
$$

Since $C_{Fe} \ll C_L$,

$$\frac{d[\text{FeL}^{2+}]}{dt} = \frac{k_1 + k_2 K_a^L[\text{H}^+]^{-1} + k_3 K_a^{Fe}[\text{H}^+]^{-1} + k_4 K_a^{Fe}[\text{H}^+]^{-2}}{(1 + K_a^L[\text{H}^+]^{-1})(1 + K_a^{Fe}[\text{H}^+]^{-1})} \tag{11.35}$$

$$\times C_L(C_{Fe} - [\text{FeL}^{2+}])$$

This equation can be integrated because it contains only one concentration variable ($[\text{FeL}^{2+}]$). However, integration is not necessary at this point for the following reasons.

When the rate of complex formation is first-order with respect to the total concentration of uncomplexed Fe(III) ions ($C_{Fe} - [\text{FeL}^{2+}] = [\text{Fe}^{3+}] + [\text{Fe(OH)}^{2+}]$) under $C_{Fe} \ll C_L$, the rate equation is given by eqn (11.36).

$$\frac{d[\text{FeL}^{2+}]}{dt} = k_{obs}(C_{Fe} - [\text{FeL}^{2+}]) \tag{11.36}$$

A comparison of eqn (11.35) and (11.36) gives k_{obs} (eqn (11.37)); integration is unnecessary. This is always true, as described by Jordan.[9]

$$k_{obs} = \frac{k_1 + k_2 K_a^L[\text{H}^+]^{-1} + k_3 K_a^{Fe}[\text{H}^+]^{-1} + k_4 K_a^{Fe}[\text{H}^+]^{-2}}{(1 + K_a^L[\text{H}^+]^{-1})(1 + K_a^{Fe}[\text{H}^+]^{-1})} C_L \tag{11.37}$$

When the plot of k_{obs} measured at different C_L and constant $[\text{H}^+]$ vs. C_L yields a straight line with zero intercept, k_{obs} is given by eqn (11.38), where k_f is the second-order rate constant.

$$k_{obs} = k_f C_L \tag{11.38}$$

Therefore, the following equation is obtained from eqn (11.37) and (11.38).

$$\left(1 + \frac{K_a^L}{[\text{H}^+]}\right)\left(1 + \frac{K_a^{Fe}}{[\text{H}^+]}\right)k_f = k_1 + \frac{k_2 K_a^L}{[\text{H}^+]} + \frac{k_3 K_a^{Fe}}{[\text{H}^+]} + \frac{k_4 K_a^{Fe}}{[\text{H}^+]^2} \tag{11.39}$$

If the plot of the left-hand side values in eqn (11.39) vs. $1/[\text{H}^+]$ yields a straight line with an intercept, the contribution of the k_4 pathway is negligible and the k_1 value and sum of $k_2 K_a^L$ and $k_3 K_a^{Fe}$ are obtained from the intercept and slope, respectively. However, only the sum of $k_2 K_a^L$ and $k_3 K_a^{Fe}$ is obtained and each value cannot be obtained independently. This implies that the k_2 and k_3 pathways in eqn (11.28) are kinetically indistinguishable. This is often called "proton ambiguity".[3] However, even in such a case, at least one of the reaction paths of k_2 and k_3 is involved in the reaction because the slope is not zero. Therefore, the upper limit values of k_2 and k_3 can be estimated, provided that only one of the two pathways contributes to the reaction. The pathway involved in the reaction can be determined chemically from the estimated upper limit values.

Table 11.4 shows the values of the rate constants for the complex formation reactions of Fe^{3+} and $Fe(OH)^{2+}$, as previously described. This table also includes the results of reaction systems without "proton ambiguity"; "proton ambiguity" can be avoided when the conjugate bases of strong acids (Br^-, Cl^-, SCN^-, *etc.*) are used as ligands.

Table 11.4 shows that the rate constant of Fe^{3+} varies considerably depending on the type of ligand, but the change is quite small in the case of $Fe(OH)^{2+}$, indicating that the former occurs *via* an I_a mechanism and the latter *via* an I_d mechanism.

11.2.2 Complex Formation Reaction with a Bidentate Ligand

We now consider the chelate formation reaction of an aqua-metal ion with a bidentate ligand, represented by the following equation:

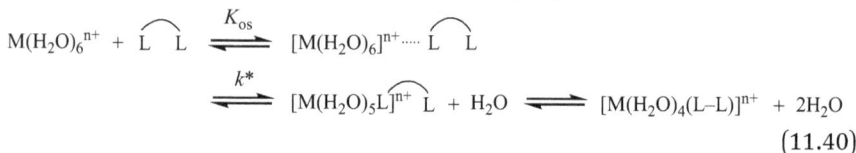

$$M(H_2O)_6{}^{n+} + \overset{\frown}{L \quad L} \underset{k^*}{\overset{K_{os}}{\rightleftharpoons}} [M(H_2O)_6]^{n+} \cdots \overset{\frown}{L \quad L}$$

$$\rightleftharpoons [M(H_2O)_5L]^{n+} \overset{\frown}{\underset{}{\quad L}} + H_2O \rightleftharpoons [M(H_2O)_4(L\text{–}L)]^{n+} + 2H_2O$$

$$(11.40)$$

The first process is similar to the complex formation reaction with a monodentate ligand shown in eqn (11.17). The aqua-metal ion forms an outer-sphere complex with the bidentate ligand, and subsequently a solvent molecule coordinated to the metal ion in the inner-coordination sphere of the aqua-metal ion is replaced by the coordination atom of the bidentate ligand to form an unchelated inner-sphere complex, in which the bidentate ligand coordinates in a monodentate manner. In the next step, a second coordinating solvent molecule is replaced by the remaining coordination atom in the bidentate ligand and chelate-ring closure occurs. Using an aqua-metal ion that reacts *via* a dissociative mechanism as an example, at the time

Table 11.4 The complexation rate constants of Fe^{3+} and $Fe(OH)^{2+}$ (25 °C).[8]

Ligand	$k_4/M^{-1} s^{-1}$ for $Fe(OH)^{2+}$	$k_2/M^{-1} s^{-1}$ for Fe^{3+}
$SO_4{}^{2-}$	1.1×10^5	2.3×10^3
Cl^-	5.5×10^3	4.8
Br^-	2.8×10^3	1.6
NCS^-	5.1×10^3	9.0×10^1
$Cl_3CCO_2{}^-$	7.8×10^3	6.3×10^1
$Cl_2HCCO_2{}^-$	1.9×10^4	1.2×10^2
$ClH_2CCO_2{}^-$	4.1×10^4	1.5×10^3
$HC_3CO_2{}^-$	$\leqq 2.8 \times 10^3$	2.7×10^1
C_6H_5OH	1.5×10^3	
$C_{10}H_{12}(=O)(OH)^a$	6.3×10^3	2.2×10^1
$H_3CC(O)NH(OH)$	2.0×10^3	1.2
H_2O (in s^{-1})	1.2×10^5	1.6×10^2

^{a} 4-Isopropyltropolone.

of ring closure, the dissociated solvent molecule (or another solvent molecule) competes with the remaining coordination atom in the bidentate ligand for the empty coordination site formed *via* the dissociation of a second solvent molecule. For aqua-metal ions with a relatively slow solvent exchange or complexation, such as Ni^{2+} ions, the rate-determining step is not chelate-ring closure. Because there is another coordination atom in the immediate vicinity of the open coordination site, the local concentration is high and, as a result, the reaction probability becomes high. In this case, the rate of the reaction is almost the same as that of a similar monodentate ligand. On the other hand, the recombination of solvent molecules is more advantageous in the case of aqua metal ions such as Co^{2+}, which have a considerably fast solvent exchange rate, or when ring closure is hindered, such as when a large chelate ring is formed. Thus, the rate of chelate closure is decreased. In the case of aqua-metal ions such as Cu^{2+} and Cr^{2+}, in which the solvent exchange reaction is extremely fast, the time required for the recombination of solvent molecules is much shorter than the time required for the remaining coordination atoms in the bidentate ligand to move to a position where the chelate can be closed, so the chelate ring closure becomes rate-limiting. In this case, the complex formation rate with the bidentate ligand is much slower than that with the monodentate ligand and the degree of rate reduction depends significantly on the structure of the ligand. Such a reaction mechanism in which the chelate ring closure becomes rate-limiting is called a sterically controlled substitution (SCS) mechanism.[1]

11.2.3 Complexation Reaction of Porphyrins

Figure 11.4 shows porphyrins which are aromatic heterocyclic macrocycle compounds that serve as tetradentate ligands and have planar molecular structures. Two of the four nitrogen atoms in the center of a porphyrin have hydrogen atoms attached to them, but these hydrogen atoms dissociate as hydrogen ions during the complexation reaction, so the porphyrin ligand has a 2− charge in the metal complex. The distance between the nuclei of the two facing nitrogen atoms in the porphyrin is ∼400 pm,[10] and relatively

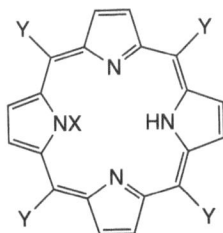

Figure 11.4 Porphyrin structure and abbreviations. TPP (H_2tpp): X=H, Y=phenyl; N-MeTPP: X=CH$_3$, Y=phenyl; TPPS (H_2tpps): X=H, Y=4-sulfonatophenyl.

small transition metal ions, such as Cu^{2+} (ionic radius: 73 pm), can enter this central cavity. However, larger metal ions such as Cd^{2+} (95 pm) and Hg^{2+} (102 pm) cannot fully enter the cavity and are placed slightly above the porphyrin plane.

The complex formation reaction of the porphyrin can be expressed using the following equation:

$$H_2\,porphyrin + M^{2+} \rightleftharpoons [M(porphyrin)] + 2H^+ \qquad (11.41)$$

The rate of the complexation reaction of porphyrins is much slower than that of acyclic ligands or the solvent exchange reaction. Figure 11.5 shows the rate of the complexation reaction of tetraphenylporphyrin (TPP) in DMF.[11,12] The rate of the porphyrin complexation reaction is several orders of magnitude slower than that of the solvent exchange reaction. This slow formation of metal porphyrin complexes is thought to be due to the planar molecular structure of porphyrins. Because the lone pairs of electrons on the nitrogen atoms of porphyrins are oriented toward the porphyrin center, the porphyrin molecule must be distorted and the lone pairs of electrons must face out of the porphyrin plane to interact with the metal ion during complexation, which may have a kinetic disadvantage in the porphyrin complexation reaction. In addition to the usual Eigen–Wilkins mechanism, an equilibrium in which the porphyrin molecule is distorted from its planar structure is assumed before the rate-determining step of the reaction. Assuming that the equilibrium constant for the deformation of the porphyrin

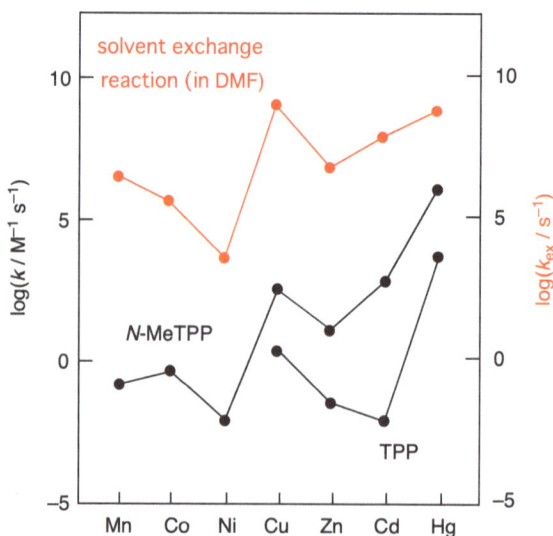

Figure 11.5 A comparison of the rates of the complexation reaction of porphyrins and the solvent exchange reaction of divalent metal ions in DMF ($T = 298$ K).[11,12]

molecule is K_D and that the rate-determining step of the porphyrin complexation reaction is a bond formation with the first nitrogen atom, the second-order rate constant for the formation of the metal porphyrin complex can be given using the following equation:

$$k_f = K_D K_{os} k_S \tag{11.42}$$

where k_S is the rate constant of the reaction in which the coordinated solvent molecule dissociates from the metal ion. The equilibrium constant for the reaction that distorts the porphyrin molecule from its planar structure should be very small $(K_D \ll 1)$, leading to a decrease in the rate of the complexation reaction.

On the other hand, in N-MeTPP, one of the hydrogen atoms bonded to the pyrrole nitrogen atoms of the porphyrin is replaced by a methyl group, the porphyrin molecule is deformed from a plane due to the steric hindrance of the methyl group, and the lone pairs of electrons on the nitrogen atom face out-of-plane. Therefore, its interaction with metal ions occurs more easily than in TPP, and the complexation reaction of N-MeTPP is faster than that of TPP. In both cases, there is a relationship between the rate of the solvent exchange reaction of metal ions and the rate of the complexation reaction of porphyrins, as shown in Figure 11.5.

The complexation reactions of porphyrins have also been investigated in aqueous solution. The rate equations for the forward and reverse reactions of the complexation reaction of TPPS with Zn(ɪɪ) and Cd(ɪɪ) ions can be expressed as follows:[13]

$$\text{Rate of forward reaction} = k_f [H_2 \text{tpps}] [M^{2+}] \tag{11.43}$$

$$\text{Rate of reverse reaction} = k_d [M(\text{tpps})] [H^+]^2 \tag{11.44}$$

The second-order rate constants for the forward reaction are $k_f = 1.13 \text{ M}^{-1}\text{s}^{-1}$ for Zn^{2+} and $5.21 \times 10^2 \text{ M}^{-1}\text{s}^{-1}$ for Cd^{2+} at $T = 298$ K, which are orders of magnitude lower than the rate constants of ordinary complexation reactions. The rate of the reverse reaction quadratically depends on the hydrogen ion concentration: $k_d = 9.37 \text{ M}^{-2}\text{s}^{-1}$ for the Zn(ɪɪ) complex and $5.49 \times 10^{12} \text{ M}^{-2}\text{s}^{-1}$ for the Cd(ɪɪ) complex at $T = 298$ K. This indicates that, in the reverse reaction, two hydrogen ions bind to the porphyrin complex prior to the rate-determining step. The values of the formation constants estimated using the rate constants are $K = 0.12$ M for Zn(ɪɪ) and $K = 9.5 \times 10^{-11}$ M for Cd(ɪɪ). The formation constant for Cd(ɪɪ) is much smaller than that for Zn(ɪɪ), indicating that the Cd(ɪɪ) ion is too large to enter the central porphyrin cavity. This difference can be mainly attributed to the difference in the rate of the reverse reaction and the reactivity of the metal complex with the hydrogen ion is much higher in the Cd(ɪɪ) complex, reflecting its molecular structure.

11.2.4 Acid Dissociation Reaction

The simplest reaction of an aqua-metal ion is perhaps the acid dissociation (hydrolysis, proton transfer) reaction to form a hydroxo-complex (eqn (11.45)). When a water molecule coordinates to a metal ion, deprotonation from the water molecule becomes electrostatically easier (Figure 11.6). Therefore, aqua-metal ions are stronger Brønsted acids than water.

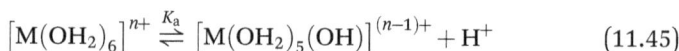

$$[M(OH_2)_6]^{n+} \overset{K_a}{\rightleftharpoons} [M(OH_2)_5(OH)]^{(n-1)+} + H^+ \qquad (11.45)$$

The greater the positive charge on the metal ion, the easier the proton dissociation from the bound water molecule, as shown in Table 11.5. The data in Table 11.5 also show that the pK_a values depend on the ionic radius, as well as the ionic charge.

For three- and four-valent aqua-metal ions with a low pK_a, the solution must be sufficiently acidic to make the amount of hydroxo-complexes negligible, especially for most 4+ aqua-metal ions, there exists a small but significant amount of hydroxo-complexes even in strongly acidic aqueous solutions.

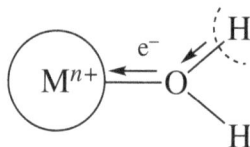

Figure 11.6 Deprotonation of aqua-metal ions.

Table 11.5 Acid dissociation constants of aqua-metal ions.[14]

Charge	pK_a						
4+	Ce^{4+}	Zr^{4+}	Hf^{4+}	Th^{4+}	U^{4+}	Np^{4+}	Pu^{4+}
	0.7	0.22	−0.12	4.31	1.68	2.3	1.6
3+	Sc^{3+}	Y^{3+}	La^{3+}	Ce^{3+}	Pr^{3+}	Nd^{3+}	Lu^{3+}
	5.11	9.1	10.1	9.13	8.5	8.5	6.6
	V^{3+}	Cr^{3+}	Pu^{3+}	Fe^{3+}	Co^{3+}	Al^{3+}	Ga^{3+}
	2.7	3.95	6.95	2.74	1.78	5.0	2.9
	In^{3+}	Tl^{3+}					
	4.41	1.18					
2+	Be^{2+}	Mg^{2+}	Ca^{2+}	Sr^{2+}	Ba^{2+}	Mn^{2+}	Fe^{2+}
	5.7	12.8	12.63	13.18	13.15	10.6	6.74
	Co^{2+}	Ni^{2+}					
	9.85	9.76					
1+	Tl^+	Li^+	Na^+, K^+, Cs^+				
	13.5	14.4	>14				

A second proton can dissociate from a second water molecule in the inner sphere of the metal ion, which may be followed by further dissociation (eqn (11.46)).

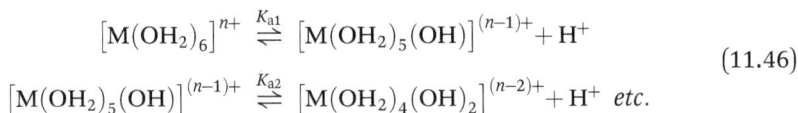

$$[M(OH_2)_6]^{n+} \overset{K_{a1}}{\rightleftharpoons} [M(OH_2)_5(OH)]^{(n-1)+} + H^+$$

$$[M(OH_2)_5(OH)]^{(n-1)+} \overset{K_{a2}}{\rightleftharpoons} [M(OH_2)_4(OH)_2]^{(n-2)+} + H^+ \ etc.$$

(11.46)

The hydroxo-complexes of metal ions, such as $Al(OH)^{2+}$ and $Fe(OH)^{2+}$, have a strong tendency to form hydroxo- and oxo-bridged species (eqn (11.47)), which leads to the formation of miscellaneous polynuclear species.

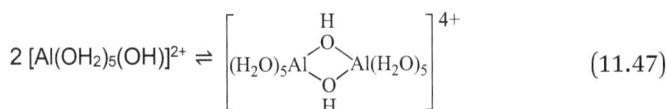

$$2 \ [Al(OH_2)_5(OH)]^{2+} \rightleftharpoons \left[(H_2O)_5Al \underset{\underset{H}{O}}{\overset{\overset{H}{O}}{<}} Al(H_2O)_5 \right]^{4+}$$

(11.47)

For aqua-metal ions with a charge of 4+ and higher, it is also possible for a single coordinated water molecule to dissociate two protons to form an oxo-aqua complex (eqn (11.48)).

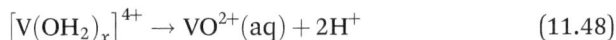

$$[V(OH_2)_x]^{4+} \rightarrow VO^{2+}(aq) + 2H^+$$

(11.48)

In general, proton transfer reactions are very fast and have large rate constants (k_r), as shown in Table 11.6. These rate constants were measured by Eigen *et al.* using various relaxation methods. The reaction between H^+ and OH^- is the fastest and is a diffusion-controlled process in an aqueous solution. Table 11.6 includes reactions that are considerably slower than this reaction. In these slower reactions, the protons to be released are involved in intramolecular hydrogen bonding or the structural changes that occur during deprotonation.

The reactions between the hydrolyzed species (for example, $Al(OH)^{2+}$ and $Cu(OH)^+$) and H^+ (eqn (11.49)) are also diffusion-controlled processes. The

Table 11.6 Proton transfer reactions and their rate constants (k_r).[15]

Reaction system	k_r (25 °C)/$M^{-1}\,s^{-1}$
$H^+ + OH^- \rightarrow H_2O$	1.4×10^{11}
$H^+ + HS^- \rightarrow H_2S$	7.5×10^{10}
$H^+ + CH_3COO^- \rightarrow CH_3COOH$	4.5×10^{10}
$H^+ + NH_3 \rightarrow NH_4^+$	4.3×10^{10}
$H^+ + Al(OH)^{2+} \rightarrow Al^{3+}$	4.4×10^9
$H^+ + Cu(OH)^+ \rightarrow Cu^{2+}$	1×10^{10}
$H^+ + HCO_3^- \rightarrow CO_2 + H_2O$	5.6×10^4
$OH^- + NH_4^+ \rightarrow NH_3 + H_2O$	3.4×10^{10}
$OH^- + C_6H_5OH \rightarrow C_6H_5O^- + H_2O$	1×10^{10}
$OH^- + {}^+HN(CH_2COO^-)_3 \rightarrow N(CH_2COO^-)_3 + H_2O$	1.4×10^7

forward rate constant (k_r) (protonation) and reverse rate constant (k_{OH}) (deprotonation, acid dissociation, or hydrolysis) are related to the acid dissociation constant (K_a) as $K_a = k_{OH}/k_r$.

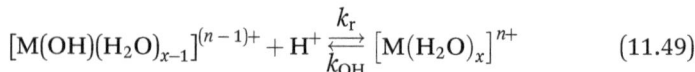

$$[M(OH)(H_2O)_{x-1}]^{(n-1)+} + H^+ \underset{k_{OH}}{\overset{k_r}{\rightleftharpoons}} [M(H_2O)_x]^{n+} \tag{11.49}$$

Table 11.6 shows $k_r \sim 10^{11}\ M^{-1}\ s^{-1}$ and k_r does not vary greatly depending on the hydrolyzed species, so the difference in the acid dissociation constant $(K_a = k_{OH}/k_r)$ mainly reflects the difference in k_{OH}. For example, in the case of $M^{n+} = Ni^{2+}$, $k_{OH} \sim 20\ s^{-1}$ (p$K_a = 9.76$). Under the conditions that the total concentration of Ni^{2+} is 1.0×10^{-3} M and $[H^+] = 1.0 \times 10^{-5}$ M, the rate of hydrolysis reaction is much slower than that of the protonation reaction. However, the hydrolysis reaction does not occur alone, but always occurs concurrently with the protonation reaction. In other words, the rate constant (k_{obs}/s^{-1}) observed under the present conditions is equal to $k_r[H^+] + k_{OH}$. Therefore, the equilibrium of the hydrolysis reaction of Ni^{2+} ions will be established sufficiently fast when compared to the ligand substitution reactions, including solvent exchange and complex formation reactions. However, in the reaction systems where hydrolysis proceeds further to form polynuclear hydrolyzed species (*e.g.*, $Fe_3(OH)_4{}^{5+}$ and $Pb_4(OH)_4{}^{4+}$), the rates of formation of these hydrolyzed species are very low and may take several days or more to reach equilibrium. Therefore, care must be taken when discussing such reaction systems.

11.2.5 Mechanism of the Ligand Substitution Reaction and Rate Law

The mechanism of the ligand substitution reaction can be determined based on a kinetic study in which the nature of the reactants is changed in a systematic manner. In the case of the A and I mechanisms, the reaction rate will be first-order in both the metal complex and entering ligand. The kinetics show sensitivity to the nature of the entering ligand and the leaving ligand for associative and dissociative activation, respectively. Even in the case of the D mechanism, where the rate-determining step is the dissociation of the leaving ligand, the reaction rate may depend on the concentration of the entering ligand, which depends on the reaction conditions. In general, it is difficult to determine the reaction mechanism based on the rate law.

The reaction mechanism that can be most clearly attributed is the one that proceeds *via* reaction intermediates that can be observed experimentally. An example of such a case is the D mechanism, which includes an intermediate with a reduced coordination number. The reaction mechanism can be expressed using the following equations:

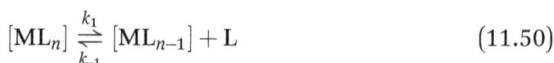

$$[ML_n] \underset{k_{-1}}{\overset{k_1}{\rightleftharpoons}} [ML_{n-1}] + L \tag{11.50}$$

$$[ML_{n-1}] + X \underset{k_{-2}}{\overset{k_2}{\rightleftharpoons}} [ML_{n-1}X] \tag{11.51}$$

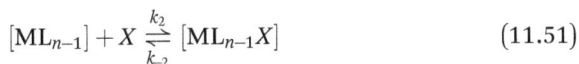

In this mechanism, a reaction intermediate $[ML_{n-1}]$ is generated *via* the dissociation of the leaving ligand, which reacts with the entering ligand to give the final product, $[ML_{n-1}X]$. The application of steady-state approximation to this intermediate gives the pseudo-first-order rate constant, as expressed by eqn (11.52) under conditions in which the concentrations of both the entering and leaving ligands are in large excess when compared to the metal complex.

$$k_0 = \frac{k_1 k_2 [X] + k_{-1} k_{-2}[L]}{k_2 [X] + k_{-1}[L]} = \frac{k_1 [X] + k_{-2}(k_{-1}/k_2)[L]}{[X] + (k_{-1}/k_2)[L]} \tag{11.52}$$

An example of these systems is the axial ligand substitution reaction of the Cr(III) tetraphenylporphyrin complex, [Cr(TPP)(Cl)(L)], as expressed by eqn (11.53).[16]

$$[Cr(TPP)(Cl)(L)] + X \rightarrow [Cr(TPP)(Cl)(X)] + L \tag{11.53}$$

where L and X represent uncharged ligands such as pyridine and 1-methylimidazole, respectively. In a non-coordinating solvent, such as toluene, the pseudo-first-order rate constant can be expressed using eqn (11.52). The rate of the substitution reaction increases upon increasing the concentration of the entering ligand but eventually converges to a constant value (k_1). The values of k_{-1}/k_2 represent the relative reactivity of the ligands toward the reaction intermediate. The reaction of the intermediate with an external ligand (X) was studied using a photochemical technique, and it was revealed that the intermediate was highly reactive and the conditions for the D mechanism were satisfied.[17]

It should be mentioned that there are other reaction mechanisms that provide the same rate law as the D mechanism. An example of the ligand substitution reaction, which involves the pre-association of metal complexes with an entering ligand, is shown in eqn (11.54).

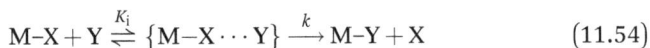

$$M{-}X + Y \overset{K_i}{\rightleftharpoons} \{M{-}X\cdots Y\} \overset{k}{\longrightarrow} M{-}Y + X \tag{11.54}$$

The pre-association complex, $\{M{-}X\cdots Y\}$, is formed in a fast pre-equilibrium and the pseudo-first-order rate constant is given by

$$k_0 = \frac{k K_i [Y]}{K_i [Y] + 1} \tag{11.55}$$

under the conditions of $[Y] \gg [MX]$. This equation predicts the same dependence of k_0 on [Y] as that in the dissociative mechanism.

Another example of a reaction system in which the reaction is considered to proceed *via* the A or I_a mechanism is the ligand substitution reaction of square-planar complexes containing metal ions with a d^8 electron configuration, such as Pd(II) and Pt(II).[18] The pseudo-first-order rate constant (k_0) under pseudo-first-order conditions in which the concentration of the entering ligand (Y) is in large excess can be expressed as follows:

$$k_0 = k_1 + k_2[Y] \tag{11.56}$$

If the solvent can coordinate to the metal ion, k_1 will be constant regardless of the entering ligand Y in the same solvent, whereas k_2 will be highly dependent on the nature of the entering ligand. Thus, k_1 corresponds to the pathway involving the participation of the solvent, whereas k_2 corresponds to the pathway in which Y directly displaces the leaving ligand. In the k_1 pathway, ligand L bound to complex $[ML_n]$ is first replaced by a solvent molecule (S) to form the complex $[ML_{n-1}S]$ and then ligand Y replaces the coordinating solvent to give the final product $[ML_{n-1}Y]$. The rate-determining step is the step in which $[ML_{n-1}S]$ is formed.

In this ligand substitution reaction, the square-planar complex has empty coordination sites in the axial direction of the coordination plane, and it is believed that the entering ligand makes a nucleophilic attack on the central metal from that direction. In other words, an attack on the planar complex by the entering ligand (Y) from the axial direction is considered to proceed *via* the square pyramidal structure, which is transformed into a trigonal bipyramidal reaction intermediate (or transition state), in which the entering and leaving ligands are located on the top of the equatorial plane, as shown in Figure 11.7. In this mechanism, the stereochemistry of the starting complex was preserved during the substitution reaction.

In this type of reaction, for example, in the case of Pt(II) complexes, the rate of substitution reaction is in the following order when the entering ligand is a halide ion:

$$F^- \ll Cl^- < Br^- < I^-$$

The rate of substitution reactions with heavier halogens is higher and a similar trend can be observed for ligands bearing group 15 and 16 elements as the coordinating atoms. This phenomenon is thought to reflect the high polarizability of the ligand, which is related to the fact that Pt(II) is a soft

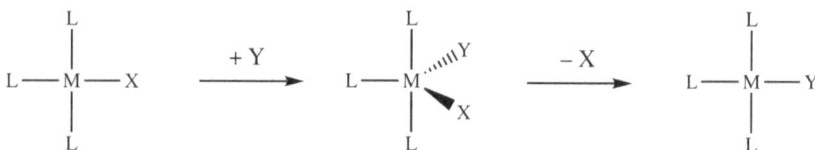

Figure 11.7   Mechanism of the substitution reaction of the square-planar complex.

metal ion and is easily polarized. The effect of the leaving ligand on the substitution rate reflects the strength of the bond between the ligand and the metal atom, and the rate of the substitution reaction of X^- by pyridine in $[Pt(dien)X]^+$ follows the order of:[19]

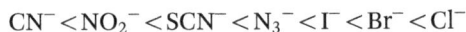

$$CN^- < NO_2^- < SCN^- < N_3^- < I^- < Br^- < Cl^-$$

Even the associative mechanism can exhibit such a leaving ligand dependence, suggesting that the Pt–X bond is cleaved at the same time the bond is formed with the entering ligand during the formation of the transition state.

The first step in discussing the mechanism of the ligand substitution reaction is to consider the rate law based on experimental results. However, there are examples where different reaction mechanisms lead to the same type of rate law, and the I mechanism, which is an intermediate between the D and A mechanisms, is by far the most common mechanism observed for the ligand substitution reaction. In addition to the rate law, kinetic parameters such as the rate constant, activation enthalpy, activation entropy, and activation volume, which can be determined from the pressure dependence of the reaction rate, should be used to elucidate the mechanism of these reactions. The chemical properties of the entering and leaving ligands are expected to have a significant effect on the rate constants and the reactivity will more strongly reflect the respective properties of the leaving ligand in the dissociative mechanism and the entering ligand in the associative mechanism. Because these ligand substitution reactions are mainly studied in solution, the entering and leaving ligands in the ligand substitution reactions are sometimes solvent molecules. In such cases, it is also necessary to consider how the solvent molecules are involved in the reaction by changing the activity of the solvent by adding an inert diluent or comparing the results obtained in different solvents. Thus, it is desirable to elucidate a reasonable reaction mechanism based on a variety of information, such as the rate constants and activation parameters, in addition to the rate law.

Problem 1

An $Fe^{2+}(aq)$ ion reacts with an oxalate ion to form the Fe(ox) complex in an aqueous solution.

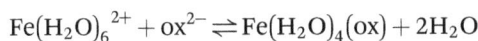

$$Fe(H_2O)_6^{2+} + ox^{2-} \rightleftharpoons Fe(H_2O)_4(ox) + 2H_2O$$

An aqueous solution at pH 4.00 contains 1.0×10^{-2} M $Fe(ClO_4)_2$ and 2.0×10^{-3} M sodium oxalate. Calculate the concentrations of the uncomplexed Fe(II) ion and Fe(ox) using the following chemical constants: $K_{FeOH} = 5.56 \times 10^6$ M^{-1} for $Fe^{2+}(aq)$ and $K_{a1} = 5.6 \times 10^{-2}$ M and $K_{a2} = 1.5 \times 10^{-4}$ M for oxalic acid, and the formation constant of the Fe(ox) complex is $K_1 = 1.1 \times 10^3$ M^{-1}.

Problem 2

Complex formation reactions of many metal ions in solution occur according to the Eigen–Wilkins mechanism shown in eqn (11.25). Under the pseudo-first-order condition, $C_M \ll C_L$, the reaction is often irreversible and the observed rate constant is given by eqn (11.26). Derive this equation.

Problem 3

Consider the following reversible reaction:

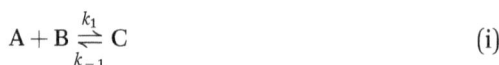

$$A + B \underset{k_{-1}}{\overset{k_1}{\rightleftharpoons}} C \tag{i}$$

If it is possible to suppress the reverse reaction by adding a substance that reacts rapidly with C, the reaction is given by eqn (ii).

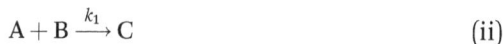

$$A + B \xrightarrow{k_1} C \tag{ii}$$

Which of eqn (i) and (ii) is faster when the change in the concentration of B or C is no longer observed under the pseudo-first-order condition of $C_A \gg C_B$?

Problem 4

In an aqueous solution of nickel(II) perchlorate containing NH_3, the Ni(II) species present are $Ni(H_2O)_6^{2+}$ and $Ni(H_2O)_{6-x}(NH_3)_x^{2+}$ ($x = 1$–6), as shown in eqn (11.2). Derive equations that express the abundance of each Ni(II) species as a function of the concentration of free NH_3. The protonation of ammonia is neglected. The abundance ratio was defined as the ratio of each Ni(II) species to the total concentration of Ni(II) species.

Problem 5

Usually, when studying the kinetics of second-order reactions, such as complex formation reactions, the conditions are set to pseudo-first-order conditions in which the concentration of one reactant is sufficiently larger than that of the other reactant. For example, under pseudo-first-order conditions, the change in the concentration of the reactant with time can be examined from the change in the absorbance of the solution with time. Show the merits of pseudo-first-order conditions for estimating the rate constants of such reactions.

Problem 6

Explain how to determine rate constants k_1 and k_{-1} for the following ligand substitution reaction under appropriate conditions.

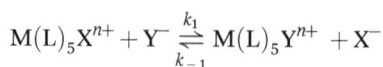

$$M(L)_5X^{n+} + Y^- \underset{k_{-1}}{\overset{k_1}{\rightleftharpoons}} M(L)_5Y^{n+} + X^-$$

Problem 7

The complexation reaction of a metal ion (M^{n+}) with a ligand (L) is expressed as follows. Derive the equation of the pseudo-first-order rate constant (k_{obs}) obtained under the condition $C_M \ll C_L$.

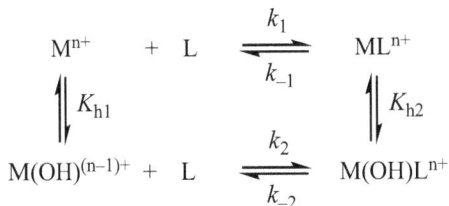

$$
\begin{array}{ccccc}
M^{n+} & + \; L & \underset{k_{-1}}{\overset{k_1}{\rightleftharpoons}} & & ML^{n+} \\
\Big\updownarrow K_{h1} & & & & \Big\updownarrow K_{h2} \\
M(OH)^{(n-1)+} & + \; L & \underset{k_{-2}}{\overset{k_2}{\rightleftharpoons}} & & M(OH)L^{n+}
\end{array}
$$

Problem 8

The D mechanism for ligand substitution reactions of metal complexes is expressed using eqn (11.50) and (11.51), and the pseudo-first-order rate constant (k_0) is given by eqn (11.52). Derive this equation.

Problem 9

The axial ligand substitution reaction of a certain binuclear complex with a ligand (L) proceeds through an $A \to B \to C$ pathway *via* intermediate B, in addition to the direct substitution pathway ($A \to C$). Derive the equation of the pseudo-first-order rate constant under the condition C_D (total concentration of the binuclear complex) $\ll C_L$, when the steady-state approximation is applicable to intermediate B.

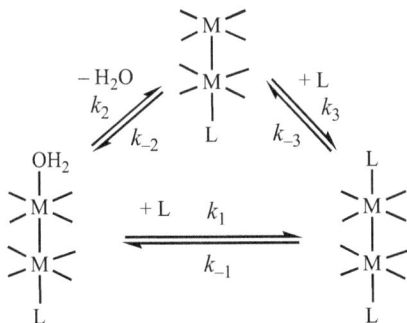

Answer to Problem 1

$$[Fe'] = [Fe^{2+}] + [Fe(OH)^+], \quad [ox'] = [H_2ox] + [Hox^-] + [ox^{2-}]$$

$$\alpha_H = 1 + (1/K_{a2}) \cdot [H_3O^+] + (1/K_{a2}) \cdot (1/K_{a1}) \cdot [H_3O^+]^2$$

$$= 1 + 6.67 \times 10^3 \times (1.0 \times 10^{-4}) + 6.67 \times 10^3 \times 17.9 \times (1.0 \times 10^{-4})^2$$

$$= 1.67$$

$$\alpha_A = 1 + K_{FeOH}[OH^-] = 1 + 5.56 \times 10^6 \times 1.010^{-10} = 1.00$$

$$K_E = K/(\alpha_H \cdot \alpha_A) = 1.1 \times 10^3/(1.67 \times 1.00) = 660 \text{ M}^{-1}$$

$$[Fe(ox)] = 1.7 \times 10^{-3} \text{ M}, \quad [Fe'] = 8.3 \times 10^{-3} \text{ M}$$

Answer to Problem 2

$$C_M \ll C_L, K_{os} = \frac{[M, L]}{[M][L]}$$

$$C_M = [M] + [M, L] + [ML] = [M](1 + K_{os}[L]) + [ML]$$

$$\frac{d[ML]}{dt} = k_{H_2O}[M, L] = k_{H_2O}K_{os}[M][L] = \frac{k_{H_2O}K_{os}[L]}{1 + K_{os}[L]}(C_M - [ML])$$

Since the formation rate of ML is first order with respect to the total concentration of uncomplexed metal ions, $\dfrac{d[ML]}{dt} = k_{obs}(C_M - [ML])$

Therefore,

$$k_{obs} = \frac{k_{H_2O}K_{os}[L]}{1 + K_{os}[L]} = \frac{k_{H_2O}K_{os}C_L}{1 + K_{os}C_L}(C_M \ll C_L)$$

Answer to Problem 3

Under pseudo-first-order conditions, the observed rate constant (k_{obs}) is equal to $k_1 C_A + k_{-1}$ for eqn (i), whereas $k_{obs} = k_1 C_A$ for eqn (ii). Therefore the answer is (i).

Answer to Problem 4

The abundance ratio R_i is expressed as follows:

$$R_i = [Ni(H_2O)_{6-i}(NH_3)_i^{2+}] \Big/ \sum_{i=0}^{6} [Ni(H_2O)_{6-i}(NH_3)_i^{2+}]$$

The denominator of this equation is the total concentration of Ni(II) species. Using the overall formation constant β_i, R_i can be expressed as follows:

$$R_0 = 1 \Big/ \left(1 + \sum_{i=1}^{6} \beta_i[NH_3]^i\right) \quad (\text{for } Ni(H_2O)_6^{2+})$$

$$R_i = \sum_i [NH_3]^i \Big/ \left(1 + \sum_{i=1}^{6} \beta_i[NH_3]^i\right) \quad (i = 1-6)$$

Answer to Problem 5

When we study the kinetics of the second-order reaction $(A + B \rightarrow C)$ under strict second-order conditions, the rate law can be expressed as

$$[A]^{-1} - [A]_0^{-1} = kt$$

It is necessary to know the exact concentration of the reactants at $t = 0$, as well as at any time during the reaction. However, under pseudo-first-order conditions, the rate law can be expressed as

$$[A] = [A]_0 \exp(-kt)$$

Due to the characteristics of the exponential function, it is not necessary to know $[A]_0$ to obtain the pseudo-first-order rate constant (k).

Answer to Problem 6

The pseudo-first-order rate constant (k_{obs}) measured by following the change in the concentration of the reactant complex under the conditions $C_X \gg C_{MX}$ and $C_Y \gg C_{MX}$ is equal to $k_1 C_Y + k_{-1} C_X$. The plot of the k_{obs} values obtained at different C_X (or C_Y) vs. C_X (or C_Y) gives a straight line with slope k_{-1} (or k_1) and intercept k_1 (or k_{-1}).

Answer to Problem 7

Because the products ML and M(OH)L are in fast equilibrium, the rate is given by the following equation:

$$\frac{d([ML] + [M(OH)L])}{dt} = k_1[M][L] - k_{-1}[ML] + k_2[M(OH)][L] - k_{-2}[M(OH)L] \quad \text{(i)}$$

Substitution of $K_{h1} = [M(OH)][H^+]/[M]$ and $K_{h2} = [M(OH)L][H^+]/[ML]$ into eqn (i) yields the following equation:

$$\frac{d([ML] + [M(OH)L])}{dt} = \left(k_1 + \frac{k_2 K_{h1}}{[H^+]}\right)[M][L] - \left(k_{-1} + \frac{k_{-2}K_{h2}}{[H^+]}\right)[ML] \quad \text{(ii)}$$

$$= \alpha_1[M][L] - \alpha_2[ML]$$

Since

$$C_M = [M] + [M(OH)] + [ML] + [M(OH)L]$$

$$= \left(1 + \frac{K_{h1}}{[H^+]}\right)[M] + \left(1 + \frac{K_{h2}}{[H^+]}\right)[ML] = \beta_1[M] + \beta_2[ML]$$

under the conditions

$$C_M \ll C_L, \quad [M] = \frac{C_M - \beta_2[ML]}{\beta_1} \tag{iii}$$

substituting this equation into eqn (ii) yields eqn (iv).

$$\frac{d([ML] + [M(OH)L])}{dt} = \frac{d[ML]}{dt}\left(1 + \frac{K_{h2}}{[H^+]}\right) = \frac{d[ML]}{dt}\beta_2$$

$$= \frac{\alpha_1(C_M - \beta_2[ML])[L]}{\beta_1} - \alpha_2[ML] \tag{iv}$$

Therefore,

$$\frac{d[ML]}{dt} = \frac{\alpha_1 C_M}{\beta_1\beta_2}[L] - \left(\frac{\alpha_1}{\beta_1}[L] + \frac{\alpha_2}{\beta_2}\right)[ML] \tag{v}$$

This equation can be integrated because it involves only one variable under the condition $C_M \ll C_L$ ([L]-C_L). Therefore, k_{obs} is equal to the coefficient of [ML]:

$$k_{obs} = \frac{k_1 + \dfrac{k_2 K_{h1}}{[H^+]}}{1 + \dfrac{K_{h1}}{[H^+]}}[L] + \frac{k_{-1} + \dfrac{k_{-2}K_{h2}}{[H^+]}}{1 + \dfrac{K_{h2}}{[H^+]}}$$

Answer to Problem 8

The rate laws of eqn (11.50) and (11.51) can be expressed as follows:

$$-d[ML_n]/dt = k_1[ML_n] - k_{-1}[ML_{n-1}][L]$$

$$d[ML_{n-1}]/dt = k_1[ML_n] - k_{-1}[ML_{n-1}][L] - k_2[ML_{n-1}][X] + k_{-2}[ML_{n-1}X]$$

By applying steady-state approximation to intermediate $[ML_{n-1}]$, the second equation can be set to zero. From these equations, the following equation can be derived:

$$-d[ML_n]/dt = (k_1 k_2[X] + k_{-1}k_{-2}[L])(k_2[X] + k_{-1}[L])^{-1}$$

$$\times [ML_n] - k_{-1}k_{-2}[L](k_2[X] + k_{-1}[L])^{-1}C_0$$

where C_0 is the total metal complex concentration. The concentration of $[ML_n]$ decreases according to the following exponential function:

$$[ML_n] = ([ML_n]_0 - q/p)\exp(-pt) + q/p$$

$$p = (k_1 k_2 [\mathbf{X}] + k_{-1} k_{-2} [\mathbf{L}])(k_2 [\mathbf{X}] + k_{-1} [\mathbf{L}])^{-1}$$

$$q = k_{-1} k_{-2} [\mathbf{L}](k_2 [\mathbf{X}] + k_{-1} [\mathbf{L}])^{-1} C_0$$

Therefore, the pseudo-first-order rate constants of reactions (eqn (11.50) and (11.51)) are given by eqn (11.52).

Answer to Problem 9

B is in a steady state; therefore, [B] is kept low and does not change with time.

$$\frac{d[\mathbf{B}]}{dt} = k_2 [\mathbf{A}] - k_{-2} [\mathbf{B}] + k_{-3} [\mathbf{C}] - k_3 [\mathbf{B}][\mathbf{L}] \approx 0$$

Therefore,

$$[\mathbf{B}] = \frac{k_2 [\mathbf{A}] + k_{-3} [\mathbf{C}]}{k_{-2} + k_3 [\mathbf{L}]} \tag{i}$$

Substituting this equation into the rate equation yields the following one-variable equation:

$$\frac{d[\mathbf{C}]}{dt} = k_1 [\mathbf{A}][\mathbf{L}] - k_{-1} [\mathbf{C}] + k_3 [\mathbf{B}][\mathbf{L}] - k_{-3} [\mathbf{C}]$$

$$= k_1 [\mathbf{A}][\mathbf{L}] - k_{-1} [\mathbf{C}] + \frac{k_3 [\mathbf{L}](k_2 [\mathbf{A}] + k_{-3} [\mathbf{C}])}{k_{-2} + k_3 [\mathbf{L}]} - k_{-3} [\mathbf{C}]$$

$$= k_1 [\mathbf{L}] C_M + \frac{k_2 k_3 [\mathbf{L}] C_M}{k_{-2} + k_3 [\mathbf{L}]} - \left(k_1 [\mathbf{L}] + k_{-1} + \frac{k_2 k_3 [\mathbf{L}] + k_2 k_{-3}}{k_{-2} + k_3 [\mathbf{L}]} \right) [\mathbf{C}]$$

The k_{obs} is equal to the coefficient of [C], i.e.,

$$k_{obs} = k_1 [\mathbf{L}] + k_{-1} + \frac{k_2 k_3 [\mathbf{L}] + k_2 k_{-3}}{k_{-2} + k_3 [\mathbf{L}]}$$

References

1. J. Burgess, *Ions in Solution, Basic Principles of Chemical Interactions*, Horwood Pub. Ltd., Chichester, 2nd edn, 1999.
2. A. Ringbom, *Complexation in Analytical Chemistry*, John Wiley & Sons, New York/London, 1963.
3. C. H. Langford and H. B. Gray, *Ligand Substitution Processes*, W. A. Benjamin, New York, 1965.
4. (a) T. J. Swift and R. E. Connick, *J. Chem. Phys.*, 1962, **37**, 307; (b) T. J. Swift and R. E. Connick, *J. Chem. Phys.*, 1964, **41**, 2553.
5. L. Helm and A. E. Merbach, *Chimia*, 2019, **73**, 179.

6. F. Basolo and and R. G. Pearson, *Mechanisms of Inorganic Reactions*, John Wiley and Son, New York, 2nd edn, 1967.

7. M. Eigen and K. Tamm, *Z. Elektrochem.*, 1962, **66**, 107.

8. (a) K. Ishihara, S. Funahashi and M. Tanaka, *Inorg. Chem.*, 1983, **22**, 194; (b) S. Funahashi, K. Ishihara and M. Tanaka, *Inorg. Chem.*, 1983, **22**, 2070.

9. R. B. Jordan, *Reaction Mechanisms of Inorganic and Organometallic Systems*, Oxford Univ. Press, 3rd edn, 2007, p. 34.

10. K. M. Smith, *Porphyrins and Metalloporphyrins*, Elsevier, Amsterdam, 1975, pp. 317–380.

11. S. Funahashi, Y. Yamaguchi and M. Tanaka, *Inorg. Chem.*, 1984, **23**, 2249.

12. S. Funahashi, Y. Yamaguchi and M. Tanaka, *Bull. Chem. Soc. Jpn.*, 1984, **57**, 204.

13. M. Inamo, A. Tomita, Y. Inagaki, N. Asano, K. Suenaga, M. Tabata and S. Funahashi, *Inorg. Chim. Acta*, 1997, **256**, 77.

14. R. M. Smith, A. E. Martell and R. J. Motekaitis, NIST Critical Stability Constants of Metal Complexes Database 46.

15. (a) M. Eigen and J. Schoen, *Z. Elektrochem.*, 1955, **59**, 483; (b) M. T. Emerson, E. Grunwald and R. A. Kromhout, *J. Chem. Phys.*, 1960, **33**, 547; (c) M. Eigen, W. Kruse, G. Maass and L. De Maeyer, *Prog. React. Kinet.*, 1964, **2**, 286.

16. P. O'Brien and D. A. Sweigart, *Inorg. Chem.*, 1982, **21**, 2094.

17. M. Inamo, M. Hoshino, K. Nakajima, S. Aizawa and S. Funahashi, *Bull. Chem. Soc. Jpn.*, 1995, **68**, 2293.

18. R. J. Cross, *Adv. Inorg. Chem.*, 1989, **34**, 219.

19. F. Basolo, H. B. Gray and R. G. Pearson, *J. Am. Chem. Soc.*, 1960, **82**, 4200.

CHAPTER 12

Electron Transfer Reactions in Solution

HIDEO D. TAKAGI

Research Center for Materials Science (RCMS), Nagoya University, Furo-cho, Chikusa-ku, Nagoya, 464-8602, Japan
Email: h.d.takagi@nagoya-u.jp

12.1 Introduction

Electron transfer reactions are concerted reactions. The term "concerted" indicates there is no intermediate and the reaction takes place *via* a unique transition state. Since almost all reactions in chemistry involve electron transfer processes to some extent, understanding of chemical reactions requires proper knowledge of the theories for electron transfer reactions. Although most examples referred to in this chapter are the electron transfer reactions between metal complexes, the theory is readily extendable to the areas of organic chemistry.[1]

Historically, electron transfer reactions have been categorized into two broad types of reactions: outer-sphere electron transfer reactions and inner-sphere electron transfer reactions.[2] In outer-sphere electron transfer reactions, an electron is transferred *via* the direct coupling between relevant metal orbitals (frequently d-orbitals in inorganic chemistry) without changes in the co-ordination geometry (coordination numbers and coordinated atoms) of metal complexes. On the other hand, in inner-sphere electron transfer reactions the coordination geometry is changed after the electron transfer process. However, the change in the coordination geometry after an electron transfer reaction is not essential for the inner-sphere reactions since transfer of the bridging lig-and/atom/ion is governed by the difference in the lability of the central metal

Coordination Chemistry Fundamentals Series No. 2
Metal Ions and Complexes in Solution
Edited by Toshio Yamaguchi and Ingmar Persson
© Japan Society of Coordination Chemistry 2024
Published by the Royal Society of Chemistry, www.rsc.org

ion before and after the electron transfer process, and the lability of the metal complex depends essentially on the electronic configurations of central metal ions.[3] With this in mind, the definition of inner-sphere electron transfer reactions may have to be re-considered on the basis of the different types of electronic coupling other than the direct coupling between metal orbitals: super-exchange reactions and sequential electron transfer reactions that involve the third orbital on the mediating ligand(s) may be more important compared with the apparent transfer of the coordinated atom or molecule to explain the acceleration of the electron transfer processes by the inner-sphere mechanism. In the super-exchange reactions, participation of an orbital on the mediating ligand lifts the degeneracy of crossing orbitals at the transition state (= the crossing point between the reactant and product surfaces along the reaction coordinate) and the energy of the newly emerged lower adiabatic surface is further lowered to ensure rapid electron transfer. In sequential transfer, the LUMO (lowest unoccupied molecular orbital) on the mediating ligand(s) assists the electron transfer from one metal to the other. In both cases, electron transfer reactions do not take place *via* the simple direct coupling between the metal orbitals, and the electron transfer reactions are generally accelerated.

In this chapter, theories for the outer-sphere electron transfer reactions are explained at first, and generalized theoretical treatments for the inner-sphere reactions such as super-exchange/sequential transfer processes will be referred to in a later section.

There are two types of outer-sphere electron transfer reactions: (a) electron self-exchange reactions and (b) cross reactions. Electron self-exchange reactions are the reactions between the same metal complexes in different oxidation states: no apparent change in the ratio of $[Fe(ttcn)_2^{3+}]/[Fe(ttcn)_2^{2+}]$ with time is observed for a mixture of $[Fe(ttcn)_2^{3+}]$ and $[Fe(ttcn)_2^{2+}]$ complexes in solution (ttcn = 1,4,7-trithiacyclononane), although electrons are perpetually exchanged between $[Fe(ttcn)_2^{3+}]$ and $[Fe(ttcn)_2^{2+}]$. On the other hand, cross reactions are the reactions between two different metal complexes, such as the reactions between $[Ni(tacn)_3]^{3+}$ and $[Cu(dmp)_2]^+$ (tacn = 1,4,7-triazacyclononane and dmp = 2,9-dimethyl-1,10-phenanthroline) and between $[Cu(dmp)_2]^+$ and $[Cu(ttcn)_2^{2+}]$.

Until the 1950s, the rates of electron transfer reactions between metal complexes in solution were believed to be governed by the size and properties of ligands such as electrical conductivity. Therefore, the electron transfer reactions between small metal complexes were thought to be rapid. However, the use of isotopes such as ^{60}Co for kinetic investigations revealed that the electron self-exchange reaction between $[Co(en)_3]^{3+}$ and $[Co(en)_3]^{2+}$ (en = ethylenediamine) in solution is very slow, despite the small size of the ligand.[4] Libby[5] pointed out that the rate of electron transfer reactions in solution is influenced by the slow rotational motion and the rapid fluctuation of the electron clouds on solvent molecules in the vicinity of the metal complexes. In 1956, the Marcus–Hush theory was publicized,[6–10] and the activation barrier for an electron transfer process was explained by the solvent properties such as the dielectric constant (related to the slow rotation of solvent molecules) and refractive index (related to the rapid response of electron clouds on the solvent molecule). Solvent molecules in the bulk exist outside the coordination sphere of metal complexes (more specifically, outside the contact pair of two metal complexes, in the Marcus theory), and therefore the Marcus–Hush theory is related to the outer-sphere contribution to the activation barrier.

Since the electron transfer reactions are accompanied by the inner-sphere structural changes (changes in the bond lengths between the coordinated atom and central metal ions in most cases) of reactants, Norman Sutin[11] introduced the idea of the inner-sphere contributions to the activation barrier: in the case of electron self-exchange reactions, the distances between the metal ion and coordinated atoms are different for reduced and oxidized forms of metals, and the structures (bond lengths between the coordinated atom and central metal ions) of each metal complex in the transition state are expected to be in between the ground-state structures of each metal complex in different oxidation states. The energy to form such a transition state from each ground-state structure of reactants is the inner-sphere contribution to the activation energy.

Rotation of solvent molecules takes place in the time scale of 10^{-10} to 10^{-11} s in general, and the vibration frequency related to the inner-sphere reorganization is in the time scale of 10^{-13} s. Note that the reaction co-ordinate of an outer-sphere electron transfer reaction is made up of the motions of all nuclei: the distance between the metal and coordinated atoms (which is related to inner-sphere motions) and the positions of all atoms in the rotating solvent molecules in the outer-sphere. Electron transfer occurs very rapidly when both inner- and outer-sphere motions concertedly form the optimum inner- and outer-sphere structures ready for the electron transfer. Since both outer- and inner-sphere contributions to the activation barriers can be expressed by mathematical equations, activation energies as well as the rate constants for outer-sphere electron transfer reactions are theoretically predicted. Preciseness of the theory has been verified for a number of reactions by many researchers to date, including the experiments at elevated pressures[12] and at electrodes.[13]

12.2 Classical and Semi-classical Marcus Theory for Outer-sphere Electron Self-exchange Reactions in Solution

When electron self-exchange reactions take place in solution, where two reactants are identical metal complexes in two different oxidation states, firstly the two reactants approach to form an encounter complex. It should be noted that both reactants are solvated, while one or both reactants are paired with counter ions depending on the formal charges of the metal complexes and on the type of the solvent (when the relative dielectric constant of the solvent is small, charged metal ions/complexes tend to form ion pairs with the counter ion(s)). In the outer-sphere electron transfer reactions in solution,

corresponding   12.2.1   12.2.2 &

sections   12.2.3

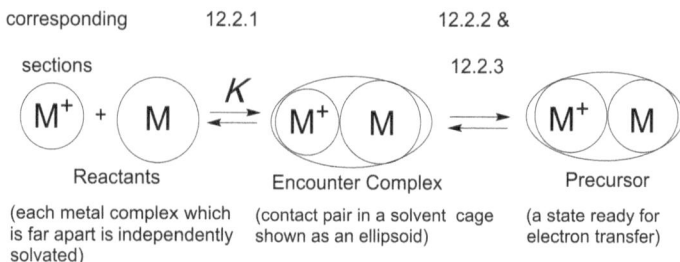

Reactants   Encounter Complex   Precursor

(each metal complex which   (contact pair in a solvent cage   (a state ready for
is far apart is independently shown as an ellipsoid)   electron transfer)
solvated)

Scheme 12.1 Formation of the encounter complex: K corresponds to the formation constant. The inner- and outer-sphere rearrangements take place inside and outside of the ellipsoid, respectively.

(1) solvated metal complexes (reactants) form an encounter complex at first, and (2) the electron transfer takes place within the encounter complex after necessary inner- and outer-sphere rearrangements. The encounter complex is a contact pair of the two metal complexes, with or without counter ions. The encounter complex is surrounded by solvent molecules, while it does not include solvent molecules in between the two metal complexes. In some cases, a counter ion may exist between two metal complexes. In such a case, the distance between two metal complexes is longer than that for the contact pair without a counter ion and the overlap between two metal orbitals decreases and the reaction is slower. On the other hand, existence of a counter ion(s) between metal complexes within the encounter complex sometimes enhances the probability of electron transfer through the super-exchange coupling when the counter ion has an appropriate mediating orbital (Scheme 12.1).

Transfer of electron(s) takes place *via* the direct overlap of the metal orbitals on the two metal complexes. The energy required to form the encounter complex, ΔG_{coul}, corresponds to the coulombic work necessary for the two charged ions to approach from the infinite distance.

12.2.1 Coulombic Work to Bring Two Reactants to Form the Encounter Complex

The equilibrium constant, K, in Figure 12.1 corresponds to the coulombic work to bring two reactants (metal complexes M^+ and M) to the contact distance from the infinite distance in a solvent with a relative dielectric constant, ε_r (solvents are treated as a dielectric continuum with a macroscopic relative dielectric constant, ε_r, in the classical Marcus theory).

The equilibrium constant K and corresponding energy ΔG^*_{coul} are given by the following equations:[14]

$$K = \frac{4\pi(r_1 + r_2)^3 N_A}{3000} \exp\left[\frac{z_1 z_2 e^2}{4\pi\varepsilon_0\varepsilon_r(r_1 + r_2)k_B T\left\{1 + B(r_1 + r_2)\sqrt{I}\right\}}\right] \quad (12.1)$$

$$\Delta G_{coul} = -RT\ln K \quad (12.2)$$

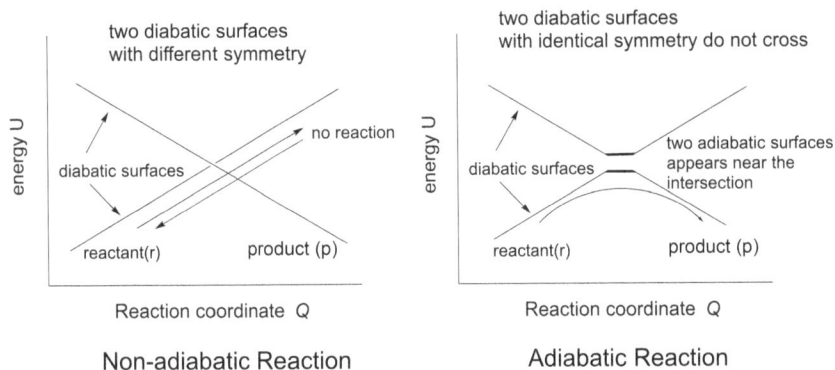

Figure 12.1 General explanation of non-adiabatic and adiabatic reactions.

where k_B is the Boltzmann constant, e is the charge of an electron, ε_0 is the dielectric constant of vacuum, I is the ionic strength of the solvent, and B and N_A are the B-parameter of the Debye–Hückel equation and the Avogadro constant. The energy terms in this section refer to the Gibbs energy unless stated otherwise. Parameters r_1, r_2, z_1, and z_2 are the radii and formal charges of the reactants. This equation is valid only when the ionic strength of the solution is small: when the experiment is carried out at reasonably high ionic strength, $0.1 < I < 1.0$, proper correction by considering the linear contribution of I to the ordinary Debye–Hückel equation should be made. When electron transfer reactions are examined in aqueous solutions, we may not have to consider the effect of ion pairing, especially when the charges on the ion z and I are small. However, when the reactions are examined in organic solvents with small dielectric constants, care must be taken since most charged ions are significantly paired with counter ion(s). As a result, K may be larger for the reactions of metal complexes with like charges in organic solvents: ion-pair formation in solvents with small dielectric constants enhances the probability of formation of the encounter complex. This term expressed by eqn (12.1) together with the nuclear frequency $k_B T/h$ that destroys the transition state may be approximated by the collision frequency (or rate constant at the diffusion limit) by using the Stokes–Einstein equation. This treatment eliminates the uncertainty in the Debye–Hückel term in eqn (12.1) for the reactions in organic solvents with a low relative dielectric constant where the ionic strength is uncertain, while the degree of ion-pair formation is high: it is calculated that more than 99% of dissolved ions with $z_1 z_2 = 1$ are ion-paired when ε_r and $r_1 + r_2$ are set to 32 and 5×10^{-8} cm at 298 K.

12.2.2 Outer-sphere Reorganization Energy for Electron Self-exchange Reactions in Solution

In electron self-exchange reactions, the formal charges of the metals in the precursor (encounter complex immediately before the electron transfer step = reactive reactants) and successor complexes (encounter complex immediately after the electron transfer step = nascent products) are exchanged rapidly: precursor and successor states are regarded as the states ready for electron transfer. Both the solvation structure around the encounter complex and the coordination structure of each metal complex in the encounter complex have to be rearranged to form the precursor state. The outer-sphere and inner-sphere activation energies correspond to the formation of such a precursor state within the encounter complex.

Molecular vibrations are in the time scale of about 10^{-13} seconds in general, while the rotational motion of solvent molecules around the encounter complex that responds to the polarity change of the encounter complex is slower, 10^{-10} to 10^{-11} s for most solvents. Slow rotational reorganization motion of solvent molecules in the outer sphere is related to the macroscopic (relative) dielectric constant of solvents, while the response of electron clouds on the solvent molecules is much faster and related to the refractive index of the solvent. Marcus assumed that the latter labile response and the former inert response create the outer-sphere activation barrier for electron transfer reactions in solution. Since most electron transfer reactions take place with insufficient overlap between the d-orbitals of metal complexes (or orbitals relevant to the electron transfer process), electron transfer is a probability process (as described later, the rate constant/activation energy depends on the degree of overlap between the relevant orbitals). Marcus assumed that there are many manifolds with different labile and inert orientations, and the electron is transferred only when the most suitable orientation is achieved.

$$\Delta G_{\text{outer-sphere}} = \frac{N_A e^2}{16\pi\varepsilon_0} \left(\frac{1}{\varepsilon_{\text{op}}} - \frac{1}{\varepsilon_{\text{r}}} \right) \left(\frac{1}{2r_1} + \frac{1}{2r_2} - \frac{1}{r_1 + r_2} \right) \tag{12.3}$$

where ε_{op} and ε_{r} are the optical dielectric constant (= squares of the refractive index, n^2) and the relative dielectric constant of the solvent, and N_A, e and ε_0 are the Avogadro number, the electron charge (1.6×10^{-19} C) and the permittivity of vacuum. The term in the first bracket of eqn (12.3) is called the Pekar factor which represents the solvent property. In the Marcus theory, two spherical complex ions are located close enough to weakly interact but not too far apart that the coulombic forces can be neglected.

Since eqn (12.3) corresponds to the single electron transfer process, the number of electrons to be transferred should be multiplied for multiple electron transfer reactions. Hush[15-17] reached the same conclusion by assuming that the intermediate state of the electron transfer reaction is given by the linear combination of the wave functions for the precursor and successor states.

Levich and Dogonadze[18,19] independently reached the same conclusion from the semi-classical treatment of non-radiative electronic transition in ionic crystals.

12.2.3 Inner-sphere Reorganization Energy for Electron Self-exchange Reactions

As noted above, metal complexes in the encounter complex optimize the structures to achieve the transition state for the electron transfer process: both the reduced and oxidized forms of metal complexes in the contact pair adjust the structures so as to achieve identical bond lengths between the central metal ion and the coordinated atoms on the ligand for the oxidized and reduced complexes. Such a structural change also optimizes the orbital overlap between two metals by adjusting the ligand field. Inner-sphere structural changes together with the outer-sphere reorganization concertedly provide the optimum condition for the electron transfer, *i.e.* the precursor (reactive state), which is identical in energy to the successor (nascent state) for self-exchange reactions. Note that the reaction coordinate for electron transfer processes is the nuclear coordinates of the atoms which are related to the inner- and outer-sphere reorganization.

The inner-sphere activation energies can be estimated by the following treatment.[20] Structural changes shown in the following diagram (Scheme 12.2) take place within the encounter complex.

$$U_p = U_0 + \frac{1}{2}N_A f_A^+ \left(r_A^+ - r_1\right)^2 + \frac{1}{2}N_A f_A (r_A - r_2)^2 \qquad (12.4)$$

$$U_s = U_0 + \frac{1}{2}N_A f_A^+ \left(r_A^+ - r_2\right)^2 + \frac{1}{2}N_A f_A (r_A - r_1)^2 \qquad (12.5)$$

The energies of the precursor and successor states are identical to each other for self-exchange reactions, $U_s = U_p$, and therefore

$$r^* = \frac{f_A^+ r_A^+ + f_A r_A}{f_A^+ + f_A} \qquad (12.6)$$

where r^* is the bond length between the metal ion and the coordinated atom of the ligands in the precursor/successor states. The inner-sphere reorganization energy ($=$ inner-sphere contribution to the activation energy) can be obtained using eqn (12.7). It should be noted here that the precursor (reactive) state and successor (nascent) state are on the adiabatic surface and these states are regarded as the transition state. Therefore, the energy to reach these states from each ground state is regarded as the activation energy.

$$\Delta G_{\text{inner-sphere}} = \frac{N_A f_A^+ f_A}{2(f_A^+ + f_A)} (r_A - r_A^+)^2 \qquad (12.7)$$

activation state
(precursor / successor) $r^* = \dfrac{f_A^+ r_A^+ + f_A r_A}{f_A^+ + f_A}$

$A^+(r_{A^+}, f_{A^+})$ + $A(r_A, f_A)$ $A(r_A, f_A)$ + $A^+(r_{A^+}, f_{A^+})$

contact pair before electron transfer   contact pair after electron transfer

$A^+(r_1)$ + $A(r_2)$ Ground state energy for
                                independent reactants / products
                                $= U_0$

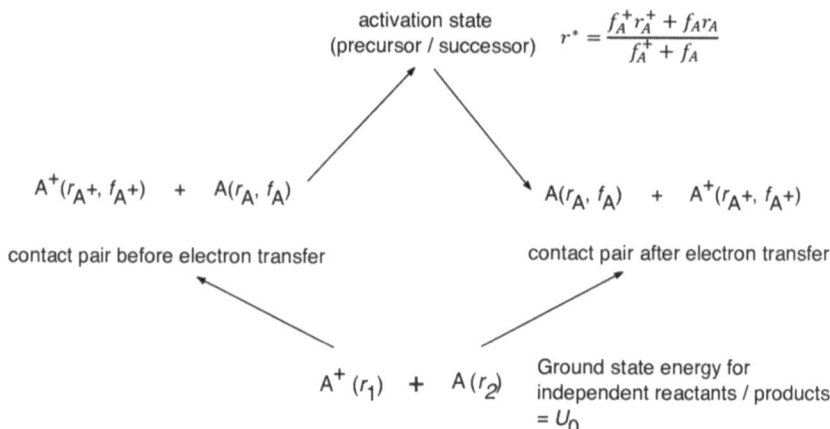

Scheme 12.2   Structural parameters related to the inner-sphere reorganization. *f*s are the force constants between the metal and the coordinated atoms (symmetric stretching mode), and therefore U_p and U_s are the sum of all contributions from each coordination bond.

$\Delta G_{\text{inner-sphere}}$ is the sum of all contributions from each coordination bond. Therefore, the contribution of each coordination bond in the reactant has to be summed up to calculate the inner-sphere contributions to the activation energy.

12.2.4   Semi-classical Expression of Outer-sphere Electron Transfer Reactions

Combining eqn (12.1), (12.3) and (12.7), one can calculate the activation energy and the corresponding rate constant for electron self-exchange reactions in solution. Generally, the following equation is used for the calculation of the rate constant, with considerations of the non-adiabaticity of the reaction and the tunneling effect:

$$k_{\text{calc}} = \kappa_{\text{el}} \Gamma_n \nu_n \exp\left(- \frac{\Delta G_{\text{coul}}^* + \Delta G_{\text{inner-sphere}}^* + \Delta G_{\text{outer-sphere}}^*}{RT} \right) \qquad (12.8)$$

$$\nu_n^2 = \frac{\nu_{\text{outer-sphere}}^2 \, \Delta G_{\text{outer-sphere}}^* + \nu_{\text{inner-sphere}}^2 \, \Delta G_{\text{inner-sphere}}^*}{\Delta G_{\text{outer-sphere}}^* + \Delta G_{\text{inner-sphere}}^*} \qquad (12.9)$$

where κ_{el}, Γ_n, and ν_n are the factors that account for the non-adiabaticity of the reaction, contribution of nuclear tunneling, and the nuclear frequency, respectively.

$\text{Exp}[-(\Delta G_{\text{inner-sphere}}^* + \Delta G_{\text{outer-sphere}}^*)]$ in eqn (12.8) is called the nuclear factor since these terms are related to the nuclear motions along the reaction coordinate, and Γ_n is the ratio of the nuclear factor of the reaction against that at the high temperature limit (HT). The nuclear frequency that corresponds to the nuclear vibration mode is expressed by $k_B T/h$ in general in the absolute rate theory, since the contribution of the inner-sphere vibration frequency overwhelms the slower outer-sphere rotational frequency in most cases. Nuclear tunneling is important only at low temperatures.

From the saddle point method by Holstein, $\Delta G^*_{inner\text{-}sphere}(HT)$, Γ_n, and κ_n are given by the following equations:

$$\Delta G^*_{inner\text{-}sphere}(HT) = \lambda_{inner\text{-}sphere}\left(\frac{kT}{h\nu_{inner\text{-}sphere}}\right)\tanh\left(\frac{h\nu_{inner\text{-}sphere}}{4kT}\right) \quad (12.10)$$

$$\Gamma_n = \exp\left\{-\frac{\lambda_{inner\text{-}sphere}}{h\nu_{inner\text{-}sphere}}\left[\tanh\left(\frac{h\nu_{inner\text{-}sphere}}{4kT}\right) - \frac{h\nu_{inner\text{-}sphere}}{4kT}\right]\right\} \quad (12.11)$$

$$\kappa_n = \exp\left\{-\left[\frac{\lambda_{outer\text{-}sphere}}{4kT} + \frac{\lambda_{inner\text{-}sphere}}{h\nu_{inner\text{-}sphere}}\tanh\left(\frac{h\nu_{inner\text{-}sphere}}{4kT}\right)\right]\right\} \quad (12.12)$$

where $\Delta G_{inner\text{-}sphere} = \dfrac{\lambda_{inner\text{-}sphere}}{4}$ and $\Delta G_{outer\text{-}sphere} = \dfrac{\lambda_{outer\text{-}sphere}}{4}$.

Note that Γ_n is taken to be 1 for most reactions at room temperature.

For electron transfer reactions with no contributions of nuclear tunneling, the rate constant is expressed by

$$k = K\kappa_{el}\nu_n\kappa_n \quad (12.13)$$

where ν_n is the nuclear frequency and κ_n is the nuclear factor. The coulombic work term, corresponding to K in eqn (12.8), described in Section 12.2.1 is not included in the nuclear factor.

The (electron) transmission coefficient, κ_{el}, is given by eqn (12.14):

$$\kappa_{el} = \frac{2\left[1 - \exp\left(\dfrac{-\nu_{el}}{2\nu_n}\right)\right]}{2 - \exp\left(\dfrac{-\nu_{el}}{2\nu_n}\right)} \quad (12.14)$$

where ν_{el} is the electron hopping frequency in the vicinity of the transition state. When κ_{el} is approximately 1, the reaction is adiabatic. Note that the reaction is taken to be adiabatic when $10^{-3} < \kappa_{el} \leq 1$ since the Marcus theory assumes medium overlap (insufficient overlap) between relevant orbitals. The electron hopping frequency ν_{el} is related to the electronic coupling element defined by $H_{rp} = \langle\psi_r|H_1|\psi_p\rangle = \langle\psi_r|\partial U/\partial Q|\psi_p\rangle Q$, where U is the total energy of the reaction system and Q is the reaction coordinate ($=$ nuclear coordinate).

When two diabatic energy surfaces of the same symmetry meet at the intersection (the crossing point of the two diabatic surfaces corresponds to the transition state), the perturbation theory for degenerate systems indicates that the degeneracy of the diabatic surfaces at the crossing point is resolved and two adiabatic surfaces which are $2H_{rp}$ apart emerge at the intersection: this is so called the non-crossing rule or avoided-crossing rule.[21,22]

When the symmetry of the two crossing orbitals (diabatic surfaces) are identical to each other, H_{rp} has a non-zero value since the reaction coordinate is always totally symmetric at the transition state (the reaction coordinate is always included either in the primary axis or in the primary plane of symmetry for the specific point group to which the precursor and successor complexes belong).[20] On the other hand, when the symmetries of two crossing orbitals (surfaces) are different, $H_{rp} = 0$ and two diabatic surfaces cross (remain degenerate) without creation of two adiabatic surfaces: the reaction is completely non-adiabatic in such a case. The non-crossing rule is valid for all types of chemical reactions including the electron transfer reaction. As the adiabaticity of a reaction depends on the symmetries of the crossing orbitals, adiabatic reactions are also called symmetry-allowed reactions, while non-adiabatic reactions are symmetry-forbidden reactions. Note that consideration of orbital symmetry is not necessary for outer-sphere electron transfer reactions, since the related orbitals somehow overlap, irrespective of the geometry of the precursor/successor complex. Therefore, nonadiabaticity of electron transfer reactions solely depends on the degree of overlap of orbitals on the electron donor and acceptor atoms.

The electron hopping frequency ν_{el} is also related to the electronic coupling element, H_{rp}, by eqn (12.15):

$$\nu_{el} = \frac{2H_{rp}^2}{h} \left(\frac{\pi^3}{\lambda RT} \right)^{\frac{1}{2}} \qquad (12.15)$$

where λ is the intrinsic energy barrier: ν_{el} decreases significantly with decreasing H_{rp}. For a series of reactions with a gradual change in the distance between two coupled sites, r, eqn (12.16) may be used to explain the decrease in the rate constant with distance.

$$H_{rp} = H_{rp}^0 \exp\left[-\frac{\beta(r - r_0)}{2} \right] \qquad (12.16)$$

The parameter β is a scaling factor and r_0 is the distance where an adiabatic electron transfer reaction is expected to occur: long range electron transfer reactions involving large proteins can be concerted due to this relationship.[23] The distance between the two coupled sites is not the only factor that reduces the value of H_{rp}: the energy gap between the two interacting orbitals is also related to the coupling element. It has been pointed out that the two orbitals with the energy difference larger than 5 eV hardly interact with each other for the second-order Jahn–Teller interactions,[22,24] indicating that two orbitals more than 5 eV apart hardly interact with each other even in the same molecule. Therefore, electron transfer reactions where two interacting orbitals are more than 5 eV apart may not take place smoothly because of significant non-adiabaticity, $\kappa_{el} \ll 1$ (the 5 eV rule). However, we never observe electron self-exchange reactions regulated by this rule since the two interacting orbitals for the self-exchange reactions are similar in energy. On the other hand, this 5 eV rule regulates the electron transfer processes of some cross reactions. Examples are shown in Section 12.8.2.

12.3   Two-state Model and Inverted Region of Cross Reactions

In this section, considered are the cross reactions in which oxidizing and reducing reagents are different metal complexes. The potential curve (total energy U_R as a function of reaction coordinate x) for the reactant is approximated by a harmonic function for electron transfer reactions.

$$U_R = A_R x^2 \tag{12.17}$$

Similarly, the potential curve for the product is given by eqn (12.18):

$$U_P = A_P(1 - x)^2 + \Delta G^\circ \tag{12.18}$$

where ΔG° is the driving force of the reaction (ΔG° is usually negative), and A_R and A_P are the coefficients related to the parabolic curvature. Positions on the reaction coordinates $x = 0$ and $x = 1$ correspond to the ground states for the reactant and product, respectively. See Figures 12.2 and 12.3.

The curvatures corresponding to the reactant and product states are expected to be similar to each other, $A_R = A_P = A$, for a series of electron transfer reactions. Since two parabolas cross when $U_R = U_P$, the transition state is positioned at $x^* = \dfrac{1}{2}\left(1 + \dfrac{\Delta G^\circ}{A}\right)$ and therefore

$$\Delta G^* = \frac{1}{4}A\left(1 + \frac{\Delta G^\circ}{A}\right)^2 \tag{12.19}$$

where A is the intrinsic energy barrier, which is $4\Delta G^*$ for electron self-exchange reactions. Electron transfer reactions with $\Delta G^\circ/A > -1$ are regarded as reactions in the normal region where the activation energy decreases with increasing absolute value of ΔG°. Note that the intrinsic energy barrier A is positive and ΔG° is negative in this chapter. On the other hand, when $\Delta G^\circ/A < -1$ the reactions are in the inverted region where the activation energy for the cross reaction increases with increasing absolute value of ΔG°. Such a relation may be examined by moving the parabola for the product

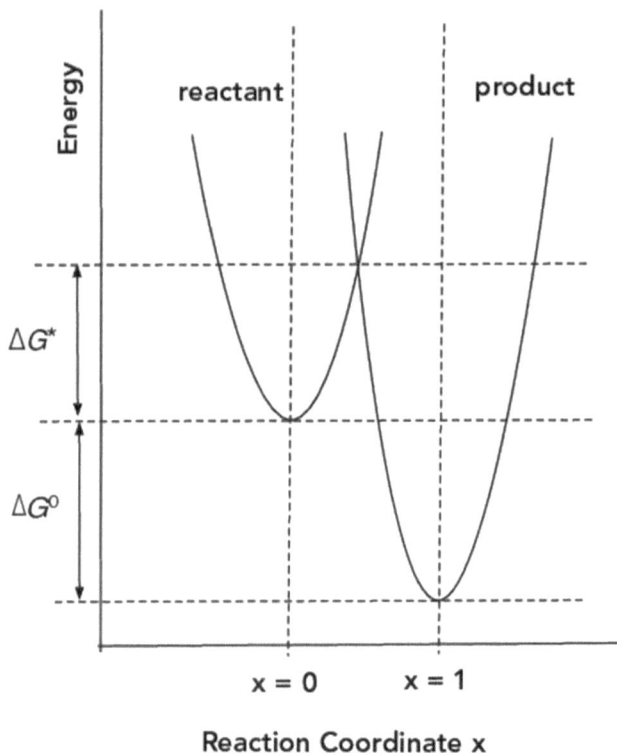

Figure 12.2   The two-state model for outer-sphere electron transfer reactions. Note that Figures 12.2–12.4 are also applicable to the series of inner-sphere electron transfer reactions as far as the activation processes of each reactant are the same irrespective of the counter reagents for a series of reactions.

state up- and downward while the parabola for the reactant state being fixed in Figure 12.2. Some electron transfer reactions in the inverted region $(\Delta G^\circ/A < -1)$ have been reported.[25]

12.4   Brønsted Coefficient

The position of the transition state on the reaction coordinate was obtained as

$$x^* = \frac{1}{2}\left(1 + \frac{\Delta G^\circ}{A}\right). \tag{12.20}$$

From this relation, it is obvious that the position of the transition state (crossing point of the reactant and product surfaces) varies as a function of the driving force ΔG°. A value defined by $-\partial\Delta G^*/\partial\Delta G^\circ$ is known as the Brønsted coefficient (since ΔG° is negative a minus sign was added).

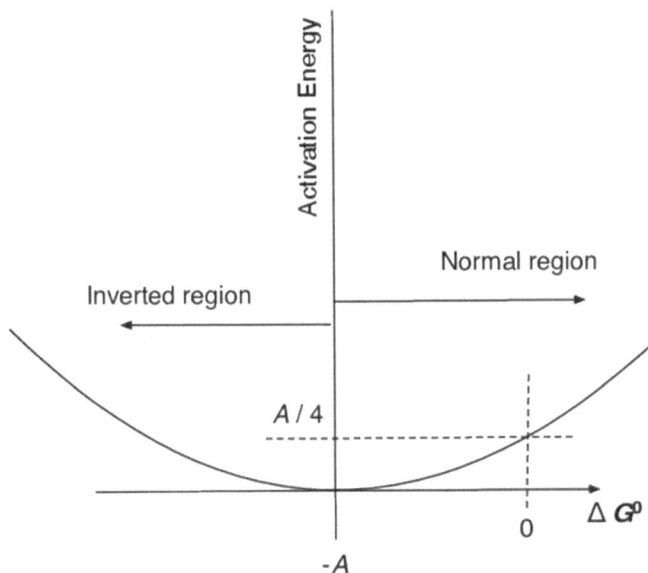

Figure 12.3 Normal and inverted regions of outer-sphere electron transfer reactions. Since the rates of bimolecular reactions in solution are regulated by diffusion, we generally observe a diffusion-controlled reaction in the inverted region (flattening of the plot in the inverted region). The inverted region is observed in solution only for unimolecular electron transfer reactions, such as the electron transfer process between the charge-separated sites in a solvent cage.

The transition state is right in the middle of the reaction coordinate $(x^* = 1/2)$ for self-exchange reactions $(\Delta G° = 0)$, and $x^* = 0$ when $-\Delta G° = A$. In the latter case, electron transfer occurs immediately after formation of the encounter complex: the inner- and outer-sphere structures at the transition state are hardly different from those of the encounter complex. Relaxation to the final products occurs after the electron transfer process.

12.5 Cross Relations

The rate constant and activation energy of a cross reaction, $A_{red} + B_{ox}$ (corresponding to reaction (12.23)), are related to those for two independent self-exchange reactions $A_{red} + A_{ox}$ and $B_{red} + B_{ox}$ (eqn (12.21) and (12.22)).

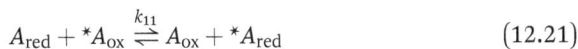
$$A_{red} + {}^*A_{ox} \overset{k_{11}}{\rightleftharpoons} A_{ox} + {}^*A_{red} \tag{12.21}$$

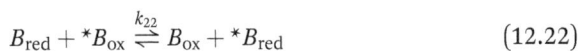
$$B_{red} + {}^*B_{ox} \overset{k_{22}}{\rightleftharpoons} B_{ox} + {}^*B_{red} \tag{12.22}$$

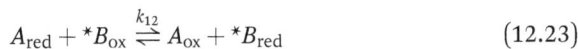
$$A_{red} + {}^*B_{ox} \overset{k_{12}}{\rightleftharpoons} A_{ox} + {}^*B_{red} \tag{12.23}$$

Eqn (12.24) was derived by using the formal redox potentials for the A_{red}/A_{ox} and B_{red}/B_{ox} couples: $\Delta G_{12}^{\circ} = -RT\ln K_{12} = -F(E_1^{\circ} - E_2^{\circ})$. Each rate constant is defined by $k = \nu_n \exp(-\Delta G^*/RT)$ from the absolute rate theory $(\nu_n (= k_B T/h) = $ *ca.* 6.6×10^{13} s^{-1}).[26]

Energy correction for coulombic work terms (w_{ij}) were introduced by Espenson and co-workers[27] for the practical use of this equation, especially for the applications to a series of reactions involving small inorganic molecules such as O_2 in water.

$$k_{12} = \sqrt{k_{11}k_{22}K_{12}f}\,W_{12} \tag{12.24}$$

$$\ln f = \frac{[\ln K_{12} + (w_{12} - w_{21})/RT]^2}{4\left[\ln\left(\dfrac{k_{11}k_{22}}{Z_{11}Z_{22}}\right) + \dfrac{w_{11} + w_{22}}{RT}\right]} \tag{12.25}$$

$$W_{12} = \exp\left(-\frac{w_{12} + w_{21} - w_{11} - w_{22}}{2RT}\right) \tag{12.26}$$

$$w_{ij} = \frac{z_1 z_2 e^2}{4\pi\varepsilon_0\varepsilon_r(r_1 + r_2)[1 + B(r_1 + r_2)\sqrt{I}]} \tag{12.27}$$

However, it has been known that the importance of the work terms may be limited to reactions in aqueous solution, since it is hardly possible to precisely estimate some essential parameters such as I and B for the reactions in organic solvents.

In 1980, Ratner and Levine derived a similar cross relation to eqn (12.24) on the basis of statistical mechanics and thermodynamic considerations.[28,29] The energy difference in the precursor and successor states is the energy difference in the electronic configuration: we ascribe the difference to only one transferring electron which is presumed to move independently from all the other electrons in the system (the principle of least motion). The energies of the states are expressed by the following integral with Hamiltonians of all coordinates for nuclei and electrons in the system, and therefore the energy of the states is depicted by the two-dimensional space as shown in Figure 12.4.

$$E_{precursor} = \int \phi_p H \phi_p^* dV \quad \text{and} \quad E_{successor} = \int \phi_s H \phi_s^* dV \tag{12.28}$$

It was theoretically shown by Levine[30] that the Brønsted coefficient monotonically changes with ΔG° when the activation energies for a series of reactions are 10 kJ mol^{-1} or larger (most cross reactions fall in this category), and the energies of the precursor (reactive reactant) and successor (nascent) states are identical to each other irrespective of ΔG°. Therefore, both precursor $\left(\text{reactive reactant; } G^{\circ}\left(A_{red}^{\ddagger}\right) + G^{\circ}\left(B_{ox}^{\ddagger}\right)\right)$ and successor $\left(\text{nascent; } G^{\circ}\left(A_{ox}^{\ddagger}\right) + G^{\circ}\left(B_{red}^{\ddagger}\right)\right)$

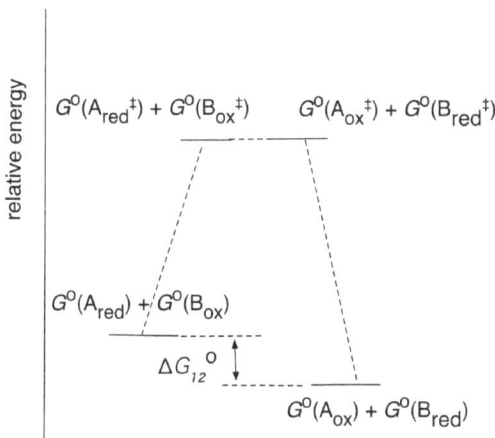

Figure 12.4   Energy diagram for chemical species involved in a cross reaction.

states in the following discussion have to be taken as the states immediately before and immediately after the electron transfer on the adiabatic surface.

$$G^\circ\left(A_{red}^\ddagger\right) + G^\circ\left(B_{ox}^\ddagger\right) = G^\circ\left(A_{ox}^\ddagger\right) + G^\circ\left(B_{red}^\ddagger\right) \tag{12.29}$$

It is important to note that the processes to reach each transition state (the precursor and the successor state in Figure 12.4) from the ground states of the reactant and product are diabatic processes. Ratner and Levine[29,30] derived the cross relation by assuming that each ground state species reaches the precursor (reactive reactant) and successor (nascent) states independently for the self-exchange and cross reactions without being affected by the nature of the counter reagents.

When the precursor and successor states for each reaction are distinguished by $\ddagger$, the activation energies for each reaction are expressed by eqn (12.30)–(12.32). $\Delta G^\ddagger$ corresponds to the activation energy for each reaction, and G° corresponds to the relative energy of each species.

$$\Delta G_{11}^\ddagger = G^\circ\left(A_{red}^\ddagger\right) + G^\circ\left(A_{ox}^\ddagger\right) - G^\circ(A_{red}) - G^\circ(A_{ox}) \tag{12.30}$$

$$\Delta G_{22}^\ddagger = G^\circ\left(B_{red}^\ddagger\right) + G^\circ\left(B_{ox}^\ddagger\right) - G^\circ(B_{red}) - G^\circ(B_{ox}) \tag{12.31}$$

$$\Delta G_{12}^\ddagger = G^\circ\left(A_{red}^\ddagger\right) + G^\circ\left(B_{ox}^\ddagger\right) - G^\circ(A_{red}) - G^\circ(B_{ox}) \tag{12.32}$$

Since

$$\Delta G_{12}^\circ = G^\circ(A_{ox}) + G^\circ(B_{red}) - G^\circ(A_{red}) - G^\circ(B_{ox}), \ 2\Delta G_{12}^\ddagger$$

$$= G^\circ\left(A_{red}^\ddagger\right) + G^\circ\left(B_{ox}^\ddagger\right) + G^\circ\left(A_{ox}^\ddagger\right) + G^\circ\left(B_{red}^\ddagger\right) - 2G^\circ(A_{red}) - 2G^\circ(B_{ox}).$$

As a result, we obtain the final eqn (12.33).

$$\Delta G_{12}^{\ddagger} = 1/2\left(\Delta G_{11}^{\ddagger} + \Delta G_{22}^{\ddagger} + \Delta G_{12}^{\circ}\right) \tag{12.33}$$

Therefore, the cross relation is given by eqn (12.34).

$$k_{AB} = \sqrt{k_{AA} k_{BB} K_{AB} \frac{Z_{AB}^2}{Z_{AA} Z_{BB}}} \tag{12.34}$$

The rate constants in eqn (12.34) are expressed by eqn (12.35) by using the diffusion rate constant (or the second-order collision frequency), Z_{ij}.

$$k_{ij} = Z_{ij} \exp(-\Delta G_{ij}^*/RT) \tag{12.35}$$

where Z_{ij} is given by the following expression using the Stokes–Einstein equation. Since eqn (12.33) and (12.34) were derived without using exotic assumption such as $A = (A_p + A_s)/2$, it seems appropriate to describe the rate constants by using the collision frequency and the probability of electron transfer by the ratio of the reactive reactant, in a sense of the trajectory theory.

$$Z_{ij} = \frac{2RT(r_i + r_j)^2}{3000\eta r_i r_j}\left(\frac{U}{\exp(U) - 1}\right) \tag{12.36}$$

$$U = \frac{z_i z_j e^2}{4\pi\varepsilon_0 \varepsilon_s (r_i + r_j) k_B T} \quad R \text{ in} = \text{erg mol}^{-1} \text{ deg}^{-1} \tag{12.37}$$

where η is the viscosity of the solvent (in poise) and r_i and r_j are the radii of the reactants (cm).

We do not have to worry about the effect of ionic strength and Debye–Hückel parameters for the use of this equation when experiments are carried out in organic solvents containing sufficient amounts of indifferent salts: when most of the ionic reactants are ion-paired, $Z_{AB}^2 \sim Z_{AA} Z_{BB}$ and $k_{AB} = \sqrt{k_{AA} k_{BB} K_{AB}}$. This relation is exploited for detailed analyses of the reactions in non-aqueous solutions with small dielectric constants. Note that Z_{ij} is the product of $k_B T/h$, K (formation constant of the encounter complex), and κ_{el}.

12.6   Verification of the Theory

There have been many experimental results[31] reported to date that verify the validity and accuracy of the Marcus theory and cross relations. Only a few of such examples are shown in this section.

Rate constants for rapid electron self-exchange reactions in solution are usually measured by the NMR line broadening technique.[32,33] However, slow reactions are usually observed either by the polarization transfer method using NMR or the direct measurement using stable isotopes. Some of the data have been reported by monitoring the change in the circular dichroism with time by using optically active isomers.[34]

It has been known that the Marcus theory predicts the electron self-exchange rate constants for a number of redox couples very well. The discrepancies between the calculated and measured results for $[Ru(OH_2)_6]^{3+/2+}$ and $[Ru(NH_3)_6]^{3+/2+}$ couples were attributed to the smaller electron transmission coefficients, κ_{el}, since $10^{-3} < \kappa_{el} < 1$ even for adiabatic reactions.

The cross relation was extensively examined and it was concluded that the self-exchange rate constants calculated using the cross relation are satisfactorily consistent with the directly observed rate constants.[35] A few examples for a series of ruthenium and manganese complexes are shown in Table 12.2.

Discrepancies of the reported self-exchange rate constants for the $[Fe(OH_2)_6]^{3+/2+}$ couple in Tables 12.1 and 12.2 have been attributed to the difference in the mechanism: cross reactions involving this couple may proceed through the inner-sphere mechanism, depending on the counter reagent. However, it is certain that the use of the cross relation is not limited only to the reactions that take place *via* the outer-sphere mechanism: the cross relation is valid as far as each ground state species reaches the precursor and successor states independently for the cross reaction and for the self-exchange reaction, irrespective of the counter reagents, which is the only assumption used to derive the Ratner–Levine cross relation.[29,30] Therefore, the rate constants estimated by the cross relation reflect the activation process of each reactant: the discrepancy in the calculated rate constants for a reaction by using the cross relation, therefore, tells us that one or both reactant(s) took different activation processes for the particular cross reaction.

12.7 Examinations of the Marcus Theory: Measurements at Elevated Pressures

In the gaseous state, the equilibrium constant K for a reaction is dependent only on T, and therefore no dependence of K on P is expected. However, an

Table 12.1 Some examples of the observed and calculated self-exchange rate constants.[a]

Redox couples	$[Fe(OH)_6]^{3+/2+}$	$[Ru(OH_2)_6]^{3+/2+}$	$[Ru(NH_3)_6]^{3+/2+}$	$[Co(NH_3)_6]^{3+/2+}$
$k_{calcd}/dm^3\ mol^{-1}\ s^{-1}$	2.8	1.5×10^5	2.2×10^5	1.1×10^{-6}
$k_{obsd}/dm^3\ mol^{-1}\ s^{-1}$	4.2	20	6.7×10^3	8×10^{-6}

[a] Data were obtained from the following articles: B. S. Brunshwig, C. Creutz, D. H. Macartney, T. K. Sham and N. Sutin, *Discuss. Faraday Soc.*, 1982, **74**, 113; P. Bernhard, L. Helm, A. Ludi and A. E. Merbach, *J. Am. Chem. Soc.*, 1985, **107**, 312; P. J. Smolenaers and J. K. Beattie, *Inorg. Chem.*, 1986, **25**, 2259, and A. Hammershoi, D. Gaselowitz and H. Taube, *Inorg. Chem.*, 1984, **23**, 979. Data at 25 °C: some of the data were obtained by extrapolating the value from the reported temperatures. The rate constant for $[Co(NH_3)_6]^{3+/2+}$ was observed at 40 °C.

Table 12.2 Some results for the verification of the Marcus cross relation (25 °C).[a]

Counter reagent	$E°$/V	k_{ex}	$k_{cross}^{measured}$	k_{cross}^{calcd}
Oxidation of $[Ru(sar)]^{2+}$ $k_{ex}=1.2 \times 10^5$ $E°$/V$=0.290$				
$[Ru(NH_3)_5py]^{3+}$	0.302	1.1×10^5	1.1×10^5	1.4×10^5
$[Ru(NH_3)_5nic]^{3+}$	0.362	1.1×10^5	2.8×10^5	4.4×10^5
$[Ru(NH_3)_5isn]^{3+}$	0.384	1.1×10^5	5.2×10^5	6.6×10^5
$[Ru(tacn)_2]^{3+}$	0.366	5.4×10^4	7.3×10^5	3.4×10^5
$[Mn(sar)]^{3+}$	0.519	170	1.7×10^5	0.9×10^5
$[Fe(OH_2)_6]^{3+}$	0.740	6.2×10^{-3}	7.2	6.7
Reduction of $[Mn(sar)]^{3+}$ $k_{ex}=170$ $E°$/V$=0.519$				
$[Ru(NH_3)_5py]^{2+}$	0.302	1.1×10^5	3.7×10^5	7.2×10^5
$[Ru(NH_3)_5isn]^{2+}$	0.384	1.1×10^5	1.4×10^5	1.7×10^5
$[Ru(tacn)]^{2+}$	0.366	1.2×10^5	2.9×10^5	1.6×10^5
$[Fe(OH_2)_6]^{2+}$	0.740	6.2×10^{-3}	12	20

[a] Data from P. Bernhard and A. M. Sargeson, *Inorg. Chem.*, 1988, **27**, 2582. Rate constants are in $dm^3 \, mol^{-1} \, s^{-1}$. sar = 3,6,10,13,16,19-hexaazabicyclo[6,6,6]eicosane, tacn = 1,4,7-triazacyclono-nane, nic = nicotinamide, isn = isonicotinamide, and py = pyridine. The ionic strength of the solution is 0.1 at 25 °C.

equilibrium constant in solution is a function of both T and P. The pressure dependence of the equilibrium constant K is expected since the isothermal compressibility of fluids is small (the compressibility of gasses is defined by pV/RT and it is constant for ideal gases). For reaction (12.38), a reaction volume, ΔV, is defined by eqn (12.39).

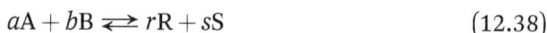

$$aA + bB \rightleftarrows rR + sS \tag{12.38}$$

$$-RT\left(\frac{\partial \ln K}{\partial P}\right)_T = rV_R + sV_S - aV_A - bV_B = \Delta V \tag{12.39}$$

where V_i is the partial molar volume of species i and ΔV is the reaction volume. The rate constant for a reaction is given by the product of $k_B T/h$ and the equilibrium constant K^* between the ground state species and the species in the transition state according to the absolute rate theory.

$$aA + bB \overset{K^*}{\rightleftarrows} [aA \bigstar \bigstar bB]^* \rightleftarrows rR + sS \tag{12.40}$$

Therefore, eqn (12.41) is obtained as a relation between the rate constant k and pressure.

$$-RT\left(\frac{\partial \ln k}{\partial P}\right)_T = \Delta V^* \tag{12.41}$$

where ΔV^* is the activation volume which describes the change in the partial molar volume during the activation process. It should be noted that eqn (12.41) can be used only when k is expressed by either molality $(mol \, kg^{-1})$ or mole fraction. When k is expressed by molarity (mol dm^{-3}), the right-hand side of the equation should be $\Delta V^* - RT\kappa(1 - a - b)$ or $\Delta V^* + RT\kappa(n - 1)$ where n is the order of the reaction and κ is the compressibility of the solution which is generally approximated by the compressibility of the solvent.

Organic chemists have applied high pressure techniques for more than a hundred years.[36] In those cases, most applications are practical and oriented either to accelerate the reactions or to achieve higher yields of products. Recent applications used to distinguish some reaction mechanisms, however, are somewhat questionable since most of the criteria for the distinction of reaction mechanisms have only a slim theoretical basis, especially in the research field of ligand substitution/solvent exchange reactions in inorganic chemistry:[12,36,37] results of a large number of studies concerning the reaction mechanisms for ligand substitution/solvent exchange reactions have been reported on the basis of the observed activation volumes. The categorization of the reaction mechanisms, such as A, D, I_a and I_d, as well as the explanations of the volume profiles corresponding to each mechanism may be wrong. Ligand exchange or substitution reactions certainly involve the movement of electron density in the activation process, and therefore these reactions may also be governed by the probability that originates from the degree of coupling between two related orbitals, the metal orbital and the ligand orbital. As a result, the thermal activation parameters reported for this type of reactions may not directly reflect the macroscopic molecular properties such as molecular sizes and volumes. On the other hand, studies related to the investigations of electron transfer reactions at elevated pressures[12] are far more reliable since such analyses have a solid theoretical basis, as described in the previous sections.

12.7.1 Electron Self-exchange Reactions in Solution at Elevated Pressures

Stranks and Swaddle developed the following equations on the basis of the Marcus theory for electron transfer reactions. By differentiating energy terms such as the work necessary to form the encounter complex and the outer-sphere activation energy, the following general equations were derived:[38–40]

$$\Delta V^*_{\text{outer-sphere}} = \frac{Ne^2}{16\pi\varepsilon_0}\left[\left(\frac{1}{2r_1} + \frac{1}{2r_2} - \frac{1}{r_1 + r_2}\right)\left\{\frac{\partial}{\partial P}\left(\frac{1}{n^2} - \frac{1}{\varepsilon}\right)\right\} - \frac{\frac{1}{n^2} - \frac{1}{\varepsilon}}{3(r_1 + r_2)}\beta\right]$$

(12.42)

$$\Delta V^*_{\text{coul}} = \frac{Nz_1z_2e^2}{16\pi\varepsilon_0}\left[\frac{\partial}{\partial P}\left(\frac{1}{\varepsilon}\right) + \frac{\beta}{3(r_1 + r_2)}\right] + \frac{RTz_1z_2C\sqrt{\mu}}{(1 + Ba\sqrt{\mu})^2}$$
$$\times \left[\left(\frac{\partial \ln\varepsilon}{\partial P}\right)(3 + 2Ba\sqrt{\mu} - \beta)\right]$$

(12.43)

(the second term of eqn (12.43) corresponds to the Debye–Hückel correction for the reactions in water).

where n, ε, β, r_i, μ, and a, B, and C are the refractive index, relative dielectric constant and the isothermal compressibility of the solvent, the radii of metal complexes, the ionic strength of the solution, and Debye–Hückel parameters, respectively. The inner-sphere contribution to the activation volume is expected to be very small and generally ignored since the internal pressure of the coordination sphere (inner-sphere) is very high and therefore the distances between the metal center and coordinated atoms hardly change with the applied external pressure (generally <200 MPa in most experiments). In most cases, the contribution of ΔV^*_{coul} is also small, and the activation volumes for electron self-exchange reactions are well described only by $\Delta V^*_{\text{outer-sphere}}$.

When the reaction is non-adiabatic, the activation volume for the self-exchange reaction is expressed by eqn (12.44) as the coupling element H_{rp} is independent of P.

$$\Delta V^{\text{NA}}_{*\,\text{outer-sphere}} = \beta RT - \left[\frac{2\alpha\beta(r_1+r_2)RT}{3}\right] + \left(\frac{RT\Delta V^*_{\text{outer-sphere}}}{2\Delta G^*_{\text{outer-sphere}}}\right) + \Delta V^*_{\text{outer-sphere}}$$

$$(12.44)$$

Since the scaling factor α (in distance^{-1}) is generally expected to be in the range of 6 to 25 nm^{-1}, large negative values (-8 to -23 cm^{-3}) are predicted as the activation volume for non-adiabatic electron self-exchange reactions. On the other hand, activation volumes for adiabatic self-exchange reactions are expected to be in the range of -2 to -5 cm^3 depending on the solvent. Activation entropy for electron transfer reactions is parallel to the activation volume: a large negative value of the activation entropy generally indicates a limited number of manifolds for the electron transfer and therefore the reaction is expected to be non-adiabatic.

A number of studies of self-exchange reactions in various solvents have been carried out and the validity of the Marcus theory has been confirmed. Even a more practical model of the outer-sphere reorganization was examined on the basis of the mean spherical approximation (MSA)[41,42] theory, and it was concluded that the classical expression of the Marcus equation reproduced the experimental results better than the MSA theory.[43]

Equipment and solvent parameters necessary for the calculations of the activation volumes are summarized elsewhere.[12,44,45] A number of reaction volumes for some redox couples which are necessary for the evaluation of volume profiles for cross reactions have also been reported.[46–48] Some of the results are summarized in Table 12.3.

12.7.2 Electron Self-exchange Reactions at Electrodes

Marcus predicted the relation between the activation energies for homogeneous and heterogeneous electron self-exchange reactions since the

Table 12.3 Selected outer-sphere self-exchange rate constants reported to date.[a]

Redox couple	Solvent	Ionic strength	Redox potential/ V	k_{ex}/ $M^{-1}s^{-1}$	ΔV^*/ $cm^3\,mol^{-1}$	Ref.
$MnO_4^{-/2-}$	Water	1.1	0.52	2.4×10^{-3}		61
$[Fe(OH_2)_6]^{3+/2+}$		0.1	0.74	1.1	−11	62, 63
$[Fe([9]aneS_3)_2]^{3+/2+}$		2.0	1.4	3.2×10^6	−21	64
$[Fe(CN)_6]^{3-/4-}$		0.1	0.42	1.9×10^4	−22	65–67
$[Co(en)]^{3+/2+}$		0.98	−0.18	7.7×10^{-5}		68
		0.5			−20	69
$[Co(sep)]^{3+/2+}$		0.2	−0.3	5.1		70, 71
$[Co([9]aneN_3)_2]^{3+/2+}$		0.1	−0.41	0.19		72
$[Co([9]aneS_3)_2]^{3+/2+}$		0.2	0.42	1.6×10^5		73
		0.1	0.42	1.3×10^5		74, 75
$[Co(phen)_3]^{3+/2+}$		0.1	0.36	12		76
$[Co(bipy)_3]^{3+/2+}$		0.1	0.31	5.7		77
$[Co(edta)]^{-/2-}$		0.2	0.37	1.4×10^{-4}		78
		0.1	0.37	2×10^{-7}		79
$[Co(edta)]^-/$ $[CoHedta]^-$		0.5	−3.2			80
$[Ni([9]aneN_3)]^{3+/2+}$		1.0	0.94	2.7×10^3		81
$[Ni([9]aneN_3)]^{3+/2+}$	AN	0.1	0.555	1.4×10^3	−8.2	82
$[Cu(dmp)_2]^{2+/+}$	AN	0.002	0.59	2.0×10^5		83
	AN	0.1	0.280	4.9×10^3	−3.4	83
$[Cu(dmbp)_2]^{2+}$	AN	0.1	0.346	5.5×10^3		84
$[Fe(Cp)_2]^{+/0}$	AN		0.000	5.3×10^6		85, 86
$[Fe(Cp^*)_2]^{+/0}$	AN		−0.500		3.8×10^7	86
$[Co(Cp)_2]^{+/0}$	AN		−1.33		3.8×10^7	86
$[Co(Cp^*)_2]^{+/0}$	AN		−1.94		4.3×10^8	86
$[Mn(bipyO_2)_3]^{3+/2+}$	AN		0.435		80	87
$[Ru(hfac)_3]^{0/1-}$	AN	0.04	0.357	5.0×10^6		88

[a] $bipyO_2 = N,N'$-dioxo-2,2'-bipyridine, AN = acetonitrile, rate constant in $kg\,mol^{-1}s^{-1}$ in this solvent.

reaction at electrodes is the mirror image of the homogeneous self-exchange reaction.[49,50]

$$\Delta G^*_{\text{heterogeneous}} = \frac{1}{2}\Delta G^*_{\text{homogeneous}} \qquad (12.45)$$

Accordingly, the activation volumes for the homogeneous and heterogeneous electron self-exchange reactions are given by eqn (12.46).

$$\Delta V^*_{\text{heterogeneous}} = \frac{1}{2}\Delta V^*_{\text{homogeneous}} \qquad (12.46)$$

As the heterogeneous electron self-exchange rate constant, k_{el}, is very sensitive to the surface conditions of the electrodes, the relation given by eqn (12.45) has been demonstrated only for a limited number of redox couples. In addition, design of the electrochemical cell is essential since the reference electrode as well as the solution in the cell is easily contaminated upon pressurizing the cell.[51,52] Examination of eqn (12.45) is not an easy task since $\Delta G^*_{heterogeneous}$ cannot be obtained unless we know the heterogeneous k^0_{el} value that corresponds to the diffusion-controlled rate constant for homogeneous reactions. The calculated value of k^0_{el} by using eqn (12.48) is about 3.2×10^2 cm s^{-1} when $D = 2 \times 10^{-5}$ cm^2 s^{-1} and $R = 4 \times 10^{-10}$ m, which is too large compared with the measured values.[53]

$$k_{el} = k^0_{el} \exp\left(-\Delta G^*_{heterogeneous}/RT\right) \tag{12.47}$$

$$k^0_{el} = (2D/\pi R)^{1/2} \tag{12.48}$$

However, the dependence of k^0_{el} on P is expected to be canceled upon application of eqn (12.46) since the collision frequency for the second-order reactions in solution also depends on the viscosity: $D \propto 1/\eta$ from the Stokes–Einstein equation.

$$Z_{ij} = \frac{2RT(r_i + r_j)^2}{3000\eta r_i r_j} \left(\frac{U}{\exp(U) - 1}\right) \tag{12.49}$$

$$U = \frac{z_i z_j e^2}{4\pi\varepsilon_0\varepsilon_s(r_i + r_j)k_B T} \tag{12.50}$$

where D is the diffusion coefficient and $R (= r_i + r_j)$ is the collision diameter. Therefore, examination of the relation in eqn (12.46), which is obtained by differentiation of eqn (12.45) with respect to P, seems quite accurate as has been shown in Figure 12.6 of ref. 52 compared with the previous results obtained using eqn (12.45) in which plots were only roughly linear and the slope of the plot was rarely 0.5.[54-60]

Self-exchange rate constants, redox potentials, and activation volumes for some redox couples are listed in Table 12.3.

12.8 Some Other Important Physical Concepts for Further Understanding of Electron Transfer Reactions

12.8.1 Pekar Factor as a Tool to Analyze the Behavior of Solvent Molecules Around the Contact Pair During the Activation Process

As described above, the theory of outer-sphere electron transfer reactions in solution was developed in the 1950s and has been verified to date. When one of the reactants is substitution inert, redox reactions are expected to proceed through the simple outer-sphere mechanism unless the substitution-inert metal complex contains ligands with orbitals that induce the sequential/ super-exchange mechanism.[89]

It is certain that the inner-sphere reorganization energy is hardly influenced by the difference of the solvents or by the counter reagents. Therefore, the rate constant of an electron self-exchange reaction in different solvents depends essentially on the Pekar factor, which is expressed by eqn (12.51), of each solvent:

$$\frac{1}{\varepsilon_{op}} - \frac{1}{\varepsilon_r} \tag{12.51}$$

Such a relation may be used to analyze the specific solvation structure around the contact pair through observations of the rate constants for electron transfer reactions: the behavior of ionic liquids around the ions or polarizing species is sometimes very different from that in the bulk.[90–93]

It has been known that the electron transfer reactions of the ferrocene/ ferricinium couple at electrodes indicated dissociation of ionic liquids around the electrode.[90] In 2003, Nakamura and Shikata reported that the dielectric response of an ionic liquid exhibited more than two different phases.[91] They explained the observed results by the existence of undissociated and dissociated ionic liquids: they concluded partial dissociation of ionic liquids in the bulk. On the other hand, Israelachvili and co-workers[92] clearly showed from the surface force measurements that ionic liquids near the surface of electrodes are completely dissociated, while no dissociation was observed in the bulk. It seems that the observation by Nakamura and Shikata included the dielectric responses of both at the electrode and in the bulk because of the standard geometry of their equipment.

The Marcus theory for outer-sphere electron transfer reactions made it possible to directly observe the behavior of ionic liquids as solvents around the reacting ions (not at the electrode surfaces) through observations of the rate constants for homogeneous electron transfer reactions. It was shown that an ionic liquid is certainly dissociated around the metal complexes with high charge density.[93] Experimental results clearly showed that the rotational correlation time of ionic liquids around the contact pair in its activation process became longer and the relative dielectric constant became larger than those for undissociated ionic liquids because of the elongated distance between the cationic and anionic components of the ionic liquid.[93,94]

12.8.2 Electron Transfer Reactions in Which Inner- and Outer-sphere Reorganizations Appear to Take Place Separately: The Gated Phenomena

Another application of the theory for outer-sphere electron transfer reactions is related to the examination of the degree of non-adiabaticity of the reaction. Non-adiabaticity of outer-sphere electron transfer reactions is not only caused by the distance between the two reactive sites but also by the large differences between the energy levels of interacting orbitals. For example, only copper complexes exhibit gated electron transfer phenomena in which the rate constant is regulated by the rate of the structural change prior to the electron transfer process.[95-103] The structural change is essential for the succeeding outer-sphere electron transfer reaction that proceeds through the super-exchange mechanism in these reactions. Note that the structural change and the electron transfer process do not necessarily share the same reaction coordinate in these reactions, and therefore they take place independently. The reason why the gated reactions are observed only for the reactions of copper complexes was explained by the fact that copper is the only metal element which has the lowest energy d-orbital of all metal elements: copper 3d-orbital is more than 7 eV lower than the d-orbital levels of any other metal elements and therefore direct coupling with the d-orbitals on other metal elements is hardly possible (the "5 eV rule"). In addition, it has been reported that the slow structural change prior to the electron transfer process is essential to form the copper species that assures the low-energy electron transfer reaction by the super-exchange mechanism: the reaction coordinate for this slow structural change (dissociation of a coordinated solvent molecule) is irrelevant to the reaction coordinate of the succeeding outer-sphere electron transfer process.

The super-exchange mechanism has been explained by the reaction in which more than three diabatic surfaces of the same orbital symmetry cross at the transition state. In such a case, degeneracy is resolved and a very low adiabatic surface appears for the thermal reaction. M. D. Newton used the name "super-exchange coupling" for the ligand-mediated electron self-exchange reaction of the ferrocene/ferricenium couple.[89] On the other hand, the name "sequential transfer" is used for the mechanism in which electron(s) are sequentially transferred from one metal site to the other through conjugated bridging ligand(s). However, when the sequential transfer process takes place at a unique point on the reaction coordinate (at the transition state), the overall reaction is regarded as a super-exchange process with the effective coupling element of H_{RP}^{eff}. In this context, gated electron transfer is taken as a super-exchange process mediated by the excited-state orbital(s) of coordinated ligand(s), and the reaction proceeds as if the electron is transferred sequentially through the orbital(s) of mediating ligand(s). The super-exchange coupling takes place within the encounter complex, and the preceding structural change (such as dissociation of a coordinated solvent molecule) necessary for the succeeding electron transfer reaction through the super-exchange coupling has to take place before formation of the encounter complex: the reaction coordinate for the preceding structural change is irrelevant to the reaction coordinate for the succeeding outer-sphere electron transfer process.

12.8.3 Concerted but Very Slow Outer-sphere Reactions

There are some concerted electron transfer reactions regulated by structural changes with high energy, such as reactions involving low-spin/high-spin configuration changes. Such reactions are generally very slow:[89,104,105] in such reactions both reduction and oxidation reactions are regulated by the spin activation process. However, this type of reaction has been successfully explained by the Marcus theory/cross relation: the high-energy inner-sphere reorganization coupled with the spin-equilibrium and the outer-sphere re-arrangement takes place concertedly in these reactions. Changes in the spin state are generally accompanied by large changes in the bond lengths be-tween the coordinated atoms and the central metal ion. Therefore, reactions involving low-spin/high-spin Co(III)/(II) couples may be regarded as gated reactions when the reaction coordinate for the inner-sphere reorganization is not coupled with the outer-sphere nuclear motions for electron transfer. However, Newton showed that reactions involving low-spin/high-spin Co(III)/(II) couples are slow but concerted outer-sphere electron transfer reactions: reactions involving low-spin/high-spin interconversion, probably through spin–orbit coupling, are considered to take place within the Frank–Condon regime.

Co(III) complexes have a low-spin electronic configuration, except for CoF_6^{3-}, and most Co(II) complexes with ordinary ligands have a high-spin configuration. The electron self-exchange rate constant between low-spin $[Co(1,4,7\text{-trithiacyclononane})_2]^{3+}$ and low-spin $[Co(1,4,7\text{-trithiacyclono-nane})_2]^{2+}$ is as large as 10^5 $kg\,mol^{-1}\,s^{-1}$ (25 °C), while the self-exchange rate constant for the reaction between low-spin $[Co(NH_3)_6]^{3+}$ and high-spin $[Co(NH_3)_6]^{2+}$ is about 10^{-6} $dm^3\,mol^{-1}\,s^{-1}$ at 40 °C. Self-exchange rate con-stants for the cobalt complexes with phenanthroline and bipyridine ligands are around 1 $kg\,mol^{-1}\,s^{-1}$. Although the basicity of phenanthroline and bi-pyridine ligands is smaller than that of amine ligands, these ligands tend to stabilize the lower oxidation state: Co(II) complexes with phenanthroline and bipyridine ligands exhibit a larger ligand field and tend to exhibit a small low-spin/high-spin equilibrium. The slightly larger self-exchange rate con-stants for these Co(III)/(II) couples originate from the ease of Co(II) to attain the low-spin states. The self-exchange rate constant for the $[Co(sep)]^{3+/2+}$ couple is also larger than those for the $[Co(en)_3]^{3+/2+}$ and $[Co(NH_3)_6]^{3+/2+}$ couples. This may be caused by the more compressed structure of the $[Co(sep)]^{2+}$ complex owing to the sepulchrate ligand that induces a larger ligand field than amine and en ligands. Slow inner-sphere rearrangements in these reactions are certainly coupled with the outer-sphere rearrange-ments to achieve the transition state for the outer-sphere electron transfer process.[98]

Electron transfer reactions may have to be regarded as probability pro-cesses because the rate of a reaction is controlled by the degree of electronic coupling. Therefore, for most chemical reactions, which involve electron transfer processes to some extent, thermodynamic activation parameters such as activation energy, activation enthalpy, and activation entropy may not directly reflect the macroscopic motions/behaviors of chemical species unless electronic coupling between the relevant orbitals is sufficient to guarantee that the observed reaction is fully adiabatic. This is important when reaction mechanisms are discussed on the basis of thermodynamic activation parameters. It has been known that we sometimes observe large negative activation entropy when the reaction is non-adiabatic: a non-adiabatic reaction proceeds through a very narrow reaction manifold to reach product(s) and the probability of such a reaction to occur is small.

Appendix 1. Coordination Bond and Metal Complexes

In general chemistry courses, students learn various kinds of chemical bonds. It is understood that, in a single covalent bond and coordination bond, a pair of electrons are shared between two atoms. One electron from each atom is shared in a covalent bond, while two electrons from only one atom are shared between two atoms in a coordination bond. As shown in the reaction of ammonia with a hydrogen ion in Figure 12.5(a), the two electrons thus shared are delocalized all over the molecule or ion: we cannot distinguish the shared electrons from the other valence electrons. In metal complexes, the donor atoms, EPDs (electron pair donors), are nitrogen atoms, sulfur atoms, oxygen atoms, *etc.*, and the EPAs (electron pair acceptors) are metal or metal ions. A hydrogen ion has an empty 1s orbital and accepts a pair of electrons from an EPD atom such as NH_3. In metal complexes, the vacant d and s, p orbitals in higher energy of metals/metal ions are the acceptor orbitals of electron pairs from EPDs. As an example of a metal(0) complex, $[Ni(CO)_4]$ is famous for its application to the extraction of Ni metal from ores (Ni(0) has a d^{10} electronic configuration and the central Ni atom has no vacant d-orbital to share). Six NH_3 molecules in $[Co(NH_3)_6]^{3+}$ exhibit octahedral coordination since Co(III) has 6 electrons in the d-orbital and $[Co(NH_3)_6]^{3+}$ uses two vacant 3d-orbitals, one higher 4s orbital and three 4p orbitals as the accepter orbitals. The coordination number is determined by (1) the ratio of the radii of metals (or metal ions) and ligands (radius ratio rule) and (2) the number of acceptor orbitals, and the energy of electronic configurations determines the coordination geometry.

Ligands with only one EPD atom are unidentate ligands. Ligands with more than two EPD atoms are bidentate, tridentate, ...multidentate ligands: 'dentate' means the ligands that can 'bite' metal/metal ions. Examples of multidentate ligands are ethylenediamine and ethylenediaminetetraacetic acid (or ethylenedinitrilotetraacetic acid (EDTA)): some examples of multidentate ligands mentioned in the text are also shown in Figure 12.5(a). Since

(a)

ammonia   ammonium ion

ethylenediamine (en)   1,4,7-triazacyclomomane (tacn)

1,4,7-trithiocyclomomane (ttcn)   2,9-dimethyl-1,10-phenanthroline (dmp)

(b)

metal-ttcn complexes: M = Fe(III)/(II),Co(II)/(III), Ni(III)/(II),etc.
M(tacn)$_2$ complexes have similar structures.

tetracarbonylnickel(0)   [Cu(dmp)$_2$]$^+$ [Cu(dmp)$_2$(OH$_2$)]$^{2+}$

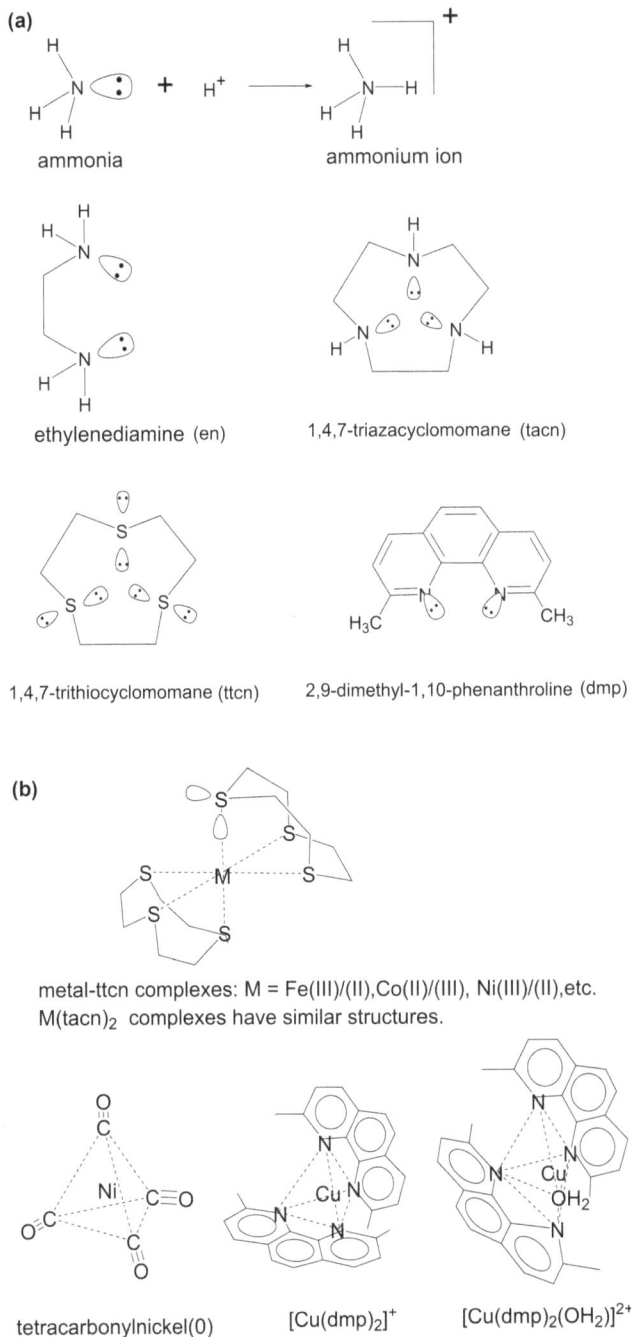

Figure 12.5 (a) Various ligands including ammonia as a unidentate ligand. (b) Structures of some metal complexes appeared in Chapter 12.

ethylenediamine has two nitrogen donor atoms, it is a bidentate ligand, and 1,4,7-triazacyclononane is a tridentate ligand. As some metals/metal ions can accept 6 pairs of electrons, complexes with 1,4,7-triazacyclononane are in the form $[M(1,4,7\text{-triazacyclononane})_2]^{n+}$. Some metal complexes with octahedral, tetrahedral and trigonal bipyramidal structures are shown in Figure 12.5(b).

In general, metal complexes which do not exchange or substitute the coordinated ligand within about a minute in solution are substitution inert: an example is Co^{3+} (low-spin d^6 configuration) complexes. Substitution-inert octahedral metal complexes do not have electrons in the e_g orbitals since e_g orbitals have an anti-bonding character in octahedral metal complexes. On the other hand, some metal complexes are substitution labile: Co^{2+} (high-spin d^7 configuration in most cases) complexes are generally labile and even $[Co(NH_3)_6]^{2+}$ in water rapidly decomposes to $[Co(H_2O)]^{2+}$ in water. Such decompositions of metal complexes can be suppressed by addition of an extra amount of free ligand in solution. Generally, metal complexes with multidentate ligands are relatively stable and substitution inert compared with metal complexes with unidentate ligands because of the chelate effect, which is also known as a local concentration effect. Therefore, some metal complexes with tridentate and quadridentate ligands are indefinitely stable in solution. Generally, metal complexes with multidentate ligands are substitution inert and the electron transfer reactions with these metal complexes proceed *via* the outer-sphere mechanism.

Appendix 2. Electronic Coupling

The term 'coupling' is defined as (1) the interaction between two systems or matter, or simply as (2) bonding; the former is generally used in physics and the latter in chemistry. An example for (2) is homo- and cross-coupling: as an example of a cross coupling reaction, A–X + B–Y produces A–B. A magnetic interaction such as the L–S coupling is an example of (1). In this section, the term 'coupling' corresponds to (1). 'Electronic coupling' means the interaction between two orbitals, and there is no electron transfer reaction without electronic coupling. Note that we cannot define an orbital or the corresponding orbital energy without electron(s) in the orbital. A coupling element as the consequence of 'coupling' is given by the integration of the product of two orbital functions, $\langle \psi_a | H_1 | \psi_b \rangle = \int \psi_a H_1 \psi_b dV$ (dV means integration over the space), in which H_1 is the perturbation term for a specific interaction. We usually do not care much about the mathematical form of H_1 unless we want to know the exact value of the integration. In general, we can know if such an interaction (= coupling) exists or not by checking the symmetry of the mathematical functions. For such purposes, the knowledge of the group theory is necessary. A 'problem' related to the group theory is given at the end of this chapter. Students may acquire some idea of how the group theory is useful by solving the problem.

Appendix 3. Physical View of Chemical Reactions

Students learn the energetics of chemical reactions by using Figure 12.6(a). This figure looks simple for understanding the meaning of 'activation energy' and the usefulness of catalysts. The vertical axis corresponds to the energy of the reaction system, while it is not clearly written what the horizontal axis stands for. Some students may consider that the horizontal axis is the degree of progress of a reaction. In fact, the horizontal axis corresponds to the 'nuclear coordinates' of all atoms involved in the reaction. The 'principle of least motion' tells us that motions of only a small number of atoms are to be considered during the activation process since the movement of a minimum number of component atoms provides the least-energy pathway for a reaction. The excitation process along a diabatic surface involves changes in the molecular vibrations, rotations, *etc.* A reaction proceeds along the diabatic surface where thermal energy is acquired from the environment (the slope of the line or surface increases with increasing nuclear coordinate). When the reaction system reaches the fully activated state, the system requires no more energy from the outside of the system (environment). This is the position of the transition state on the reaction coordinate: the reactant(s) have sufficient energy for bond rupture, cross coupling, or electron transfer, *etc.*: the slope of the energy against the nuclear coordinate is zero at the transition state which means that the process at the transition state is adiabatic. After the transition state, the energy of the reaction system decreases by releasing excess energy to the outside of the reaction system: the slope of total energy against the reaction coordinate is negative here since this corresponds to the diabatic condition. In conclusion, the reaction is adiabatic only at the transition state. The term 'adiabatic' means that the reaction system neither acquire energy from the environment nor discharge energy to the environment.

In Figure 12.6(b), dealt with is the bond rupture reaction of a diatomic A–A′ molecule (A′ is used only to distinguish two A atoms). The reaction coordinate (horizontal axis) may be taken as the distance between atoms A and A′ since the increment of energy is related to the distance between A and A′. There are two lines drawn in Figure 12.6(b): one line increases (positive slope) and the other decreases (negative slope) with increasing distance between A and A′.

The quantum mechanical explanation for the formation of a covalent bond between two A atoms is provided in Figure 12.6(c): a valence electron in the orbital on the A atom interacts (couples) with an electron in the valence orbital on another A (= A′) atom, and a pair of bonding and antibonding orbitals is created. The bonding orbital with lower energy is occupied by the two electrons from two A atoms, and the antibonding orbital with higher energy is vacant in the A–A′ molecule. Therefore, Figure 12.6(c) explains that the two A atoms are stabilized by forming the A–A′ bond. Arrows in the bonding orbital denote the two

(a)

Energy

activation energy

reactant

product

Profile of a reaction

(1)

(b)

Energy

antibonding orbital

a

a

bonding orbital

bondlength

Symmetries of bonding and antibonding orbitals are *a*

upper adiabatic surface

a

a

(2)

lower adiabatic surface

(c)

energy

antibonding orbital

Atomic orbital of Atom A

Bonding orbital

Atomic orbital of Atom A'

Molecular orbital of A -A'

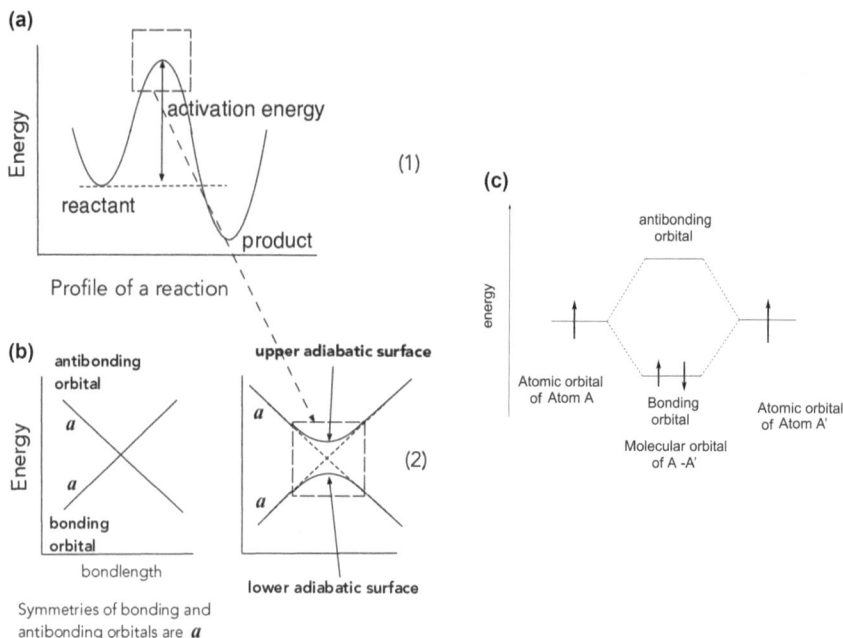

Figure 12.6   (a) Example of a reaction profile. (b) Energy–reaction coordinate relation for the bond rupture reaction of diatomic molecules. (c) Formation of a covalent bond between two A atoms.

shared electrons with different spins which are distinguished by the direction of arrows.

The lines with positive and negative slopes shown in Figure 12.6(b) correspond to the changes in the energy of the bonding and antibonding orbitals with the distance between two A atoms, respectively.

Along the diabatic surface (the straight line with a positive slope on the left), the total energy of the molecule A–A increases by acquiring energy from the environment), and the stability of the A–A bond decreases. The energy of the antibonding orbital decreases with increasing distance between two A atoms accordingly. At the intersection of the two diabatic lines (surfaces), there is no difference in energy between A–A and A + A: this is the activation state. Since the bonding and antibonding orbitals immediately before and after the bond rupture (= transition state) are in the same symmetry (both are designated to the same irreducible representation), two adiabatic surfaces (lines) appear at the upper and lower positions of the crossing point of two diabatic surfaces according to the non-crossing rule (or avoided crossing rule): it has been proven that bond-rupture reactions are always symmetry-allowed (= adiabatic). Therefore, "the principle of microscopic reversibility" confirms that the bond formation reaction is always symmetry-allowed.

The discussions for adiabatic reactions in the text and in the reference articles are not given on the basis of non-relativistic quantum mechanics.

Problem 1

Calculate the outer-sphere contribution to the activation energy for the self-exchange reactions between $Fe_{aq}^{2+}/Fe_{aq}^{3+}$ and $[Co(NH_3)_6]^{2+}/[Co(NH_3)_6]^{3+}$ in water by using the following parameters (the relative dielectric constant and refractive index for water are 80.4 and 1.33, respectively).

Reaction	r_1 (Angstrom)	r_2 (Angstrom)
$Fe_{aq}^{2+} + Fe_{aq}^{3+}$	3.33	3.19
$[Co(NH_3)_6]^{2+} + [Co(NH_3)_6]^{3+}$	3.41	3.19

Problem 2

Discuss the energetics of the outer-sphere electron transfer reaction when the electronic coupling between reactants is very large.

Problem 3

The nuclear vibration frequency, n_n, is given by eqn (12.6), and the inner-sphere vibration frequency, n_{in} (or $n_{inner\text{-}sphere}$), is approximated by using the following equation where n_1 and n_2 are the vibration frequencies between the metal and coordinated atom in the oxidized and reduced forms of the redox couple:

$$n_{in}^2 = \frac{2v_1^2 v_2^2}{v_1^2 + v_2^2}$$

The force constant f_i is given by $f_i = (2\pi v_i c)^2 m$ where v_i (cm^{-1}) is the observed vibrational frequency, m is the reduced mass of the ligand, and c is the speed of light. However, the application of these equations to estimate the inner-sphere contributions to the activation energy is not easy and therefore the calculated results are unreliable in many cases. Therefore, the inner-sphere contribution to the activation energy may roughly be estimated by subtracting the calculated outer-sphere contribution to the activation energy from the observed overall activation energy. Rate constants observed in more than two different solvents are also used to estimate the inner-sphere contribution to the activation energy since it is rarely influenced by the solvent properties, and the outer-sphere contributions in each solvent can be calculated with sufficient accuracy.

(a) The observed self-exchange rate constants for the $Fe_{aq}^{3+/2+}$ and $[Co(NH_3)_6]^{3+/2+}$ couples at 25 °C are 4.2 M^{-1}s^{-1} (ionic strength = 0.55, $K = 0.055$ M^{-1}) and 8×10^{-6} M^{-1}s^{-1} (ionic strength = 2.5, $K = 0.093$ M^{-1}) in water, respectively (K is the equilibrium constant which corresponds to the coulombic work). Roughly estimate the inner-sphere contribution to the activation energy for each self-exchange reaction.

(b) Explain why inner-sphere contributions estimated in (a) for these two self-exchange reactions are largely different.
(c) Explain why the estimation of the force constants f between the central metal ion and coordinated ligand is not an easy task.

Problem 4

(a) Draw a set of precursor and successor states expected for the following pericyclic reaction.
(b) Explain why the reaction coordinate is totally symmetric only at the transition state.

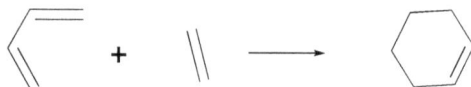

Problem 5

It is known that organic reactions are categorized into three types: acid–base reactions, radical reactions, and pericyclic reactions.

(a) Examine these 3 types of reactions involved in the electron transfer process.
(b) In general chemistry textbooks, it is written that all chemical reactions are categorized into five types: combination, decomposition, single-replacement, double-replacement, and combustion reactions. Examine these 5 types of reactions involved in electron transfer processes.

Problem 6

The group theory is particularly important upon learning chemistry. A few useful ideas are shown here to understand the concept of adiabatic reactions.

When we want to examine integrations such as the perturbed term $\langle \phi_a | H_1 | \phi_b \rangle$, we can easily tell if this integration is zero or not if we have the knowledge of the group theory. We cannot know the exact value of this integration unless we know the exact mathematical functions corresponding to ϕ_a, ϕ_b, and H_1.

One of the only two rules necessary for the judgement is that "the integration has a value only when the integrand includes 'totally symmetric representation' of the particular group in question": totally symmetric representation is the one with all characters being 1 for every symmetry operation of the group, such as the A_1 representation of the C_{2v} point group. Each mathematical function (such as the orbital function) belongs to one of the representations in each point group. The second important rule may be described as follows: only the product of mathematical functions that belong to the same irreducible representation in a point group includes at least one "totally symmetric" representation, such as $A_1 \times A_1$, $E_g \times E_g$, etc. The d_{xy}, d_{yz} and d_{zx} orbitals are assigned to t_{2g} and the s orbitals to the a_{1g} representation of the O_h point group, and A_{1g} is the totally symmetric representation of the O_h point group.

Find the condition in which $H_{rp} = \langle \psi_r | H_1 | \psi_p \rangle = \langle \psi_r | \partial U / \partial Q | \psi_p \rangle Q$ has a nonzero value. Note that U is the total energy of the system and Q is the reaction coordinate.

Problem 7

Read the following article and summarize the effect of super-exchange/sequential transfer mechanisms on the electron transfer reactions (ref. 89) M. D. Newton, K. Ohta and E. Zhong, *J. Chem. Phys.*, 1991, **95**, 2317–2326.

Answer to Problem 1

As ε_0 is 8.85×10^{-12} F m^{-1} and the Pekar factor for water is 0.553, the outer-sphere contributions to the activation energy, $\Delta G^*_{outer-sphere}$, for the self-exchange reactions of $Fe_{aq}^{3+/2+}$ and $[Co(NH_3)_6]^{3+/2+}$ couples are calculated as 29.4 and 29.1 kJ mol^{-1}, respectively.

Answer to Problem 2

When the electronic coupling between the reactants is very large, inner- and outer-sphere contributions to the activation energy are expected to be zero since these two reorganization energies are required to maximize the electronic coupling between the reactants. For almost all outer-sphere electron transfer reactions, transmission coefficients (refer to Section 12.2.4) for reactions are smaller than 1 (reactions are not fully adiabatic as seen in many reactions, $10^{-3} < \kappa_{el} < 1$), even after the inner- and outer-sphere reorganizations.

Answer to Problem 3

(a) By assuming the reactions are adiabatic, $\kappa_{el} = 1$, and there is no tunneling, the following values are calculated as the overall activation energy for the $Fe_{aq}^{3+/2+}$ and $[Co(NH_3)_6]^{3+/2+}$ couples, respectively: 51 kJ mol^{-1} ($Fe_{aq}^{3+/2+}$) and 96 kJ mol^{-1} ($[Co(NH_3)_6]^{3+/2+}$). Therefore, subtractions of the results for Problem 1 and the contribution of K from these values yield roughly estimated inner-sphere contributions of *ca.* 22 kJ mol^{-1} and 67 kJ mol^{-1} for the $Fe_{aq}^{3+/2+}$ couple and $[Co(NH_3)_6]^{3+/2+}$ couple, respectively. These values (especially the difference, 45 kJ mol^{-1}) may be compared with the directly calculated values by using the force constants and nuclear frequencies estimated from the experimentally observed vibration frequencies, 35 kJ mol^{-1} and 73 kJ mol^{-1} (the difference is 38 kJ mol^{-1}): in many cases, the difference in the inner-sphere contributions between two couples is useful for further discussions.

(b) Even rough estimations indicate that the inner-sphere reorganization energy depends largely on the difference in the metal–ligand bond lengths of the reactants. This is because most metal–ligand vibrations are observed in the far-infrared region, 200 to 800 cm^{-1}: ν_1 and ν_2 were reported as 390 and 490 cm^{-1} for $Fe_{aq}^{2+/3+}$ and 360 and 400 cm^{-1} for $[Co(NH_3)_6]^{2+/3+}$.

(c) Some of many problems for the observation and assignments of signals in the far-infrared region are (1) interferences by the signals of water vapor and intermolecular vibrations such as those by the hydrogen bonding and (2) the lack of a strong light source and more reliable detectors for the measurements. Moreover, (3) methods of the assignments of signals are not straightforward and difficult even by using a computer. Precise estimations of activation energies are also obscured by uncertainties such as the degree of adiabaticity and tunneling effect, the latter of which is sometimes expressed by using the term $\Gamma_n = 3 \times 10^{10} \nu_n$ (s^{-1}) instead of $\dfrac{k_B T}{h}$: $k = K \kappa_{el} \Gamma_n \exp\left(-\dfrac{\Delta G^*}{RT}\right)$ where ν_n is about 400 cm^{-1} (Γ_n is in the order of 10^{13} s^{-1}).

Answer to Problem 4

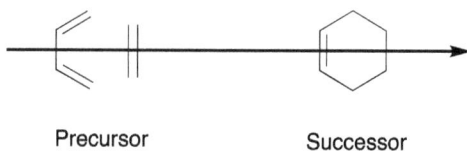

Precursor   Successor

Precursor and successor states are the reactive reactant and nascent state. Therefore, they have the structures ready for the electron transfer/back-electron transfer process, and electrons move back and forth between these states in less than a femtosecond. Although the speed of electrons is less than the speed of light, it is known that the 1s electron in mercury atoms exceeds 50% of the speed of light. Molecules included in the precursor and successor are in their excited states: bond lengths and bond angles have been changed from those for each ground-state structure. The structure of the successor is not that of the final product either: it must have a flat structure (C_{2v}) since this nascent form is also ready for the back-electron transfer. Since the only difference between the precursor and successor states is the positions of electron(s), the 'reaction coordinate at the transition state' includes the aligned precursor and successor as described in the figure according to the definition by Levine:[29] the reaction coordinate at the transition state (shown by the arrow in the figure) is therefore always totally symmetric. The C_2 axis (shown by an arrow) is retained during the electron transfer process, and it is the totally symmetric primary axis (A_1 representation of the C_{2v} point group).

Answer to Problem 5

(a) & (b) Each type of reaction may be considered as a consecutive multi-step reaction. For example, acid–base reactions may be regarded as the interaction between the electron-pair donor and acceptor according to the definition of Lewis, and based on the definition of Usanovich acid–base reactions are considered as redox reactions.

Answer to Problem 6

As the integrand in the equation $H_{rp} = \langle \psi_r | H_1 | \psi_p \rangle = \langle \psi_r | \partial U / \partial Q | \psi_p \rangle Q$ is regarded as the product of ψ_r, ∂U, ∂Q, and ψ_p, examination of each term will lead to the conclusion: (1) since U is the energy, ∂U is also totally symmetric (isotropic), and the reaction coordinate Q is totally symmetric only at (or in the vicinity of) the transition state as explained in the answer to the problem in Section 12.3. Therefore, the term $\partial U / \partial Q$ is totally symmetric at the transition state. When H_{rp} has a non-zero value, the product of ψ_r and ψ_p has to be totally symmetric. Therefore, ψ_r and ψ_p should have the identical symmetry to each other at least at the transition state.

Answer to Problem 7

It has been known that the "super-exchange-coupled sequential electron transfer mechanism through the ligand's occupied/vacant orbitals (or CT-perturbed outer-sphere electron transfer)" provides a significantly lower-energy pathway than the non-adiabatic outer-sphere process with a very small electronic coupling between the ground-state orbitals of reactants (and even ordinary outer-sphere reactions with $10^{-3} < k_{el} < 1$), when a convenient mediating ligand's orbital is available.[24,89] The basic concept of this mechanism may be summarized as follows.

The matrix element H_{0,CT_j}^{ii} corresponds to the charge transfer (CT) interaction on a metal site where 0 and CT_j indicate that the corresponding wave functions are the unperturbed functions and the matrix element corresponds to the ML (metal-to-ligand)/LMCT (ligand-to-metal charge transfer) interaction. Superscripts i and f indicate the place where the CT interaction takes place: i corresponds to one of the reactants and f to the counter re-agent. Therefore, the superscript if indicates the coupling between the two CT-excited levels of the reductant and oxidant when both reactants exhibit CT interactions.

$$H_{RP}^{eff} = \sum_j \frac{H_{0,CT_j}^{ii} H_{CT_j,CT_j}^{if} H_{0,CT_j}^{ff}}{\left(\Delta E_{CT_j}^0 \right)^2}$$

where $E_{CT_j}^0$ is the energy gap between the diabatic thermal activation energy and the corresponding CT energy level at the position of the thermal transition state on the reaction coordinate. Matrix elements, Hs, are given by the ordinary bra ($\langle|$) and ket ($|\rangle$) notation with the electronic Hamiltonian as follows:

$$H_{lm} = \langle \psi_l | H_{el} | \psi_m \rangle$$

This equation is valid for any electron transfer reaction in the Marcus–Sutin regime that takes place within the encounter complex. It is certain that such species in the CT-excited state (in the state of H_{0,CT_j}^{ii}) which is in equilibrium with the ground-state species are sometimes far more reactive compared with the ground-state species.

For example, in the cross reactions involving Cu(II)/(I) couples, consideration of the CT coupling only at the copper site is necessary since the d-orbital levels of Cu are too low for the sufficient electronic coupling with the d-orbitals of other metals. The effective electronic coupling element for the electron transfer process of copper is, therefore, given by the following equation:

$$H_{CT_jM}^{eff} = \frac{H_{0,CT_j}^{ii} H_{CT_j,0}^{jM}}{\left(\Delta E_{CT_j}^{0}\right)^2}$$

where H_{0,CT_j}^{ii} and $H_{CT_j,0}^{jM}$ are the coupling constants between the ground state and CT excited state of the copper complex, and between the CT perturbed (excited) state of the copper complex and the ground state d-orbital of the counter reagent, respectively. The increase in the rate constant by this type of coupling is about 160% for the electron self-exchange reaction of the ferricenium/ferrocene couple compared with that by the normal direct coupling between d-orbitals on each metal: the electron transfer rate constant for the ferricenium/ferrocene couple is as large as 10^6 $M^{-1}s^{-1}$ even without the super-exchange coupling. Acceleration of less-adiabatic reactions by the super-exchange coupling is expected to be much larger than this.

Acknowledgements

The author of this section wishes to express his thanks to Professors Roderick D. Cannon (University of East Anglia, UK; deceased in 2015) and Thomas W. Swaddle (University of Calgary, Canada) for guiding the author to the theoretical fields of chemistry. They were excellent mentors in the author's early carrier as a scientist. Dr Swaddle also kindly read over the original manuscript and gave the author several important remarks.

References

1. L. Eberson, *Electron transfers in Organic Chemistry*, Springer-Verlag, Heidelberg, Germany, 1987.
2. A. G. Lappin, *Redox Mechanisms in Inorganic Chemistry*, Ellis Horwood, West Sussex, UK, 1994.

3. S. F. A. Kettle, *Physical Inorganic Chemistry*, Oxford University Press, Oxford, UK, 1998.

4. R. L. Platzman and J. Franck, *Z. Phys.*, 1954, **138**, 411.

5. W. F. Libby, *J. Phys. Chem.*, 1952, **56**, 863.

6. R. A. Marcus, *J. Chem. Phys.*, 1956, **24**, 966.

7. R. A. Marcus, *J. Chem. Phys.*, 1956, **24**, 979.

8. R. A. Marcus, *J. Chem. Phys.*, 1957, **26**, 867.

9. R. A. Marcus, *J. Chem. Phys.*, 1957, **26**, 872.

10. R. D. Cannon, *Electron Transfer Reactions*, Butterworth, UK, 1980.

11. *Progress in Inorganic Chemistry*, ed. S. J. Lippard, 1983, vol. 30, p. 441.

12. *Inorganic High Pressure Chemistry*, ed. R. van Eldik, Elsevier, Amsterdam, 1986.

13. A. J. Bard and L. R. Faulkner, *Electrochemical Methods*, Wiley, USA, 1980.

14. R. M. Fuoss, *J. Am. Chem. Soc.*, 1958, **80**, 5059.

15. N. S. Hush, *Trans. Faraday Soc.*, 1961, **57**, 557.

16. N. S. Hush, *Prog. Inorg. Chem.*, 1967, **8**, 391.

17. N. S. Hush, *Chem. Phys.*, 1975, **10**, 361.

18. V. G. Levich and R. R. Dogonadze, *Colln Czech. Chem. Commun., Engl. Ed.*, 1961, **26**, 193.

19. V. G. Levich, *Adv. Electrochem. Electrochem. Eng. Ed.*, 1966, **4**, 249.

20. N. Sutin, *Annu. Rev. Nucl. Sci.*, 1962, **12**, 285.

21. L. D. Landau and E. M. Lifshitz, *Quantum Mechanics*, English/Japanese edn, 1980/1983.

22. R. G. Pearson, *Symmetry Rules for Chemical Reactions*, John Wiley and Sons, Inc., USA, 1976.

23. *Electron Transfer in Biology and the Solid State*, ed. M. K. Johnson *et al.*, American Chemical Society, 1990.

24. H. D. Takagi, *Inorganic Chemistry on the basis of Quantum Theory, Japanese Ed.*, Nagoya Univ. Press, Nagoya Japan, 2nd edn, 2018.

25. I. R. Gould, J. E. Moser, B. Armitage, S. Farid and M. S. Herman, *J. Am. Chem. Soc.*, 1989, **111**, 1917, and references therein.

26. H. Eyring, *J. Chem. Phys.*, 1935, **3**, 107.

27. K. Zahir, J. H. Espenson and A. Bakac, *J. Am. Chem. Soc.*, 1988, **110**, 5059.

28. N. Sutin, *Prog. Inorg. Chem.*, 1983, **30**, 441.

29. M. A. Ratner and R. D. Levine, *J. Am. Chem. Soc.*, 1980, **102**, 4898.

30. R. D. Levine, *J. Phys. Chem.*, 1979, **83**, 159.

31. R. A. Marcus, *Pure Appl. Chem.*, 1997, **69**, 13.

32. J. Sandstrom, *Dynamic NMR Spectroscopy*, Academic Press, London, 1982.

33. S. F. Lincoln, *Kinetic Application of NMR Spectroscopy, Progress in Inorganic Chemistry*, Pergamon Press, 1977, vol. 9

34. H. Doine and T. W. Swaddle, *Inorg. Chem.*, 1997, **30**, 1858.

35. R. B. Jordan, *Reaction Mechanisms of Inorganic and Organometallic Systems*, Oxford Univ. Press, 1998.

36. T. Asano and J. Le Noble, *Chem. Rev.*, 1978, **78**, 407.

37. T. Kowall, P. Caravan, H. Bourgeois, L. Helm, F. P. Rotzinger and A. E. Merbach, *J. Am. Chem. Soc.*, 1998, **120**, 6569.
38. D. R. Stranks, *Pure Appl. Chem.*, 1974, **38**, 303.
39. T. W. Swaddle, *Inorg. Chem.*, 1990, **29**, 5017.
40. T. W. Swaddle, *Can. J. Chem.*, 1996, **74**, 631.
41. P. G. Wolynes, *J. Chem. Phys.*, 1987, **86**, 5133.
42. I. Rips, J. Klafter and J. Jortner, *J. Chem. Phys.*, 1988, **88**, 3246.
43. H. D. Takagi and T. W. Swaddle, *Chem. Phys. Lett.*, 1996, **248**, 207.
44. J. F. Skinner, E. L. Cossler and R. M. Fuoss, *J. Am. Chem. Soc.*, 1968, **72**, 1057.
45. K. R. Srinivasan and R. L. Kay, *J. Solution Chem.*, 1977, **6**, 357.
46. M. Matsumoto, T. Tarumi, I. Takahashi, T. Noda and H. D. Takagi, *Z. Naturforch.*, 1997, **52b**, 1087.
47. J. I. Sachinidis, R. D. Shalders and P. A. Tregloan, *J. Electroanal. Chem.*, 1992, **327**, 219.
48. J. I. Sachinidis, R. D. Shalders and P. A. Tregloan, *Inorg. Chem.*, 1994, **33**, 6180.
49. R. A. Marcus, *J. Chem. Phys.*, 1965, **43**, 679.
50. R. A. Marcus, *Electrochim. Acta*, 1968, **13**, 995.
51. R. D. Cannon, *Elentron Transfer Reactions*, Butterworths and Co. Ltd., UK, 1980, ch. 6.8.
52. T. W. Swaddle and P. A. Tregloan, *Coord. Chem. Rev.*, 1999, **187**, 255 and references therein.
53. T. Saji, T. Yamada and S. Aoyagi, *J. Electroanal. Chem.*, 1975, **61**, 147.
54. T. Saji, Y. Maruyama and S. Aoyagui, *J. Electroanal. Chem.*, 1978, **86**, 219.
55. M. J. Weaver, *J. Phys. Chem.*, 1980, **84**, 568.
56. *Mechanistic Aspects of Inorganic Reactions*, ed. J. T. Hupp, M. J. Weaver, D. B. Rorabacher and J. F. Endicott, ACS Symposium, 1982, vol. 198, p. 181.
57. J. T. Hupp and M. J. Weaver, *Inorg. Chem.*, 1983, **22**, 2557.
58. J. T. Hupp and M. J. Weaver, *J. Phys. Chem.*, 1985, **89**, 2795.
59. M. J. Weaver, *J. Phys. Chem.*, 1990, **94**, 8608.
60. P. W. Crawford and F. A. Schultz, *Inorg. Chem.*, 1994, **33**, 4344.
61. L. Spiccia and T. W. Swaddle, *Inorg. Chem.*, 1987, **26**, 2265.
62. B. S. Brunschwig, C. Creutz, D. H. Macartney, T. K. Sham and N. Sutin, *Faraday Discuss. Chem. Soc.*, 1982, **74**, 113.
63. W. H. Jolley, D. R. Stranks and T. W. Swaddle, *Inorg. Chem.*, 1990, **29**, 1948.
64. H. Doine and T. W. Swaddle, *Can. J. Chem.*, 1990, **68**, 2228.
65. A. Heim and N. Sutin, *Inorg. Chem.*, 1976, **15**, 476.
66. R. J. Campion, C. F. Deck, P. King and A. C. Wahl, *Inorg. Chem.*, 1967, **6**, 672.
67. H. D. Takagi and T. W. Swaddle, *Inorg. Chem.*, 1992, **31**, 4669.
68. F. P. Dwyer and A. M. Sergeson, *J. Phys. Chem.*, 1961, **65**, 1892.

69. W. H. Jolley, D. R. Stranks and T. W. Swaddle, *Inorg. Chem.*, 1990, **29**, 385.
70. I. I. Creaser, J. M. Harrowfield, A. J. Herlt, A. M. Sergeson, J. Sprinborg, R. J. Geue and M. R. Snow, *J. Am. Chem. Soc.*, 1982, **104**, 6016.
71. H. Doine (Takagi) and T. W. Swaddle, *Inorg. Chem.*, 1991, **30**, 1858.
72. H. J. Kuppers, A. Neves, C. Pomp, D. Ventur, K. Wieghardt, B. Nuber and J. Weiss, *Inorg. Chem.*, 1986, **25**, 2400.
73. H. J. Kuppers, K. Wieghardt, S. Steenken, B. Nuber and J. Weiss, *Z. Anorg. Allg. Chem.*, 1989, **57325**, 43.
74. H. Doine (Takagi) and T. W. Swaddle, *Inorg. Chem.*, 1991, **30**, 1858.
75. S. Chandrasekhar and A. MaAuley, *Inorg. Chem.*, 1992, **32**, 482.
76. R. M. Warren, A. G. Lappin, B. D. Mehta and H. M. Neumann, *Inorg. Chem.*, 1990, **29**, 4185.
77. R. M. Warren, A. G. Lappin and A. Tatehata, *Inorg. Chem.*, 1992, **31**, 1566.
78. Y. A. Im and D. H. Busch, *J. Am. Chem. Soc.*, 1961, **83**, 3357.
79. R. G. Wilkins and R. E. Yelin, *Inorg. Chem.*, 1968, **7**, 2267.
80. W. H. Jolley, D. R. Stranks and T. W. Swaddle, *Inorg. Chem.*, 1992, **31**, 507.
81. A. McAuley and C. Xu, *Inorg. Chem.*, 1988, **27**, 1204.
82. S. Itoh, N. Koshino, M. Katsuki, T. Noda, K. Ishihara, M. Inamo and H. D. Takagi, *Dalton Trans.*, 2004, 1862.
83. H. Doine (Takagi), Y. Yano and T. W. Swaddle, *Inorg. Chem.*, 1989, **28**, 2319.
84. N. Koshino, Y. Kuchiyama, S. Funahashi and H. D. Takagi, *Chem. Phys. Lett.*, 1999, **306**, 291.
85. E. S. Yang, M. Chan and A. C. Wahl, *J. Phys. Chem.*, 1980, **84**, 3094.
86. R. M. Nielson, G. E. McManis, M. N. Golovin and M. J. Weaver, *J. Phys. Chem.*, 1988, **92**, 3441.
87. N. Koshino, Y. Kuchiyama, H. Ozaki, S. Funahashi and H. D. Takagi, *Inorg. Chem.*, 1999, **38**, 3352 and references therein.
88. M. Chan and A. C. Wahl, *J. Phys. Chem.*, 1982, **86**, 126.
89. M. D. Newton, K. Ohta and E. Zhong, *J. Chem. Phys.*, 1991, **95**, 2317.
90. W. R. Fawcett, A. Gaál and D. Misicak, *J. Electroanal. Chem.*, 2011, **660**, 230.
91. K. Nakamura and T. Shikata, *ChemPhysChem*, 2010, **11**, 285.
92. M. A. Gebbie, M. Valtiner, X. Banquy, E. T. Fox, W. A. Henderson and J. N. Israelachvili, *Proc. Natl. Acad. Sci. U. S. A.*, 2013, **110**, 9674.
93. T. Mabe, F. Doseki, T. Yagyu, K. Ishihara, M. Inamo and H. D. Takagi, *J. Solution Chem.*, 2018, **47**, 993 and references therein.
94. K. Baba, H. Ono, E. Itoh, S. Itoh, K. Noda, T. Usui, K. Ishihara, M. Inamo, H. D. Takagi and T. Asano, *Chem. – Eur. J.*, 2006, **12**, 5328.
95. J. K. Yandell, *Copper Coordination Chemistry: Biochemical and Inorganic Perspectives*, ed. K. D. Karlin and J. Zubieta, Adenine Press, N. Y., 1983.
96. D. B. Rorabacher, *Chem. Rev.*, 2004, **104**, 651.
97. B. M. Hoffman and M. A. Ratner, *J. Am. Chem. Soc.*, 1987, **109**, 6237.
98. B. C. Brunschwig and N. Sutin, *J. Am. Chem. Soc.*, 1989, **111**, 7454.

99. N. Koshino, Y. Kuchiyama, H. Ozaki, S. Funahashi and H. D. Takagi, *Inorg. Chem.*, 1999, **38**, 3352.
100. A. Yamada, T. Mabe, R. Yamane, K. Noda, Y. Wasada, M. Inamo, K. Ishihara, T. Suzuki and H. D. Takagi, *Dalton Trans.*, 2015, **44**, 13979.
101. S. Itoh, N. Kishikawa, T. Suzuki and H. D. Takagi, *Dalton Trans.*, 2005, 1066.
102. S. Itoh, S. Funahashi and H. D. Takagi, *Chem. Phys. Lett.*, 2001, **344**, 441.
103. N. Koshino, S. Itoh, Y. Kuchiyama, S. Funahashi and H. D. Takagi, *Inorg. React. Mech.*, 2000, **2**, 93.
104. *ACS Symposium Series*, ed. D. R. Salahub and M. C. Zerner, ACS Washington D. C., 1989, vol. 394, p. 378.
105. M. D. Newton, *J. Phys. Chem.*, 1991, **95**, 30.

CHAPTER 13

Metal Complexes in Supercritical Fluids

TATSUYA UMECKY

Faculty of Science and Engineering, Saga University, Honjo-machi, Saga, 840-8502, Japan
Email: umecky@cc.saga-u.ac.jp

13.1 Introduction

It is widely accepted that a substance is in a gaseous, liquid, or solid state. Figure 13.1 shows the p–T phase diagram of CO_2, an example of a phase diagram.[1] The triple point (in the phase diagram of CO_2, at 216.6 K and 0.518 MPa) is the temperature and pressure at which gas, liquid, and solid phases coexist. At temperatures lower than the triple point, the gas–solid equilibrium is represented by the sublimation curve; at higher temperatures, the gas–liquid equilibrium is represented by the evaporation curve. When the temperature increases at a pressure higher than the triple point, the substance changes from a solid to a gas through the liquid. The end of the high-temperature and high-pressure side of the evaporation curve is known as the critical point (in the phase diagram of CO_2, at 304.1 K and 7.38 MPa). At temperatures and pressures above the critical point, the substance is not in a liquid state, regardless of the gaseous substance being compressed. Supercritical fluids (SCFs) are substances at a temperature and pressure that exceed the critical temperature (T_c) and critical pressure (p_c). In this section, the features of SCFs as solvents and SCF solutions in which metal complexes are dissolved are introduced.

Coordination Chemistry Fundamentals Series No. 2
Metal Ions and Complexes in Solution
Edited by Toshio Yamaguchi and Ingmar Persson
© Japan Society of Coordination Chemistry 2024
Published by the Royal Society of Chemistry, www.rsc.org

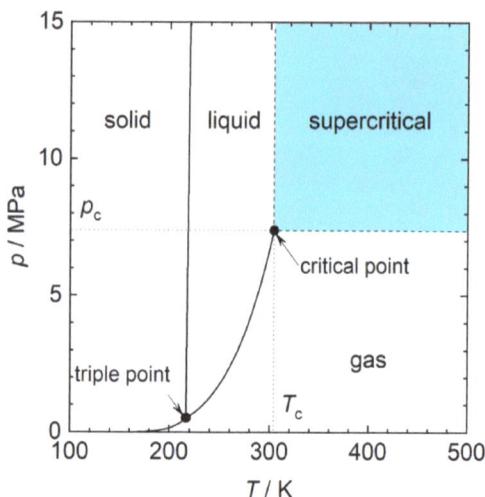

Figure 13.1   Phase diagram of CO_2.

Table 13.1   Fundamental properties of gas, liquid, and supercritical fluids.

	Gas	Supercritical	Liquid
Density/kg m^{-3}	0.6–2.0	200–900	600–2000
Viscosity/mPa s	0.01–0.03	0.01–0.1	0.2–10
Diffusion coefficient/10^{-9} m^2 s^{-1}	1000–4000	20–700	<2
Thermal conductivity/W m^{-1} K^{-1}	~0.001	0.001–0.1	~0.1

13.2   Fundamental Properties of SCFs

The fundamental physicochemical properties of substances in the super-critical, gaseous, and liquid states are listed in Table 13.1. The main features of SCFs are as follows:

(1) The thermal motion of molecules is strenuous because the sub-stance temperature exceeds the critical temperature. Therefore, SCFs have a lower viscosity, higher diffusivity, and lower surface tension than liquid solvents. Near the critical point, the thermal conduct-ivities of SCFs are higher, which indicates that the heat-transfer rate is higher.

(2) The density of SCFs can be continuously changed, widely from a dilute substance, such as an ideal gas, to a dense substance, such as a liquid, by slightly changing the temperature and pressure. This indicates that microscopic solvent properties, such as the diffusion coefficient, thermal conductivity, and dielectric constant, which depend heavily on density, can be arbitrarily tuned.

(3) In SCFs, the intermolecular repulsive forces antagonize the intermolecular attractive forces. This indicates that microscopic factors, such as solvation structures and dynamics, which control chemical reactions, can be controlled by the operating temperature and pressure, as mentioned in feature 2.

Feature 1 indicates that SCFs can easily permeate through microspaces such as pores. Such advantages of SCFs are being utilized in cleaning nanometre-scale microstructured materials, such as semiconductors, electronic parts, and foaming polymer materials. Feature 2 is a typical property specific to SCFs. This enables the optimisation of operating conditions, for example, temperature and pressure, in industrial processes. For example, the crystallisation technology for preparing nanoparticles is achieved by changing the solubility of a substance, arising from extreme changes in the state and phase of a fluid with changing temperature and pressure. Feature 3 is essential for understanding the core of SCFs. This is important from the viewpoint of solution chemistry. The solvent properties of a fluid depend on its molecules and are governed by intermolecular interactions. In SCFs, where the repulsive and attractive intermolecular forces are opposed, the density fluctuations are large, and thus heterogeneous structures are easily formed. Near the critical point, the density fluctuations of SCFs are particularly significant. For SCF solutions, it has been mentioned that the solvent molecules excessively aggregate around solute molecules (clustering).[2,3] Thus, the progress of chemical reactions near the critical point by such specific solvation phenomena is often observed.[4]

All substances have unique critical points; hence, there are several SCFs.[5] However, in numerous research and technology applications, CO_2 ($T_c = 304.1$ K, $p_c = 7.38$ MPa) and H_2O ($T_c = 647.2$ K, $p_c = 22.1$ MPa) are used as the fluids. These two substances are thermally stable and inexpensive because they are abundant in nature. They have a significant advantage in terms of industrial use. CO_2 can achieve a supercritical state at moderate temperatures and pressures. Since the 1980s, supercritical CO_2 (SC-CO_2) has been used for the extraction and separation of food, natural products, and pharmaceuticals. Currently, a wide variety of application studies on extraction, separation, as well as organic/inorganic chemical reactions, nanomaterial synthesis, waste treatment, and precision cleaning are being performed.[6] In the following section, the dissolution behaviour and solvation structure of metal complexes in SC-CO_2 are described with respect to extraction and separation.

13.3 Dissolution Characteristics of Substances in SC-CO_2

The extraction and separation of metal elements from solids and aqueous solutions using SC-CO_2 have wide-ranging applications, such as the removal of hazardous metals and recovery of useful rare metals from urban ores.

Extraction using SC-CO$_2$ has the same principle as that using organic liquid solvents. Its advantages are listed below.

(1) Unwanted substances from the extraction solutions and residual solvent in the collection of the extracts can be suppressed.
(2) The extraction rate can be improved by using SC-CO$_2$ with a lower viscosity and higher diffusivity. Moreover, the extraction equipment can be downsized because the surface tension of SC-CO$_2$ is low.
(3) The target substances can be selectively extracted *via* tuning solvent properties with temperature and pressure.
(4) In the separation between the extracts and solvents, the removal of CO$_2$ as a solvent is easier than that of organic solvents. Moreover, CO$_2$ is more chemically stable and decomposes less with H$_2$O or radiation than organic solvents.

Figure 13.2 shows the standard solubility curve for liquid or solid substances in SC-CO$_2$. At a constant temperature, the solubility of substances in SC-CO$_2$ significantly increases with increasing pressure. The target substances (component A in Figure 13.2) extracted at a higher pressure (p_A) can be collected by reducing the pressure (p_B). Simultaneously, the extracts can be separated from the CO$_2$ solvents. A large volume of CO$_2$ can be recovered and reused. Temperature swing and the addition of an inert gas are also

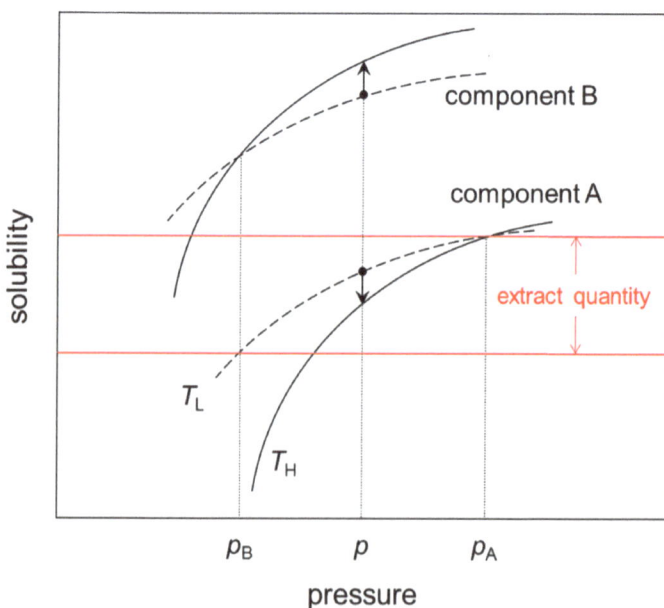

Figure 13.2 Solubility curve of substances in SC-CO$_2$. Temperatures: T_H (solid line) $> T_L$ (broken line) $> T_c$. Pressures: $p_A > p_B > p_c$. When components A and B are dissolved at a low temperature, T_L, and pressure, p $(p_B < p < p_A)$, and heated to a high temperature, T_H, only component A becomes supersaturated and precipitates.

performed. As shown in Figure 13.2, the solubility curves at the two temperatures of T_H and T_L are reported to intersect. A retrograde crystallisation method, in which pure target substances are separated from mixtures that are challenging to separate, has been proposed.[7]

CO_2 has a large quadrupole moment, but no permanent dipole moment. Because of this, the dielectric constant of SC-CO_2 is low, and, as a result, SC-CO_2 negligibly dissolves metal ions. Hence, in extraction with SC-CO_2, ligands are generally used as extractants to form neutral complex molecules and ion pairs. Thus far, various types of metal ions have been extracted with chelating ligands, such as β-diketones and dithiocarbamates, as well as macrocyclic compounds, such as crown ethers and porphyrins. Complexation between metal ions and ligands depends heavily on the charge and size of the metal ions and coexisting ligands. This is virtually the same as that of the organic solvents. In the extraction using SC-CO_2, in addition to the aforementioned physical factors such as temperature and pressure, the affinity with the CO_2 solvent, the chemical structure and stability of the formed metal complex, and the influence of coexisting substances such as polar molecules are crucial factors.[8–12]

13.4 Solvation Structure of β-Diketonato Complexes in SC-CO_2

β-Diketone ligands are one of the extractants that have often been used for the extraction of metal ions using SC-CO_2. The β-diketone molecule contains two carbonyl groups at the 2,4-(β-) positions, and keto–enol tautomerization, in which the hydrogen atom at the 3-position is transferred to the oxygen atom of the carbonyl group (Figure 13.3). The solution was acidic because of the formation of the enol forms. The keto–enol equilibrium exists also in SC-CO_2. For acetylacetone (Hacac), the enol form accounts for ~70% near 300 K.[13] The fraction of the keto form increases with increasing temperature and pressure. When an electron-withdrawing group is introduced into the β-diketone molecule, generally, the keto–enol equilibrium of β-diketone shifts to the formation of the enol form.

β-Diketone forms a monovalent anion by releasing a proton and can coordinate with metal ions as a bidentate chelate. The neutral complex, in which the ligand molecule where the number is equal to the valence of the metal ion is coordinated, is chemically stable and highly soluble in SC-CO_2. The outer surface of the metal complex is required to be hydrophobic.

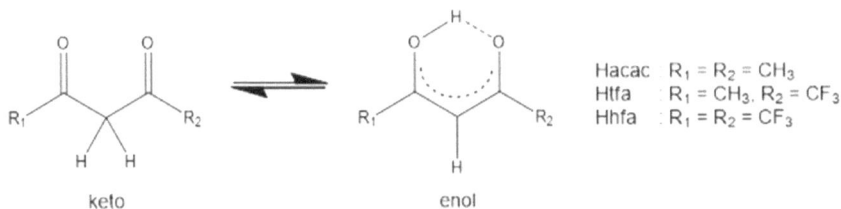

Hacac $R_1 = R_2 = CH_3$
Htfa $R_1 = CH_3, R_2 = CF_3$
Hhfa $R_1 = R_2 = CF_3$

Figure 13.3 Keto–enol tautomerization of β-diketones.

SC-CO$_2$ extraction with various β-diketones has been reported, and the introduction of fluorinated substituents to β-diketone molecules has been observed to be effective in increasing the extraction efficiency. This results from the improved solubility of the metal complexes in SC-CO$_2$. For Cr(III) complexes, the solubility in SC-CO$_2$ increases in the order of [Cr(acac)$_3$]<[Cr(tfa)$_3$]<[Cr(hfa)$_3$]. The solubility of [Cr(hfa)$_3$] is approximately 100 times higher than that of [Cr(acac)$_3$].[11] The introduction of fluorinated substituents to β-diketone molecules increases the volatility of the metal complexes, but, as described below, specific intermolecular interactions of the complexes with CO$_2$ molecules have also been reported.

To efficiently develop extractants suitable for SC-CO$_2$ extraction, a thorough understanding of the intermolecular interactions between the metal complex and the CO$_2$ solvent is important. Because the generation of SC-CO$_2$ requires high pressure, normal spectrometers cannot be used as they are; however, *in situ* spectroscopic measurements have been attempted for the SC-CO$_2$ solvent and solutions.[14] The quadrupole longitudinal relaxation times of the central metals [Be(acac)$_2$], [Al(acac)$_3$], and [Al(hfa)$_3$] in SC-CO$_2$ have been measured using the high-pressure nuclear magnetic resonance method.[15] The rotational correlation times (τ_r) were determined from the quadrupole relaxation times of these metal complexes, and the rotation dynamics of the complexes were discussed in terms of intermolecular interactions with CO$_2$ molecules based on the hydrodynamic model. As shown in Figure 13.4, the τ_r of the complexes in SC-CO$_2$ near the critical point is higher than expected from the η/T ratio, and convex shapes are observed for all complexes. The larger τ_r values near the critical point are attributable to the clustering phenomena. From the temperature at which

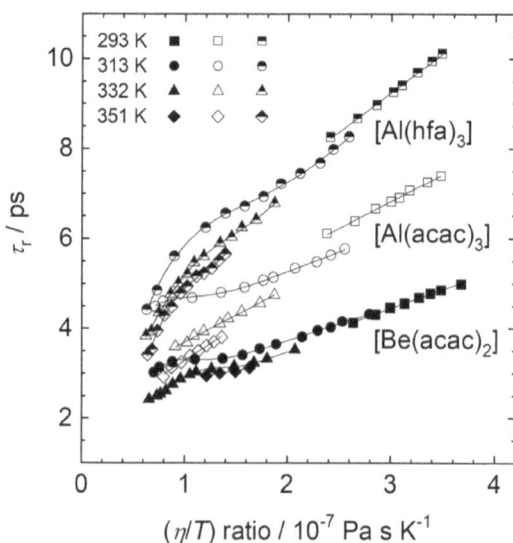

Figure 13.4   Relationship between the η/T ratio and rotational correlation times of β-diketonato complexes in SC-CO$_2$.

excess τ_r is observed, the intermolecular interactions with CO_2 are observed to be in the order of $[Al(acac)_3] < [Be(acac)_2] < [Al(hfa)_3]$. CO_2 molecules have been reported to exhibit a high affinity for aromatic compounds,[16] indicating that CO_2 can interact with pseudo-aromatic chelate rings. In β-diketonato complexes, the space around the chelate ring of the tetrahedral $[Be(acac)_2]$ is wider than that of the octahedral $[Al(acac)_3]$, and the $Be(II)$ complex is more advantageous for interaction than the nonfluorinated $Al(III)$ complex. In the intermolecular interaction of CO_2 molecules with fluorinated compounds, the Lewis acid–base interaction between the positively charged carbon atom of CO_2 and the negatively charged fluorine atom plays an important role in the dissolution of metal complexes in $SC\text{-}CO_2$. In addition, the introduction of carbonyl and ether groups was also effective in improving the affinity with CO_2.[17]

13.5   Entrainer Effects on SC-CO₂ Extraction

$SC\text{-}CO_2$ is a poorer extraction solvent than ordinary organic liquids. The solubility of highly polar compounds and high molecular weight molecules in $SC\text{-}CO_2$ is not sufficiently high. To improve the solubility of such solutes, small amounts of various substances were added to $SC\text{-}CO_2$. The added substances are commonly referred to as "entrainers" or "cosolvents". Methanol and ethanol are favourable entrainers for improving the solubility of polar substances in $SC\text{-}CO_2$.[18] The entrainer effect is also effective in the extraction of metal complexes with $SC\text{-}CO_2$. For example, Ohashi *et al.* observed that the solubility of $[Cr(acac)_3]$ significantly increases when a fluorinated alcohol, such as CF_3CH_2OH, was added to $SC\text{-}CO_2$.[19] From the results of IR spectra measurements, the increase in solubility was attributed to the hydrogen bonding of $[Cr(acac)_3]$ with the alcohol. Furthermore, there is a cooperative effect of the extractant on metal ions, such as lanthanoids and actinides. In the extraction of $U(VI)$ with β-diketones, the coexistence of tributyl phosphate (TBP) can significantly improve the extraction efficiency compared with that for each of the extractants.[8,20] TBP was additionally coordinated to the metal ion to increase its solubility in $SC\text{-}CO_2$. However, in the extraction of metal complexes using $SC\text{-}CO_2$, H_2O adversely affects the solubility of the metal complexes in $SC\text{-}CO_2$. When H_2O is present in $SC\text{-}CO_2$, the metal complexes with β-diketonato ligands such as Hhfa are irreversibly hydrolysed, thereby decreasing the extraction efficiency.[10] The extraction efficiency of metal ions decreases with increasing solubility of H_2O in $SC\text{-}CO_2$ in the high-pressure region.[12]

In the extraction of metal ions using $SC\text{-}CO_2$, as described above, the extraction and separation conditions, such as temperature and pressure, as well as using the ligands as extractants and the coexisting substances as entrainers are important. However, when extracting metal ions from aqueous solutions, the chemical environmental factors of the extraction system, such as pH, must be considered. H_2O in contact with $SC\text{-}CO_2$ shows acidity at pH ~3 owing to the chemical equilibrium of CO_2 and H_2O.

$$CO_2 + H_2O \rightleftharpoons H_2CO_3 \rightleftharpoons H^+ + HCO_3{}^- \rightleftharpoons 2H^+ + CO_3{}^{2-} \qquad (13.1)$$

Acidic extractants, such as β-diketones, do not form metal complexes in acidic regions where the pH is extremely low because the extractants do not dissociate H^+ ions in acidic solutions. The pH of the extraction system must be maintained neutral using buffer solutions or by adding bases. By adding NaOH to the aqueous solution of the extraction system, heavy metal ions, such as Cr(III), Co(II), and Au(III), can be selectively extracted to SC-CO_2.[21]

13.6   Summary

The fundamental properties of SC-CO_2 and the behaviour of metal complexes in SC-CO_2 were briefly outlined from the viewpoint of extraction and separation. Research on metal complexes in SC-CO_2 is expanding to the field of extraction and separation as well as the field of homogeneous catalysts for chemical reactions and precursors of fine metal particles. In addition, technological developments for solubilising metal ions with reverse micelles formed in SC-CO_2 are underway. Supercritical CO_2 and H_2O are environmentally friendly solvents in line with the concept of "green sustainable chemistry". SCFs are generally generated under high-temperature and high-pressure conditions. The physicochemical properties of solvents and phase behaviours can be easily controlled by tuning the operating temperature and pressure, allowing the production of special environments that are not observed in conventional liquid solvents. The fusion of SCFs and coordination chemistry will provide valuable insights into the development of science and technology.

Problem 1

Draw the p–T phase diagram of H_2O, for which the triple point is 273.2 K and 612 Pa and the critical point is 647.2 K and 22.1 MPa.

Problem 2

Draw a schematic illustration of solvent clustering around a solute near the critical point and solvation of a solute in a dense liquid.

Problem 3

In Figure 13.2, the solubility of component B in SC-CO_2 at pressure p increases with the temperature increase from T_L to T_H. Explain the increase in the solubility of component B in SC-CO_2 at p with increasing temperature.

Problem 4

In Figure 13.2, the solubility of component A in SC-CO_2 at pressure p decreases with the increase in temperature from T_L to T_H. Explain the decrease in the solubility of component A in SC-CO_2 at p with increasing temperature.

Answer to Problem 1

It is worth noting that, as p increases, the melting temperature of H_2O decreases, whereas that of CO_2 increases (Figure 13.1).

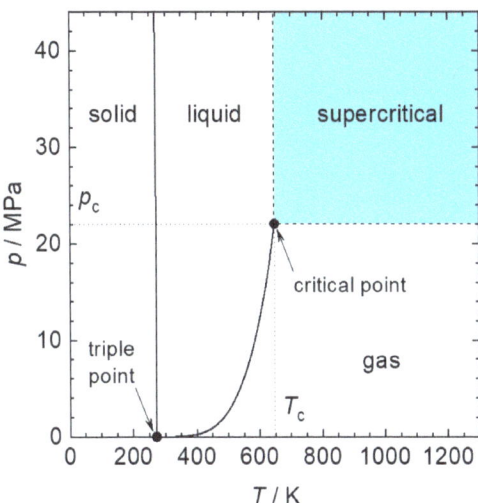

Answer to Problem 2

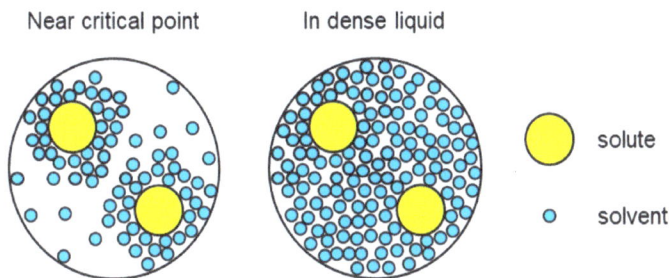

Near the critical point, solvent molecules excessively aggregate around solute molecules.

Answer to Problem 3

The major factors determining the solubility of a solute in SC-CO_2 are the vapour pressure of the solute and the solute–solvent intermolecular interactions. The increase in the solubility of component B in SC-CO_2 at p with increasing temperature can be attributed to the increase in the vapour pressure of component B. At pressures above the p_B of component B in Figure 13.2, the effect of the vapour pressure of the solute on the solubility is more significant than that of the intermolecular interactions. The important point to note is that below the p_B the solubility of component B in SC-CO_2 decreases with the temperature increase from T_L to T_H; the effect of the vapour pressure of the solute on the solubility is less significant than that of the intermolecular interactions.

Answer to Problem 4

The decrease in the solubility of component A in SC-CO_2 at p with increasing temperature can be attributed to the weakening of the intermolecular interactions between component A and CO_2.

Abbreviations

acac$^-$	2,4-Pentanedionate
hfa$^-$	1,1,1,5,5,5-Hexafluoro-2,4-pentanedionate
tfa$^-$	1,1,1-Trifluoro-2,4-pentanedionate
η	Viscosity of the solution
p	Pressure
p_c	Critical pressure
SCF	Supercritical fluid
SC-CO_2	Supercritical carbon dioxide
T	Temperature
T_c	Critical temperature
τ_r	Rotational correlation time

References

1. S. Angus, B. Armstrong and K. M. de Reuck, *International Thermodynamic Table of the Fluid state-3 Carbon Dioxide*, IUPAC, Blackwell Science, Oxford, 1976.
2. J. F. Kauffman, *Anal. Chem.*, 1996, **68**(7), 248A.
3. S. C. Tucker, *Chem. Rev.*, 1999, **99**, 391.
4. J. F. Brennecke and J. E. Chateauneuf, *Chem. Rev.*, 1999, **99**, 433.
5. *CRC Handbook of Chemistry and Physics*, ed. D. R. Lide, CRC Press, Boca Raton, FL, 86th edn, 2005, Section 6.
6. X. Ye and C. M. Wai, *J. Chem. Educ.*, 2003, **80**, 198.
7. C. Y. Tai, G.-S. You and D.-C. Wang, *Ind. Eng. Chem. Res.*, 2000, **39**, 4357.
8. C. M. Wai and S. Wang, *J. Chromatogr. A*, 1997, **785**, 369.
9. N. G. Smart, T. Carleson, T. Kast, A. A. Clifford, M. D. Burford and C. M. Wai, *Talanta*, 1997, **44**, 137.
10. J. A. Darr and M. Poliakoff, *Chem. Rev.*, 1999, **99**, 495.
11. C. Erkey, *J. Supercrit. Fluids*, 2000, **17**, 259.
12. A. Ohashi and K. Ohashi, *Solvent Extr. Res. Dev., Jpn.*, 2008, **115**, 11.
13. S. L. Wallen, C. R. Yonker, C. L. Phelps and C. M. Wai, *J. Chem. Soc., Faraday Trans.*, 1997, **93**, 2391.
14. *Chemical Synthesis Using Supercritical Fluids*, ed. P. G. Jessop and W. Leitner, Wiley-VCH, Weinheim, 1999, ch. 3.
15. T. Umecky, M. Kanakubo and Y. Ikushima, *J. Phys. Chem. B*, 2002, **106**, 11114.

16. N. Wada, M. Saito, D. Kitada, R. L. Smith, H. Inomata, K. Arai and S. Saito, *J. Phys. Chem. B*, 1997, **101**, 10918.
17. P. Raveendran, Y. Ikushima and S. L. Wallen, *Acc. Chem. Res.*, 2005, **38**, 478.
18. J. M. Dobbs, J. M. Wong, R. J. Lahiere and K. P. Johnston, *Ind. Eng. Chem. Res.*, 1987, **26**, 56.
19. A. Ohashi, H. Hoshino, J. Niida, H. Imura and K. Ohashi, *Bull. Chem. Soc. Jpn.*, 2005, **78**, 1804.
20. Y. Meguro, S. Iso and Z. Yoshida, *Anal. Chem.*, 1998, **70**, 1262.
21. J. P. Hanrahan, K. J. Ziegler, J. D. Glennon, D. C. Steytler, J. Eastoe, A. Dupont and J. D. Holmes, *Langmuir*, 2003, **19**, 3145.

CHAPTER 14

Reactions and Equilibria of Metal Complexes Under High Pressure

K. ISHIHARA

Waseda University, Okubo, Shinjuku-ku, Tokyo 169-8555, Japan
Email: ishi3719@waseda.jp

14.1 Introduction

Research on the reaction mechanism of metal complexes in solutions until around the 1980s was focused on obtaining activation parameters ($\Delta H^{\ddagger}$, $\Delta S^{\ddagger}$) from the temperature dependence of the rate constants. Furthermore, the reaction mechanism has been discussed based on these kinetic parameters. However, the reported values of activation entropy ($\Delta S^{\ddagger}$), even for the same reaction system, vary in magnitude and sign depending on the researcher, so the reliability is poor. Under such circumstances, from the end of the 1970s, it has begun to be recognized that the activation volume ($\Delta V^{\ddagger}$) obtained from the pressure dependence of the rate constant is more reliable than $\Delta S^{\ddagger}$ and was effective in estimating the reaction mechanism. As a result, the development of a high-pressure apparatus for measuring $\Delta V^{\ddagger}$ has been done. In the 1980s, most major high-pressure apparatuses for measuring reaction rates under pressure were developed. It became possible to measure the rate constants of chemical reactions under pressures up to 300 MPa.

The study of rapid inorganic reactions under high pressure originated in the study by Brower in 1968. Later, in the 1970s and 1980s, several groups developed high-pressure temperature jump (T-jump) and high-pressure

Coordination Chemistry Fundamentals Series No. 2
Metal Ions and Complexes in Solution
Edited by Toshio Yamaguchi and Ingmar Persson
© Japan Society of Coordination Chemistry 2024
Published by the Royal Society of Chemistry, www.rsc.org

pressure jump (P-jump) apparatuses.[1] Research on the complex formation reactions of metal ions began. In the late 1970s and 1980s, Merbach *et al.* and Swaddle *et al.* developed a high-pressure NMR apparatus and studied the solvent exchange reactions of metal ions.[1] On the other hand, the flow method, which is complementary to these relaxation methods, started with the development of a high-pressure stopped-flow apparatus by Heremans *et al.* in 1977. Since then, by the early 1980s, three different types of equipment were reported by Osugi *et al.*, Karan and Macey, and Tanaka *et al.* at about the same time. Later, at the end of the 1990s, Koizumi *et al.* developed a high-pressure stopped-flow apparatus[1] that had the characteristics of the apparatus of Heremans *et al.* and Tanaka *et al.*

Applying pressure gives work to the system, but the energy due to compression is less than 2 kcal mol^{-1} even when compressed to 10 000 atm. The chemistry under a pressure of 10 000 to 20 000 atm or less in which the substance exists as a liquid is essentially the same as the chemistry under ambient pressure.

14.2 Reaction Volume and Activation Volume

Consider the reaction represented by eqn (14.1). Here, k_f and k_r refer to the rate constants of the forward reaction and the reverse reaction, respectively, and K refers to the equilibrium constant. If reaction (14.1) is elementary, k_f/k_d is equal to K $(k_f/k_d = K)$, and the standard change in Gibbs free energy, $\Delta G°$, for eqn (14.1) is given by $\Delta G° = -RT \ln K$. Therefore, the pressure effect on reaction (14.1) is expressed by eqn (14.2) or eqn (14.3) at a constant temperature. Here, $\Delta V°$ is called a reaction volume and is defined as in eqn (14.4) using partial molar volumes, V_A, V_B, V_C, and V_D, for each species.

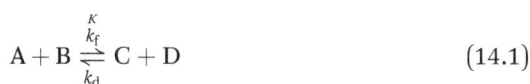

$$A + B \underset{k_d}{\overset{\overset{K}{k_f}}{\rightleftharpoons}} C + D \tag{14.1}$$

$$\left(\frac{\partial \Delta G°}{\partial P} \right)_T = \Delta V° \tag{14.2}$$

$$\left(\frac{\partial \ln K}{\partial P} \right)_T = -\frac{\Delta V°}{RT} \tag{14.3}$$

$$\Delta V° = (V_C + V_D) - (V_A + V_B) \tag{14.4}$$

Eqn (14.3) shows that the volume change that occurs when A and B react to produce C and D can be determined from the pressure dependence of the equilibrium constant. Eqn (14.4) shows that the reaction volume can be determined by directly measuring the difference between the volumes before and after mixing of the solutions of reactants A and B using a dilatometer.

On the other hand, according to the activated complex theory, the reaction in eqn (14.1) can be expressed as in eqn (14.5).

$$A + B \rightleftharpoons (AB)^{\ddagger} \underset{k_d}{\overset{k_f}{\rightleftharpoons}} C + D \qquad (14.5)$$

Here, $(AB)^{\ddagger}$ is an activated complex (the transition state), which is in equilibrium with the reactants (the initial state). When the activation free energy and the equilibrium constant for this equilibrium are expressed as $\Delta G^{\ddagger}$ and $K^{\ddagger}$, respectively, $\Delta G^{\ddagger} = -RT \ln K^{\ddagger}$, so as in eqn (14.2) and (14.3), eqn (14.6) and (14.7) hold.

$$\left(\frac{\partial \Delta G^{\ddagger}}{\partial P} \right)_T = \Delta V^{\ddagger} \qquad (14.6)$$

$$\left(\frac{\partial \ln K^{\ddagger}}{\partial P} \right)_T = -\frac{\Delta V^{\ddagger}}{RT} \qquad (14.7)$$

Since the rate constant k is related to $K^{\ddagger}$ as $k = \kappa \frac{k}{h} K^{\ddagger}$, where κ is a transmission coefficient, k is the Boltzmann constant, and h is the Planck constant, eqn (14.8) holds.

$$\left(\frac{\partial \ln k}{\partial P} \right)_T = -\frac{\Delta V^{\ddagger}}{RT} \qquad (14.8)$$

Here $\Delta V^{\ddagger}$ is called an activation volume, a difference in partial molar volume between the transition state and the initial state. Eqn (14.8) shows that the activation volume is obtained from the pressure dependence of the rate constant.

The activation volumes of the forward and the reverse reactions are expressed by the following equations using the partial molar volumes of each chemical species.

$$\Delta V_f^{\ddagger} = V_{(AB)^{\ddagger}} - (V_A + V_B) \qquad (14.9)$$

$$\Delta V_d^{\ddagger} = V_{(AB)^{\ddagger}} - (V_C + V_D) \qquad (14.10)$$

Therefore, the reaction volume is related to the activation volume by the following equation:

$$\Delta V^{\circ} = (V_C + V_D) - (V_A + V_B) = \Delta V_f^{\ddagger} - \Delta V_d^{\ddagger} \qquad (14.11)$$

The reaction volume can be calculated from this equation if the activation volumes of the forward and reverse reactions are obtained. Even if the reaction is irreversible, if the reaction volume ($\Delta V°$) and the activation volume of the forward reaction ($\Delta V_f^{\ddagger}$) are obtained, the activation volume of the reverse reaction ($\Delta V_d^{\ddagger}$) can be calculated.

14.3 Reaction Volumes and Activation Volumes

Many reaction volumes and activation volumes for inorganic and organic reactions have been compiled three times every 10 years.[2-4]

14.3.1 Reaction Volumes

Table 14.1 shows the reaction volumes for several acid dissociation reactions. The values in the table were measured using a dilatometer or a high-pressure vessel.[5]

An uncharged acid or base dissociates into ions when dissolved in water. Ions are solvated much more strongly than uncharged molecules, and solvent molecules (water molecules) are strongly attracted around the ions. As a result, solvent molecules around the ions undergo strong electrostriction. Therefore, the sum of the partial molar volumes of the final state is smaller than that of the initial state, and so the reaction volume becomes negative. Therefore, dissociation of the uncharged acid or base is promoted when pressurized, and the value of the dissociation constant increases with pressure as shown in Table 14.2 (eqn (14.3)).

It is known that the partial molar volumes ($cm^3\,mol^{-1}$, 25 °C, 1 atm) of the hydrogen ion and hydroxide ion are, respectively, given as follows:

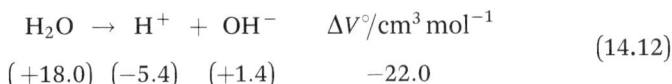

$$H_2O \rightarrow H^+ + OH^- \qquad \Delta V°/cm^3\,mol^{-1}$$
$$(+18.0)\ (-5.4)\ (+1.4) \qquad\qquad -22.0 \tag{14.12}$$

Table 14.1 Reaction volumes for some acid dissociation reactions (25 °C, 1 atm).

Compound	$\Delta V°/cm^3\,mol^{-1}$	Compound	$\Delta V°/cm^3\,mol^{-1}$
H_2O	−22.07	NH_4^+	+7.0
H_2S	−16.3	$CH_3NH_3^+$	+5.6
HCOOH	−8.5	$Fe(H_2O)_6^{3+}$	−1.2
CH_3COOH	−11.3	$Cr(H_2O)_6^{3+}$	−3.8

Table 14.2 Change in the dissociation constant with pressure (25 °C).

Compound	$K_a/10^{-5}\,mol\,kg^{-1}$ or $K_b/10^{-5}\,mol\,kg^{-1}$			
	1 atm	1000 atm	2000 atm	3000 atm
HCOOH	17.4	24.4	32.5	41.8
CH_3COOH	1.71	2.70	3.91	5.38
NH_3	1.75	5.11	12.2	24.3

$$\text{H}_2\text{S} \quad \rightarrow \quad \text{H}^+ \quad + \quad \text{SH}^- \qquad \Delta V°/\text{cm}^3\,\text{mol}^{-1}$$

$$(+35.9) \quad (-5.4) \quad (+24.6) \qquad\qquad -16.3 \tag{14.13}$$

The difference between V_{OH^-} and V_{SH^-} reflects the higher charge density of OH$^-$.

Eqn (14.14) and (14.15) are obtained from eqn (14.12) and (14.13).

$$V_{\text{H}_2\text{O}} - V_{\text{OH}^-} = 16.6 \ \text{cm}^3\,\text{mol}^{-1} \tag{14.14}$$

$$V_{\text{H}_2\text{S}} - V_{\text{SH}^-} = 11.3 \ \text{cm}^3\,\text{mol}^{-1} \tag{14.15}$$

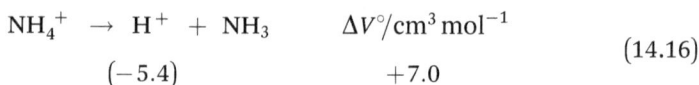

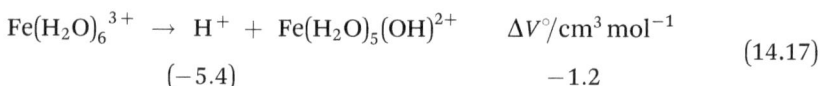

$$\text{NH}_4^+ \quad \rightarrow \quad \text{H}^+ \ + \ \text{NH}_3 \qquad \Delta V°/\text{cm}^3\,\text{mol}^{-1}$$

$$(-5.4) \qquad\qquad\qquad +7.0 \tag{14.16}$$

$$\text{Fe}(\text{H}_2\text{O})_6^{3+} \quad \rightarrow \quad \text{H}^+ \ + \ \text{Fe}(\text{H}_2\text{O})_5(\text{OH})^{2+} \qquad \Delta V°/\text{cm}^3\,\text{mol}^{-1}$$

$$(-5.4) \qquad\qquad\qquad\qquad\qquad -1.2 \tag{14.17}$$

Eqn (14.18) and (14.19) are obtained from eqn (14.16) and (14.17).

$$V_{\text{NH}_4^+} - V_{\text{NH}_3} = -12.4 \ \text{cm}^3\,\text{mol}^{-1} \tag{14.18}$$

$$V_{\text{Fe}^{3+}} - V_{\text{FeOH}^{2+}} = -4.2 \ \text{cm}^3\,\text{mol}^{-1} \tag{14.19}$$

From eqn (14.14) and (14.15), an uncharged acid has a partial molar volume larger than its conjugate base, and from eqn (14.18) and (14.19), a charged acid has a partial molar volume smaller than that of the conjugated base. In other words, the effect of electrostriction is greater for ions than for neutral molecules and is greater for ions with a high valence than for ions with a low valence.

14.3.2 Activation Volumes

The reaction mechanism of complex formation is generally expressed as follows, based on the mechanism proposed by M. Eigen (Eigen mechanism):

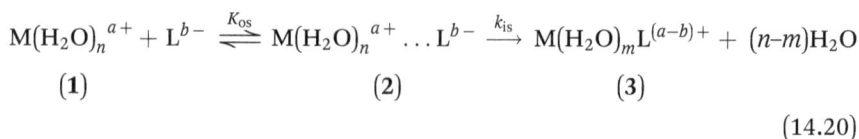

$$\text{M}(\text{H}_2\text{O})_n{}^{a+} + \text{L}^{b-} \underset{}{\overset{K_{os}}{\rightleftharpoons}} \text{M}(\text{H}_2\text{O})_n{}^{a+}\ldots\text{L}^{b-} \xrightarrow{k_{is}} \text{M}(\text{H}_2\text{O})_m\text{L}^{(a-b)+} + (n-m)\text{H}_2\text{O}$$

$$\text{(1)} \qquad\qquad\qquad\qquad \text{(2)} \qquad\qquad\qquad \text{(3)}$$

$$\tag{14.20}$$

where K_{os} is the formation constant for the diffusion-controlled process in which the metal ion (1) associates with the ligand L^{b-} to form an outer-sphere complex (or ion pair) (2) and k_{is} is the rate constant for the rate-determining step forming an inner-sphere complex (3) from (2).

The second-order rate constant k_f $(M^{-1}s^{-1})$ for the forward reaction in eqn (14.20) is expressed by the following equation:

$$k_f = K_{os}k_{is} \tag{14.21}$$

Therefore, the activation volume $\Delta V_f^{\ddagger}$ obtained from the pressure dependence of k_f is expressed by the sum of volume changes in each process in eqn (14.20) (eqn (14.22)). When the two reactants have opposite charges, the outer-sphere complex is specifically called an ion-pair, and in many cases ΔV_{os} is the dominant factor of $\Delta V_f^{\ddagger}$. The value of ΔV_{os} for ion-pair formation can be estimated by differentiating Fuoss' equation with pressure.

$$\Delta V_f^{\ddagger} = \Delta V_{os} + V_{is}^{\ddagger} \tag{14.22}$$

Table 14.3 shows some examples of the values of $\Delta V_{is}^{\ddagger}$ for the complexation of metal ions whose reaction mechanism (classification by Langford and Gray[6]) is clarified based on kinetic data under ambient pressure. As is apparent from this table, the $\Delta V_{is}^{\ddagger}$ values of $Fe(H_2O)_6^{3+}$ that reacts with an I_a mechanism are all negative, irrespective of the types of ligand, whereas the $\Delta V_{is}^{\ddagger}$ values of $Fe(H_2O)_5(OH)^{2+}$ and $Ni(H_2O)_6^{2+}$ that react with an I_d mechanism are all positive. For other metal ions, negative $\Delta V_{is}^{\ddagger}$ values are obtained for the reaction systems for which an associative mechanism is suggested, and positive $\Delta V_{is}^{\ddagger}$ values are obtained for the reaction systems for which a dissociative mechanism is proposed. This shows that the positive/negative value of activation volume is effective for discriminating the mechanism of the ligand substitution reaction.

In the solvent exchange reaction, both leaving and entering ligands are solvent molecules, so the forward reaction and the reverse reaction are completely equivalent. A. E. Merbach *et al.* measured the activation volume of the solvent exchange reaction $\Delta V_{ex}^{\ddagger}$ on many metal ions using a high-pressure NMR apparatus developed by themselves and stated that $\Delta V_{ex}^{\ddagger}$ is directly related to the reaction mechanism, as shown in Figure 14.1,[7] provided that "the bond lengths between the central metal ion and the solvent molecules not involved in the exchange reaction remain unchanged in the process from the initial state to the transition state". Although the argument based on this assumption has received many criticisms, the solvent exchange reaction is only the case where the entering ligand is a solvent molecule in the complex formation reaction, and so the solvent exchange mechanism should be consistent with the complexation mechanism. In fact, since $\Delta V_{ex}^{\ddagger}$ and $\Delta V_{is}^{\ddagger}$ for the same metal ion are very similar (Table 14.3), $\Delta V_{ex}^{\ddagger}$ is considered to be a criterion for judging the reaction mechanism.

Table 14.3 Activation volumes for complexation reactions of metal ions in H_2O at 25 °C.

Metal ion	Mechanism	Ligand	$\Delta V^{\ddagger}_{is}$/cm^3 mol^{-1}	Metal ion	Mechanism	Ligand	$\Delta V^{\ddagger}_{is}$/cm^3 mol^{-1}
Ni^{2+}	I_d	Succinate^{2-}	+6	Fe^{3+}	I_a	SCN$^-$	−6.1
		Malonate^{2-}	+8			Acetohydroxamic acid	−10.0
		Glycolate$^-$	+11			4-Isopropyltropolone	−8.7
		Glycinate$^-$	+8			Cl$^-$	−6.5
		Murexide$^-$	+9			H_2O	−5.4
		NH$_3$	+6.0	$FeOH^{2+}$	I_d	SCN$^-$	+6.9
		Iso-quinoline	+7.4			Acetohydroxamic acid	+7.7
		pada	+7.7			4-Isopropyltropolone	+4.1
		bipy	+5.5			Cl$^-$	+6.2
		terpy	+6.7			H_2O	+7.0
		H_2O	+7.2				

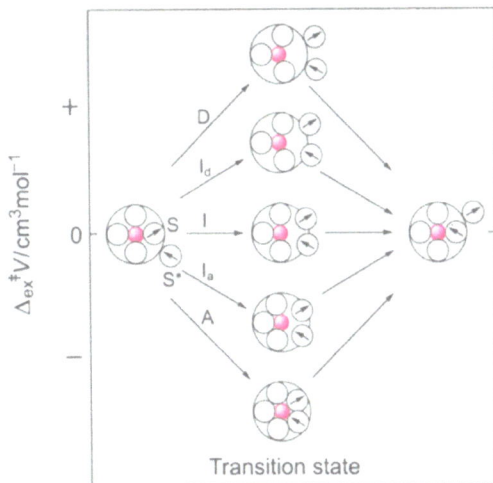

Figure 14.1 Activation volumes for the solvent exchange reaction and reaction mechanism.[8]

14.4 Volume Changes for the Reactions in Solution

The pressure effect on the reactions in solution appears only for reaction systems with non-zero reaction volumes. As described in Section 14.2, the pressure jump apparatus can measure the reaction rate only for reversible fast reaction systems with volume changes. In reactions involving ions, such as salts and acid–bases, charges are generated or neutralized during the reaction, so a large volume change occurs during the reaction. In such reactions, the solvation changes considerably due to electrostriction so the volume changes greatly and the density of the solution changes. This can be easily confirmed by, for example, the following qualitative experiments:

- When water is added up to 200 mL to a vessel containing about 8 g of sodium hydroxide pellets and the vessel is shaken to dissolve the pellets completely, a volume decrease of less than 3 mL is observed.
- When water is added up to 50 mL to a vessel containing about 15 g of sodium chloride crystal and the vessel is shaken well to dissolve the crystal completely, a volume decrease of 4 to 5 mL is observed.

Thus, ions have the property of strongly attracting polar water molecules. This is why the density of seawater containing 3.3 to 3.8% salinity (1.02 to 1.03 $g\,cm^{-3}$) is higher than that of pure water. Super-water-absorbent resins for home use take advantage of this property.

Among organic reactions, there are reactions with very large negative activation volumes of $-50\ cm^3\,mol^{-1}$ and reactions with a large solvent dependence.[4,5] Since the effect of electrostriction is larger in a solvent having a lower dielectric constant, the value of the activation volume becomes more negative in such a solvent, and so a larger acceleration effect by pressurization can be expected. Therefore, in the case of an organic reaction that causes charge separation in the transition state, the activation volume is highly dependent on the solvent used.

Assuming that the activation volume is -50 cm^3 mol^{-1}, this value is constant independent of pressure, and the reaction is first-order, it is calculated from eqn (14.23) derived from the integration of eqn (14.8) that the rate constant under 5000 atm at 25 °C is approximately 20 000 times that under atmospheric pressure. This means that even if the reaction takes 10 days to achieve 99% completion (7 half-lives) at atmospheric pressure, the reaction will be completed within 1 minute at 5000 atm.

$$\frac{k^P}{k^0} = \exp\left(-\frac{\Delta V^{\ddagger}}{RT}\right) \tag{14.23}$$

Problem 1

The ion product constant (the dissociation constant) for water is 1.00×10^{-14} at 25 °C and 1 atm. The reaction volume for the dissociation of water is -22.07 cm^3 mol^{-1} (Table 14.1). Calculate the value of the dissociation constant for water at 25 °C and 1000 atm, assuming that the reaction volume is independent of pressure.

Problem 2

The volume change $\Delta V_f^{\ddagger}$ for k_f in eqn (14.21) is expressed by eqn (14.22) using the volume changes in each process in eqn (14.20). Derive eqn (14.22).

Problem 3

Find the following values using the data in Table 14.1:

1. (1) Volume change when 200 mL of 1.00 mol L^{-1} sodium hydroxide solution is neutralized with 200 mL of 1.00 mol L^{-1} hydrochloric acid at 25 °C.
2. (2) Volume change when 100 mL of 2.00 mol L^{-1} acetic acid is neutralized with 100 mL of 2.00 mol L^{-1} sodium hydroxide solution at 25 °C.

Problem 4

Calculate the activation volumes for each complexation reaction using the values at 25 °C in the following table:

$$\text{Mn}^{2+} + \text{terpy} \xrightleftharpoons{k_f\,(\text{M}^{-1}\text{s}^{-1})} \text{Mn(terpy)}^{2+} \quad (\text{trepy} = 2, 2', 6', 2''\text{-terpyridyl})$$

$$\text{Ni}^{2+} + \text{bipy} \rightleftharpoons \text{Ni(bipy)}^{2+} \quad (\text{bipy} = 2, 2'\text{-bipyridyl})$$

	P/MPa	1	25	50	75	100
Mn^{2+} + terpy	$k_f/10^4$	8.2	8.4	8.6	8.8	9.0
Ni^{2+} + bipy	$k_f/10^3$	2.22	2.08	2.00	1.89	1.77

Problem 5

Consider a first-order chemical reaction with a half-life of 10 hours at 25 °C. Given that the activation volume for this reaction, which is independent of pressure, is -20.0 cm^3 mol^{-1}, calculate the half-life under 200 MPa.

Problem 6

The complex formation reaction of Ni^{2+} with isoquinoline (isoq) is expressed by the following equation. The reaction is first-order with respect to the concentration of uncomplexed isoquinoline under pseudo-first-order conditions in which the total concentration of Ni^{2+}, C_{Ni}, is in large excess over the total concentration of isoquinoline. Then, the pseudo-first-order rate constant k_{obs} is expressed as $k_{obs} = k_f C_{Ni} + k_d$. The values of k_f and k_d in an aqueous solution at 25 °C, 1 atm, and $I = 0.10$ M are $k_f = 2.03 \times 10^3$ kg mol^{-1} s^{-1} and $k_d = 30.9$ s^{-1}, and the activation volume values are $\Delta V_f^{\ddagger} = 7.4$ cm^3 mol^{-1} and $\Delta V_d^{\ddagger} = 8.9$ cm^3 mol^{-1}, respectively, both of which are pressure independent.

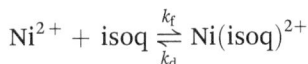

$$\text{Ni}^{2+} + \text{isoq} \underset{k_d}{\overset{k_f}{\rightleftharpoons}} \text{Ni(isoq)}^{2+}$$

(1) Calculate the reaction volume for this complex formation reaction.
(2) Calculate the values of k_f and k_d at 25 °C and 1000 atm, and find the value of k_{obs} at 25 °C, 1000 atm, and $C_{Ni} = 0.100$ mol kg^{-1}.
(3) Calculate the value of the formation constant at 25 °C and 1000 atm.

Answer to Problem 1

Because the reaction volume is independent of pressure, eqn (14.3) can be integrated to get the following equation. 1 atm $= 0.1013$ MPa, $R = 8.314$ MPa cm^3 (K^{-1} mol^{-1})

$$K^P = K^0 \cdot \exp\left(-\frac{\Delta V^\circ}{RT} P\right)$$

$$= 1.00 \times 10^{14} \cdot \exp\left(\frac{22.07}{8.314 \times 298.15} \times 1000 \times 0.1013\right) = 2.46 \times 10^{14}$$

Answer to Problem 2

$$\ln k_f = \ln K_{os} + \ln k_{is}$$

Differentiating this equation with respect to P at constant T gives the following equations:

$$\left(\frac{\partial \ln k_f}{\partial P}\right)_T = \left(\frac{\partial \ln K_{os}}{\partial P}\right)_T + \left(\frac{\partial \ln k_{is}}{\partial P}\right)_T$$

$$-\frac{\Delta V_f^{\ddagger}}{RT} = -\frac{\Delta V_{os}^{\circ}}{RT} - \frac{\Delta V_{is}^{\ddagger}}{RT}$$

$$\therefore \Delta V_f^{\ddagger} = \Delta V_{os}^{\circ} + \Delta V_{is}^{\ddagger}$$

Answer to Problem 3

(1) $H^+ + Cl^- + Na^+ + OH^- \rightarrow Na^+ + Cl^- + H_2O$

$$+\frac{22.07}{5} = +4.4 \text{ cm}^3$$

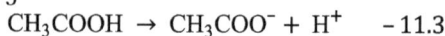

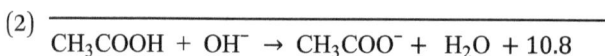

$$CH_3COOH \rightarrow CH_3COO^- + H^+ \qquad -11.3$$

$$+) \; H^+ + OH^- \rightarrow H_2O \qquad\qquad\qquad +22.07$$

(2) ──

$$CH_3COOH + OH^- \rightarrow CH_3COO^- + H_2O + 10.8$$

$$+\frac{10.8}{5} = +2.2 \text{ cm}^3$$

Answer to Problem 4

The plots of $\ln k$ vs. P for the Mn^{2+} + terpy and Ni^{2+} + bipy systems give the slopes 9.434×10^{-4} and -2.209×10^{-3}, respectively. $R = 8.314$ MPa cm^3 (K^{-1} mol^{-1}).

The slope is equal to $-\dfrac{\Delta V^{\ddagger}}{RT}$.

$$\Delta V_{Mn}^{\ddagger} = -9.434 \times 10^{-4} \times RT = -9.434 \times 10^{-4} \times 8.314 \times 298.15 = -2.3 \text{ cm}^3\text{mol}^{-1}$$

$$\Delta V_{Ni}^{\ddagger} = +2.209 \times 10^{-3} \times RT = +5.5 \text{ cm}^3\text{mol}^{-1}$$

Answer to Problem 5

The first-order rate constant $k = \dfrac{\ln 2}{t_{1/2}}$

$$k^P = k^0 \exp\left(-\frac{\Delta V^{\ddagger}}{RT}P\right) = \frac{\ln 2}{t_{1/2}}\exp\left(-\frac{\Delta V^{\ddagger}}{RT}P\right)$$

$$t_{1/2}^P = \frac{\ln 2}{k^P} = \frac{\ln 2}{\dfrac{\ln 2}{t_{1/2}}\exp\left(-\dfrac{\Delta V^{\ddagger}}{RT}P\right)} = \frac{t_{1/2}}{\exp\left(-\dfrac{\Delta V^{\ddagger}}{RT}P\right)}$$

$$= \frac{10}{\exp\left(\dfrac{20.0}{8.314 \times 298.15} \times 200\right)} \approx \frac{10}{5.0} = 2.0 \text{ h}$$

Answer to Problem 6

(1) $\Delta V^\circ = \Delta V_f^\ddagger - \Delta V_d^\ddagger = 7.4 - 8.9 = -1.5 \text{ cm}^3\text{mol}^{-1}$

(2) $k_f^P = k_f^0 \exp\left(-\dfrac{\Delta V^\ddagger}{RT}P\right) = 2.03 \times 10^3$

$$\times \exp\left(-\dfrac{7.4}{8.314 \times 298.15} \times 1000 \times 0.1013\right) = 1.50 \times 10^3$$

$$k_d^P = k_d^0 \exp\left(-\dfrac{\Delta V^\ddagger}{RT}P\right) = 30.9 \times \exp\left(-\dfrac{8.9}{8.314 \times 298.15} \times 1000 \times 0.1013\right) = 21.5$$

$$k_{obs}^P = k_f^P C_{Ni} + k_d^P = 1.50 \times 10^3 \times 0.100 + 21.5 \approx 170 \text{ s}^{-1}$$

(3) $K^P = \dfrac{k_f^P}{k_d^P} = \dfrac{1.50 \times 10^3}{21.5} = 70$

References

1. K. Ishihara, Y. Kondo and M. Koizumi, *Rev. Sci. Instrum.*, 1999, **70**, 244.
2. T. Asano and W. J. le Noble, *Chem. Rev.*, 1978, **78**, 407.
3. R. van Eldik, T. Asano and W. J. le Noble, *Chem. Rev.*, 1989, **89**, 549.
4. A. Drljaca, C. D. Hubbard, R. van Eldik, T. Asano, M. V. Basilevsky and W. J. le Noble, *Chem. Rev.*, 1998, **98**, 2167.
5. *Inorganic High Pressure Chemistry: Kinetics and Mechanisms*, ed. R. van Eldik, Elsevier, Amsterdam, 1986.
6. C. H. Langford and H. B. Gray, *Ligand Substitution Processes*, W. A. Benjamin, New York, 1965.
7. *High Pressure Chemistry: Synthetic, Mechanistic, and Supercritical Applications*, ed. R. van Eldik and F.-G. Klaerner, Wiley-VCH, Weinheim, 2002.
8. L. Helm and A. E. Merbach, *Coord. Chem. Rev.*, 1999, **187**, 151.

CHAPTER 15

Metal Complexes in Ionic Liquids

TOSHIYUKI TAKAMUKU*[a] AND MASAYASU IIDA[b]

[a] Faculty of Science and Engineering, Saga University, Honjo-machi, Saga 840-8502, Japan; [b] Emeritus Professor, Nara Women's University, Nara 630-8506, Japan
*Email: takamut@cc.saga-u.ac.jp

15.1 Metal Complexes in Ionic Liquids

Ionic liquids (ILs) are electrolytes consisting of large cations and anions, between which a strong electrostatic attraction does not act, unlike normal electrolytes like sodium chloride. In addition, IL cations, such as ammonium, phosphonium, pyrrolidinium, imidazolium, and pyridinium, usually involve a long alkyl chain. The two factors, the weak electrostatic force between the cation and anion and the entropic effect of the long alkyl chain, lead to much lower melting points of ILs than normal electrolytes. Room-temperature ILs are in the liquid state around 25.0 °C. Despite liquids, ILs have the properties as electrolytes, such as non-flammability, negligible low volatility, and electroconductivity. Moreover, the combinations of the cation and anion might be countless. Therefore, ILs attract much attention as designable solvents from various physical and chemical fields. For example, ILs are applied to novel solvents of organic syntheses and thermally stable electrolytes for secondary batteries such as lithium batteries and metal plating. Additionally, the application of ILs in CO_2 capture, dissolution of biomass such as cellulose, and lubricants is also investigated all over the world. In analytical chemistry, ILs are investigated as green solvents for extraction because of their much lower volatility than conventional solvents,

Coordination Chemistry Fundamentals Series No. 2
Metal Ions and Complexes in Solution
Edited by Toshio Yamaguchi and Ingmar Persson
© Japan Society of Coordination Chemistry 2024
Published by the Royal Society of Chemistry, www.rsc.org

such as chloroform and carbon tetrachloride. Molecular dynamics (MD) simulations have suggested that the charged moieties of the IL cation, such as the imidazolium ring, and anions aggregate with each other to form polar domains in neat ILs, while the long alkyl chains interact with themselves to form nonpolar domains.[1–4] It is interesting how the microsegregation of polar and nonpolar domains affects chemical reactions in ILs.

For metal extraction and plating using ILs, understanding of the complex formation of metal ions in ILs is essential to develop their green processes. In the extraction, metal ions move from the aqueous phase to the ionic liquid phase as a form of metal complex with ligands. Metal ions approach the surface of metallic materials as complexes in the plating. However, the complexation equilibria of metal ions with ILs themselves and other ligand molecules in ILs have not yet been well understood compared to those in conventional molecular liquids (MLs).

On the other hand, ILs whose cation or anion involves a metal ion have been synthesized. In other words, metal complexes, where halides or organic molecules such as chloride and amines coordinate to metal ions, play a role as cations or anions in ILs depending on the positive or negative charge of the complexes. Such ILs are called here as metallo-ionic liquids (metallo-ILs).

In the following sections, the complexation equilibria of transition metal ions, such as Ni^{2+} and Cu^{2+}, with ligand molecules in ILs and the properties of metallo-ILs, are explained.

15.1.1 Solvation Structure of Transition Metal Ions in ILs

In solutions of conventional MLs, such as water and organic solvents, transition metal ions are solvated by coordination of ML molecules. The electron donicity of MLs is the key property to stabilize metal ions in ML solutions. In the case of ILs, transition metal ions should be solvated by IL molecules. Usually, IL anions having electron donicity (Lewis basicity) may coordinate with metal ions in ILs. Acetate, trifluoromethanesulfonate (TfO^-), and bis(trifluoromethanesulfonyl)amide (Tf_2N^-) may coordinate to metal ions. However, the donicities of the IL anions are very low. For example, the Gutmann's donor number of Tf_2N^- ($D_N = 7$),[5,6] which is the most suitable gauge of electron donicity,[7] is much lower than those of MLs, such as water and acetonitrile ($D_N = 18.0$ and 14.1, respectively).[7] Therefore, the solvate complexes of transition metal ions in ILs may be less stable than in MLs.

Imidazolium-based ILs, such as 1-alkyl-3-methylimidazolium bis(trifluoromethanesulfonyl)amide ($C_n$mimTf$_2$N, n represents the alkyl chain length), are one of the most investigated families of ILs. In Scheme 15.1, the structures of the cations and anions of ILs are depicted.

The structure of solvate complexes for transition metal ions in ILs has been investigated using several techniques. Raman spectroscopy experiments have reported that three Tf_2N^- anions coordinate with the first transition metal (d-block element) ions from Mn^{2+} to Zn^{2+} by the oxygen atoms of Tf_2N^- to form an octahedral structure with three chelate rings.[8] Tf_2N^- may have two

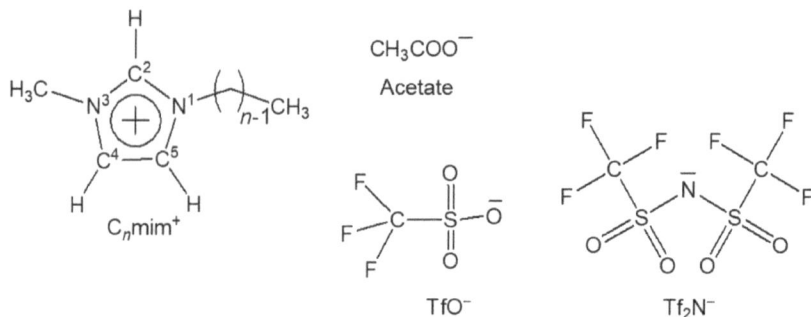

Scheme 15.1 Structures of the imidazolium cation with the alkyl chain length n and anions of ionic liquids.

conformations of the *cis* and *trans* forms against the S–N–S bond. In the solvate complex, the three Tf_2N^- anions are in the *cis* form to easily coordinate to the metal ion. The solvation structure of Zn^{2+} in Tf_2N^--based ILs with various cations, such as 1-butyl-3-methylimidazolium (C_4mim^+), 1,8-bis(3-methylimidazolium-1-yl)octane ($C_8(mim)_2^{2+}$), *N,N,N*-trimethyl-*N*-(2-hydroxyethyl)ammonium (choline$^+$), and butyltrimethylammonium (BTMA$^+$), was investigated using extended X-ray absorption fine structure (EXAFS) with the help of MD simulations.[9] Zn^{2+} ions take two structures in the ILs: a 5-fold one with one bidentate and four monodentate Tf_2N^- and a 6-fold one with only monodentate ligands. However, the IL cations do not significantly influence the solvation structure of Zn^{2+}. The other MD simulations suggested that the Zn^{2+} ion is coordinated with the six oxygen atoms of three Tf_2N^- as bidentate ligands.[10] In the case of Co^{2+} and Ni^{2+} ions, six Tf_2N^- anions coordinate with the metal ions as monodentate ligands to form an octahedral structure in ILs,[11,12] while the Co^{2+} ion is coordinated by two monodentate and two bidentate Tf_2N^- in the crystal state with the IL cation of 1-butyl-1-methylpyrrolidinium.[13] Thus, the structure of the solvate complexes of the metal ions in the Tf_2N^--based ILs is still controversial. This is because Tf_2N^- anions very weakly solvate the metal ions in the ILs for the observation of the solvate complexes. Particularly, the monodentate and bidentate ligations of Tf_2N^- cannot be easily distinguished by the radial distribution functions obtained from EXAFS experiments because of the same coordination oxygen atoms. Despite the unknown coordination manner of Tf_2N^-, monodentate or bidentate, the transition metal ions are surely coordinated by the six oxygen atoms of Tf_2N^- to form an octahedral structure according to the abovementioned investigations.

15.2 Complex Formation of Transition Metal Ions with Inorganic Anions in ILs

Transition metal ions are coordinated by the oxygen atoms of Tf_2N^- in Tf_2N^--based ILs. The coordination ability of Tf_2N^- for transition metal ions is lower due to its much lower electron donicity compared to those of other inorganic anions like halides and nitrate and neutral molecules with higher electron donicity. Thus, when such anions and MLs are added into IL solutions with metal ions and their concentration is increased, the Tf_2N^- anions that initially coordinate with the metal ion are gradually replaced by anions and MLs. The ligand exchange equilibria from Tf_2N^- to other ions or molecules in several IL systems have been investigated using ultraviolet-visible (UV-vis) spectroscopy and calorimetry.[14–16] The d–d transition spectrum of the metal ion observed by spectroscopy changes due to the replacement of the ligand with varying its concentration. Thus, the ligand concentration dependence of the spectra gives us the concentration of each complex formed in the solutions. The stability constant of each complex can be determined from the UV-vis spectra recorded as a function of the ligand concentration. Details are explained in Section 15.3. In calorimetry, the trivial temperature change arising from the ligand exchange can be exactly measured using a solid-state thermistor. The concentration dependence of the thermal change enables us to simultaneously determine the enthalpies and stability constants of the complexation.

The complexation equilibria of the uranyl cation, UO_2^{2+}, with nitrate NO_3^- in 1-butyl-3-methylimidazolium bis(trifluoromethanesulfonyl)amide, C_4mimTf_2N, have been evaluated using UV-vis spectroscopy.[14] Tf_2N^- anions bound to UO_2^{2+} in the IL are gradually replaced by NO_3^- ions added as C_4mimNO_3 with increasing NO_3^- concentration. The highest trinitrate complex of $[UO_2(NO_3)_3]^-$ finally forms in the IL. The overall stability constants $\log \beta_N$ of the mono-, di-, and trinitrato complexes, $[UO_2(NO_3)]^+$, $[UO_2(NO_3)_2]$, (here, Tf_2N^- anions are omitted) and $[UO_2(NO_3)_3]^-$, with the reaction steps of $N = 1$, 2, and 3 were estimated to be 4.81 ± 0.45, 8.31 ± 0.65, and 12.17 ± 0.81, respectively.

In the same IL, the complexation equilibria of Ni^{2+} with NO_3^- have been studied using microcalorimetry.[15] The EXAFS experiments revealed that three NO_3^- anions coordinate with Ni^{2+} by their oxygen atoms to form the octahedral structure of $[Ni(NO_3)_3]^-$ with three chelate rings. The overall stability constants $\log \beta_N$ are 4.2 ± 0.2, 7.5 ± 0.3, and 10.1 ± 0.4 for the mono-, di-, and trinitrato complexes with $N = 1$, 2, and 3, respectively. The microcalorimetry showed that the first and second steps ($N = 1$ and 2) of the ligand replacement are enthalpically driven, whereas the third step ($N = 3$) is entropy stabilised.

Another calorimetric investigation has also been conducted on the complex formation of Cu^{2+} with chloride Cl^- in 1-butyl-3-methylimidazolium trifluoromethanesulfonate, $C_4mimTfO$, by Kanzaki and co-workers.[16] The aid of quantum chemical calculations with density functional theory (DFT) showed that Cu^{2+} is solvated by the four oxygen atoms of TfO^- anions to form the square planar complex in the IL. As the concentration of Cl^- from C_4mimCl increases, the ligand exchange from TfO^- to Cl^- progresses in the IL solution. $[CuCl_4]^{2-}$ forms as the highest complex in $C_4mimTfO$. The total coordination number of Cu^{2+} is saturated to four in the IL. The number is dominated by the electrostatic hindrance of the ligand anions in the primary coordination sphere. This is a characteristic feature in the IL.

In conventional ML solvents, the coordination number of metal complexes tends to be rather dominated mainly by the steric hindrance. The 1 : 1 ligand exchange of TfO$^-$ by Cl$^-$ in the IL is an athermal process. Thus, the enthalpic gain for the coordination of Cl$^-$ to Cu^{2+} is almost compensated by the enthalpic loss for the elimination of TfO$^-$ from Cu^{2+}. The main driving force of the chloro-complexation in C$_4$mimTfO is the entropy gain of the desolvation of TfO$^-$ from Cu^{2+}.

15.3 Complex Formation of Transition Metal Ions with MLs in ILs

Most of the conventional MLs, such as dimethyl sulfoxide (DMSO), methanol (MeOH), and acetonitrile (AN), have higher electron donicities than Tf$_2$N$^-$. Thus, the ligand replacement of Tf$_2$N$^-$ anions on a transition metal by ML molecules may easily progress in ILs as the concentration of ML increases. In fact, X-ray crystallography on the single crystals of Co^{2+} and Ni^{2+} complexes grown from 1-ethyl-3-methylimidazolium bis(trifluoromethanesulfonyl)amide (C$_2$mimTf$_2$N) solutions with high concentrations of MLs, namely, DMSO and AN, revealed that six ML molecules coordinate with both metal ions through their lone pair of oxygen and nitrogen atoms, respectively, to form the octahedral structure.[17] Any Tf$_2$N$^-$ does not coordinate to the metal ions in the unit cell of the crystals. Therefore, the highest complex of [Co(ML)$_6$]$^{2+}$ and [Ni(ML)$_6$]$^{2+}$ may also be formed in IL–ML solutions.

The equilibria for the complexation of Ni^{2+} with MLs, namely, DMSO, MeOH, and AN, in C$_2$mimTf$_2$N and 1-octyl-3-methylimidazolium Tf$_2$N$^-$ (C$_8$mimTf$_2$N) have been investigated using UV-vis spectroscopy. Figure 15.1 shows the changes in the spectra for Ni^{2+}–C$_8$mimTf$_2$N–ML ternary solutions with increasing ML concentration. The ternary solutions were prepared by dissolving Ni(Tf$_2$N)$_2$ into C$_8$mimTf$_2$N–ML binary solvents. The absorption peak at $\sim$420 nm and that at $\sim$800 nm for the d–d transition gradually shift to shorter wavelengths with increasing ML content. It is suggested that Tf$_2$N$^-$ anions that are initially bound to Ni^{2+} in the IL are

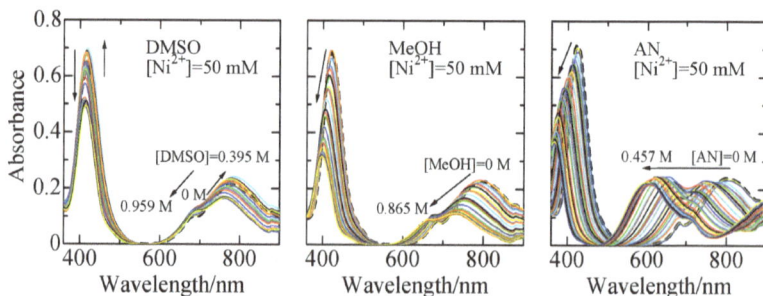

Figure 15.1   UV-vis spectra of Ni(Tf$_2$N)$_2$–C$_8$mimTf$_2$N solutions with DMSO, MeOH, and AN at 25.0 °C as a function of ML concentration.[20]

gradually replaced by ML molecules as the ML concentration increases. The most significant change in the spectra for the AN solutions against the increase in the ML concentration among the three ML systems arises from the replacement of the coordination atoms from the Tf_2N^- oxygen to the AN nitrogen atoms.

Although the coordination manner of Tf_2N^- in solvation complexes is still controversial as mentioned above, if the Tf_2N^- anion coordinates with Ni^{2+} as a bidentate ligand, the overall equilibrium for the replacement of Tf_2N^- with ML molecules on Ni^{2+} in ILs can be described by

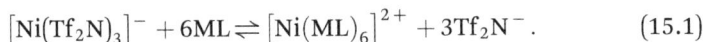

$$[Ni(Tf_2N)_3]^- + 6ML \rightleftharpoons [Ni(ML)_6]^{2+} + 3Tf_2N^-. \tag{15.1}$$

Then the conditional overall stability constant until reaction step N is explained by

$$\beta_N = \frac{[Ni(ML)_N]}{[Ni][ML]^N}. \tag{15.2}$$

In the stability constant, Tf_2N^- anions can be regarded as solvation molecules. To determine the stability constants, the UV-vis absorption spectra were fitted with the equations of the theoretical absorption of the solutions and the mass balances of the Ni^{2+} ion and ML, respectively,

$$A_{calc,\lambda} = \varepsilon_{Ni^{2+},\lambda}[Ni^{2+}] + \varepsilon_{ML,\lambda}[ML] + \sum_i^N \varepsilon_{N,\lambda}\beta_N[Ni^{2+}][ML]^N, \tag{15.3}$$

$$C_{Ni^{2+}} = [Ni^{2+}] + \sum_N \beta_N[Ni^{2+}][ML]^N, \tag{15.4}$$

$$C_{ML} = [ML] + \sum_N N\beta_N[Ni^{2+}][ML]^N, \tag{15.5}$$

using a nonlinear least squares refinement program, *e.g.* MQSPEC.[18] In the equations, ε represents the molar absorption coefficient for each species as a function of wavelength λ.

Table 15.1 shows the overall conditional stability constants $\log \beta_N$ of the ML complexes in C_2mimTf_2N and C_8mimTf_2N with different alkyl chain lengths of the ethyl and octyl groups, respectively. As shown in the table, the di-, tetra-, and hexa-ML complexes are formed in both ILs with DMSO, MeOH, and AN. In the C_8mimTf_2N solution, the mono-DMSO complex is also found. These findings suggest that one Tf_2N^- anion bound to Ni^{2+} is simultaneously replaced by two ML molecules at each step, as shown in Figure 15.2. It is plausible that Tf_2N^- coordinates to Ni^{2+} as a bidentate ligand in the ILs. In the C_2mimTf_2N solutions with the shorter ethyl chain, the stability of the highest hexa-ML complex is higher in the order of

Table 15.1 Overall conditional stability constants, $\log(\beta_N \ \mathrm{mol}^{-N}\mathrm{dm}^{3N})$, of $[\mathrm{Ni(dmso)}_N]$, $[\mathrm{Ni(meoh)}_N]$, and $[\mathrm{Ni(an)}_N]$ in $\mathrm{C_2mimTf_2N}$ and $\mathrm{C_8mimTf_2N}$ solutions at 25.0 °C.[a]

	$\mathrm{C_2mimTf_2N}$[b]		
	$[\mathrm{Ni(dmso)}_N]$ $[\mathrm{Ni^{2+}}]=66$ mM[c]	$[\mathrm{Ni(meoh)}_N]$ $[\mathrm{Ni^{2+}}]=100$ mM	$[\mathrm{Ni(an)}_N]$ $[\mathrm{Ni^{2+}}]=100$ mM
$\log \beta_1$			
$\log \beta_2$	9.48(88)	3.89(3)	5.72(2)
$\log \beta_4$	19.03(1.76)	8.25(5)	10.62(4)
$\log \beta_6$	27.26(2.64)	11.73(7)	14.40(5)
	$\mathrm{C_8mimTf_2N}$[d]		
	$[\mathrm{Ni(dmso)}_N]$ $[\mathrm{Ni^{2+}}]=50$ mM	$[\mathrm{Ni(meoh)}_N]$ $[\mathrm{Ni^{2+}}]=50$ mM	$[\mathrm{Ni(an)}_N]$ $[\mathrm{Ni^{2+}}]=50$ mM
$\log \beta_1$	5.18(18)		
$\log \beta_2$	10.11(35)	5.00(7)	7.69(16)
$\log \beta_4$	16.03(37)	10.30(15)	14.91(32)
$\log \beta_6$	19.42(39)	15.23(21)	20.82(48)

[a] Values in parentheses are standard deviations.
[b] Ref. 17.
[c] mM = mmol dm^{-3}.
[d] Ref. 20.

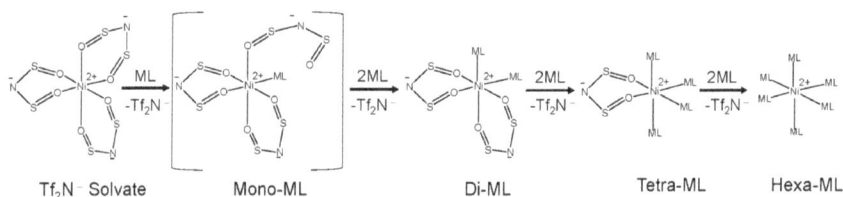

Figure 15.2 Complexation process of $\mathrm{Ni^{2+}}$ with ML molecules, if $\mathrm{Tf_2N^-}$ anions coordinate with $\mathrm{Ni^{2+}}$ as bidentate ligands. The mono-ML complex is formed only in $\mathrm{C_8mimTf_2N}$–DMSO solutions.

$[\mathrm{Ni(dmso)}_6]^{2+} > [\mathrm{Ni(an)}_6]^{2+} > [\mathrm{Ni(meoh)}_6]^{2+}$. On the other hand, the values for the $\mathrm{C_8mimTf_2N}$ solutions with the longer octyl chain are in the order of $[\mathrm{Ni(an)}_6]^{2+} > [\mathrm{Ni(dmso)}_6]^{2+} > [\mathrm{Ni(meoh)}_6]^{2+}$. These orders differ from that of the electron donicities of DMSO > MeOH > AN ($D_N = 29.8$, 19.0, and 14.1, respectively).[7] In addition, the electron acceptability of MeOH (Mayer–Gutmann's acceptor number $A_N = 41.3$) is higher than those of DMSO and AN ($A_N = 19.3$ and 19.3, respectively) due to the hydroxyl hydrogen atom of MeOH.[19] Therefore, the stability of the complexes does not depend only on the electron donicities (Lewis basicities) of ML molecules. In other words, the complexation equilibria must be affected by the microscopic interactions among the IL cation, anion, and ML on the molecular scale. Furthermore, the alkyl chain length of the IL cation of $\mathrm{C_nmim^+}$ also influences the metal complex formation in the ILs.

According to the microscopic interactions among the IL cation, anion, and ML molecules observed by Raman, IR, and NMR spectroscopy, together with the mixing state of the IL–ML binary solvents mesoscopically clarified by a small-angle neutron scattering (SANS) technique, the mechanisms of the complexation of Ni^{2+} with DMSO, MeOH, and AN in the ILs are explained as follows.[17,20] SANS experiments enable us to observe the heterogeneous mixing of binary solvents, that is, aggregation of molecules of each solvent (formation of solvent clusters).

Despite the lowest electron donicity of AN among the three MLs, the hexa-AN complex $[Ni(an)_6]^{2+}$ is comparably stable in both ILs. This is because AN molecules in the IL–AN solvents are freer compared to DMSO and MeOH molecules. The Raman, IR, and NMR spectroscopic experiments revealed that AN molecules are hardly hydrogen-bonded with the three hydrogen atoms of the imidazolium ring within $C_n mim^+$.[21] AN molecules weakly interact with Tf_2N^- *via* the dipole–dipole interaction. Furthermore, the SANS experiments suggested that AN molecules do not remarkably aggregate with themselves in the ILs because AN molecules mainly interact with them through the weak dipole–dipole interaction.[20] Therefore, the enthalpy gain for the coordination of AN molecules to Ni^{2+} is directly related to the stability of the AN complex in the ILs.

In contrast to AN, DMSO molecules can be strongly hydrogen-bonded with all the three hydrogen atoms of the imidazolium ring.[21,22] When DMSO molecules coordinate to Ni^{2+} in the ILs, the hydrogen bonds of DMSO with the hydrogen atoms of the imidazolium ring should be broken. This results in the enthalpic loss for the complex formation. In the imidazolium-based ILs, the polar domains consisting of the charged imidazolium rings and anions and the nonpolar domains where the alkyl chains aggregate with themselves may be formed, as shown by the MD simulations.[1–4] In the $C_8 mimTf_2N$ solution with the longer alkyl chain, DMSO molecules may be more strongly confined into the relatively small polar domains by the hydrogen bonding with the imidazolium ring than in the $C_2 mimTf_2N$ solution with the shorter chain. Therefore, the larger enthalpic loss for the release of DMSO molecules in the former more significantly influences the complexation than the latter. The hexa-DMSO complex in the $C_8 mimTf_2N$ solution becomes less stable than the hexa-AN complex in the same IL. In the smaller confined space of the polar domains in $C_8 mimTf_2N$ than $C_2 mimTf_2N$ due to the longer octyl chains, Tf_2N^- may less easily dissociate from Ni^{2+} in the former. This is the possible reason for the formation of the mono-DMSO complex.

The SANS experiments have shown that MeOH molecules form their clusters in $C_n mimTf_2N$,[23] while DMSO and AN molecules do not significantly aggregate to form their clusters.[20,22,24] This is attributed to the self-hydrogen bonds among MeOH molecules due to the hydroxyl group. Thus, the complex formation of MeOH with Ni^{2+} in the ILs is markedly affected by the enthalpic loss for the disruption of the self-hydrogen bonding among MeOH molecules. MeOH molecules should be eliminated from the clusters to coordinate with the Ni^{2+} ion. Consequently, the hexa-MeOH complex in both C_2mimTf_2N and C_8mimTf_2N solutions is the most unstable among the hexa-ML complexes studied.

As explained above, the complexation of the transition metal ions with ML in the ILs is surely influenced by the microscopic interactions among the IL cation, anion, and ML molecules depending on their properties. In investigations on the equilibria of metal complexation, the ionic strength should be constant to maintain the activity coefficients of species in solutions. For conventional ML systems, a supporting inert electrolyte, such as tetraalkylammonium perchlorate, is often dissolved in the solutions to generate a stronger electric field compared to that caused by the species concerning the chemical reaction. However, the activity coefficients of species in the IL solutions are not taken into account in the abovementioned investigations. This is because ILs can concurrently generate a strong electric field in the solutions. It should finally be mentioned that a new concept should be established to estimate the genuine thermodynamic parameters of metal complexation in ILs.

15.4 Ionic Liquids Having Chelate Amines

Transition-metal-containing ionic liquids (ILs) have attracted significant attention due to their unique physicochemical properties;[25] here, the metal ions are highly condensed in the electrical double-layer shell of the ILs, which can prevent the formation of large colloidal aggregates. The presence of chelating groups in ILs can stabilize the metal ions.[26] Ethylenediamine (En) is the simplest and most common chelating ligand used for the complexation of transition metals (d-block elements). ILs having amine chelating groups are further favorable for incorporating other Lewis acids, such as protons and CO_2. In this section, the properties of ILs containing amine chelates are reviewed, with an emphasis on transition-metal (d-block element) complexes with long lifetimes. The complexation of transition-metal ions in protic ionic liquids (PILs) of multidentate amines is then discussed.

15.4.1 Metal Complexes of *N*-alkylethylenediamines (Alkyl-en) and the Formation of ILs

When transition-metal ions are bound to neutral ligands to form transition-metal complex cations, the obtained salts tend to have higher melting points than conventional ILs. Accordingly, there are much fewer reported cationic metallo-ILs than anionic metallo-ILs, whereas cationic metallo-ILs have potential applications in a wide range of fields owing to their affinities to abundant anionic surfaces and to carbon surfaces with high π-electron density. For example, cationic metallo-ILs are considered as good electrolytes for electrodeposition applications.[27]

Alkylmonoamine–silver(I) cationic complexes with NO_3^- and BF_4^- have been synthesized; these complexes form liquid crystals, in which the cations have U-shaped bilayer structures below 100 °C.[28] The low melting points for such complexes are attributed to the monopositive charge and to the bulky ligands, which allowed the formation of room-temperature ionic liquids (RTILs) by simply choosing the appropriate counter anions. Dai *et al.* reported the synthesis of RTILs of monoamine complexes, *viz.* [Ag(alkyl-NH$_2$)$_2$]Tf$_2$N and [Zn(alkyl-NH$_2$)$_4$](Tf$_2$N)$_2$ (alkyl = methyl, ethyl, propyl, and *t*-butyl),[29] which were employed in the electrodeposition of silver and zinc, respectively. Furthermore, upon changing both the organic ligands and central metal ions, these complexes had tunable specific functionalities as electrolytes.

On the other hand, Iida *et al.* synthesized *N*-alkylethylenediamine (alkyl-en) silver(I) complexes, *viz.* [Ag(hexyl-en)$_2$]PF$_6$, [Ag(2-ethylhexyl-en)$_2$]-NO$_3$0.3H$_2$O, and [Ag(octyl-en)$_2$]Tf$_2$N, which formed RTILs (Scheme 15.2).[30,31] In this section, the ligands hexyl-en, 2-ethylhexyl-en, octyl-en, and dodecyl-en are abbreviated as hex-en, ethex-en, oct-en, and dod-en, respectively.

Among the three Ag(I) complexes (Scheme 15.2) of RTILs, the ethex-en complex showed a glass transition (T_g) at -54 °C, below which the X-ray scattering profile exhibited an amorphous structure.[30] Although this complex was sparingly soluble in water, it incorporated a small amount of water to form a transparent liquid. The self-diffusion coefficient (D) of the complex was measured using pulse gradient spin-echo nuclear magnetic resonance (PGSE-NMR) in the temperature range of 30–70 °C and the water content ($W_0 = [H_2O]/[Ag$-complex$]$) was in the 0.3–5.5 region. When $W_0 = 0.3$–5.0, D almost followed the Vogel–Tammann–Fulcher (VTF) equation:

$$D = D_0 \exp\left(-\frac{B}{T - T_0} \right) \tag{15.6}$$

where T is the temperature and D_0, B, and T_0 are the adjustable parameters for the given system.

As W_0 increased, the temperature dependence of D approached the Arrhenius relationship, and at $W_0 = 5.5$ the log D *vs.* T^{-1} plot was linear. Self-diffusion and electrical conductivity studies have revealed that the Ag

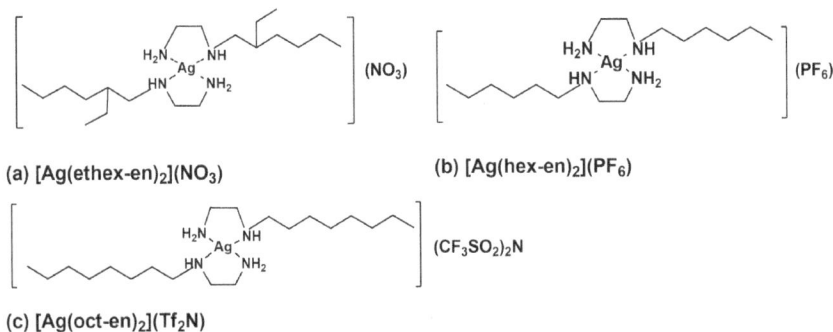

(a) [Ag(ethex-en)$_2$](NO$_3$)

(b) [Ag(hex-en)$_2$](PF$_6$)

(c) [Ag(oct-en)$_2$](Tf$_2$N)

Scheme 15.2 Structures of Ag(I) complexes of RTILs. (a) [Ag(2-ethylhexyl-en)$_2$]NO$_3$, (b) [Ag(hexyl-en)$_2$]PF$_6$, and (c) [Ag(octyl-en)$_2$]Tf$_2$N.

complexes behave like monomers in the liquid state, despite the dimer structure of the dod-en derivative in the crystal state.[32]

The ethex-en complex particularly favored the formation of well-dispersed Ag(0) nanoparticles (NPs) from their own IL *via* a reaction with an aqueous NaBH$_4$ solution, whereas the same treatment for the hex-en complex did not result in soluble Ag(0) NPs, but only in silver precipitates. This result indicates that the stabilizing ability of the ethex-en ligands prevented the aggregation of Ag(0) NPs. The metal complex acted as both a metal source and a particle protector.

Ag(I) complexes with bulky ligands favor the formation of RTILs due to the weaker intermolecular interactions in the condensed state. Zn(II) complexes also tend to have low melting points for the dipositive charge and can form RTILs, *e.g.*, [ZnY$_4$](Tf$_2$N)$_2$ (Y = hex-en, ethex-en and oct-en),[33] [Zn(alkyl-NH$_2$)$_4$](Tf$_2$N)$_2$ (mentioned above),[29] and [Zn(alkyl-im)$_4$](NO$_3$)$_2$ (alkyl = N-decyl and N-dodecyl; im = imidazolium).[25] These tetrahedral zinc(II) complexes are flexible around the zinc(II) center; consequently, their melting points are significantly lower than those of the other dipositive analogues.

The metallo-ILs described herein can be regarded as long-lifetime complexes, with all ligating groups belonging to the same molecule. Consequently, they can be classified as novel solvate (chelate) ILs, as proposed by Angell.[34]

15.4.2 PILs of Alkyl-en and of *N*-alkyldiethylenetriamines (Alkyl-dien)

The replacement of an Ag(I) ion by a proton in the [Ag(alkyl-en)$_2$]$^+$ complex yields PILs of H(alkyl-en)$^+$. Monoprotonated PILs of *N*-alkylethylenediaminium (alkyl = butyl, hexyl (Hex), 2-ethylhexyl (Ethex), octyl, decyl, and dodecyl) and *N*-alkyldiethylenetriaminium (H(alkyl-dien)$^+$, alkyl = Hex and Ethex) have been prepared using Tf$_2$N$^-$ or trifluoroacetate (TFA$^-$) counter anions, and their physical properties have been studied.[31,35–37] The melting points (T_m) of all the studied complexes were below 34 °C. Among these, HEthex-en(Tf$_2$N) and all of the dien-PILs exhibited T_g in the range from −70 to −60 °C. The detailed properties of the hexyl PILs, such as HHex-en(A) and HHex-dien(A), where A = Tf$_2$N or TFA (Scheme 15.3a–c), were investigated

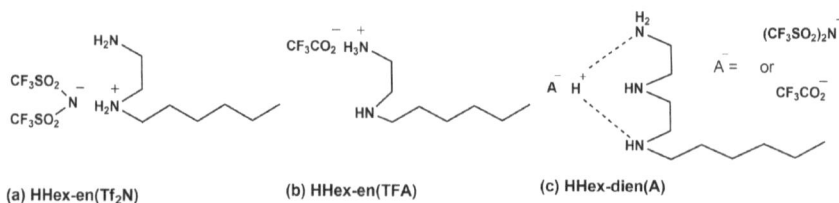

(a) HHex-en(Tf$_2$N)   (b) HHex-en(TFA)   (c) HHex-dien(A)

Scheme 15.3   Structures of PILs composed of alkyl multidentate aminium cations and anions. (a) *N*-hexylethylenediaminium (Tf$_2$N), (b) *N*-hexylethylenediaminium (TFA), and (c) *N*-hexyldiethylentriiaminium (Tf$_2$N) or *N*-hexyldiethylentriiaminium (TFA).

because they were RTILs, and their ^{13}C NMR spectra were simpler than those of the longer alkyl-chain derivatives.

H(alkyl-en)$^+$ and H(alkyl-dien)$^+$ have two and three amines, respectively, and the predominant 1:1 protonation positions were determined for HHex-en$^+$ and HHex-dien$^+$ using ^{15}N and ^{13}C NMR chemical shifts, as shown in Scheme 15.3.

15.4.3 Cu(II) Ion Complexations in the PILs of Hex-en and Hex-dien

The Cu(II) complexations in HHex-en(Tf$_2$N)[35,36] and HHex-dien(Tf$_2$N)[37] PILs have been investigated using visible absorption spectroscopy as well as paramagnetic Cu^{2+}-induced ^{13}C and ^{15}N NMR relaxations. The present polydentate amine PILs generally dissolved MX$_2$ well (M = first-row transition metal, X = monovalent anion), and for M = Ni and Cu, the solution remained stable for more than 3 weeks at 30 °C. In HHex-en(Tf$_2$N), the solubility of Cu(Tf$_2$N)$_2$ (0.1 mol kg^{-1}) was less than those of the other salts like chloride and bromide ($\sim$0.2 mol kg^{-1}) due to the common-ion effect. The complexation of Cu(Tf$_2$N)$_2$ in HHex-en(Tf$_2$N) (or in HHex-dien(Tf$_2$N)) was investigated and compared with that of Cu(Tf$_2$N)$_2$ in the monoamine-type PIL of HHex-am(Tf$_2$N) and then with that of Cu(TFA)$_2$ in HHex-en(TFA) (or in HHex-dien(TFA)) to observe the chelate and counter ion effects, respectively.

The following observations were noted. (1) In HHex-en(A) (A = Tf$_2$N or TFA), the Cu(II) ion results in N$_4$O$_2$ coordination, where the hex-en forms a bis complex, while the A anions coordinate to the Cu(II) ion at the axial sites. (2) In the HHex-am(Tf$_2$N) PIL, the Cu(II) ion is coordinated with the Tf$_2$N$^-$ anion rather than with the HHex-am$^+$ cation. (3) Upon the addition of CuX$_2$ (X = Cl$^-$ or Br$^-$) to HHex-en(Tf$_2$N), the Tf$_2$N$^-$ anion coordinated to the copper(II) ion is gradually replaced by the X anions, while in HHex-en(TFA) the TFA anion keeps coordinating to the copper(II) ion up to a CuX$_2$ concentration of 0.1 mol kg^{-1}. (4) In HHex-dien(A) (A = Tf$_2$N or TFA), the Cu(II) ion results in N$_5$O coordination, where the Hex-dien donates five nitrogen atoms, while the A anion donates one oxygen atom. In the dien PILs, the replacement of the N$_5$O complex by Cl$^-$ does not occur in either counter anion system. (5) Upon complexation, protons are released from HHex-en(Tf$_2$N) and HHex-dien(Tf$_2$N) with only slight pH changes (7.4 → 7.1 and 7.2 → 7.0, respectively) due to the buffering action of the amines. (6) The lifetimes of the Cu(II) complexes in the neat liquids with the chelate amine moieties follow at ambient temperature the order: HHex-dien(Tf$_2$N) PIL (10^{-2} s) $\gg$ HHex-en(Tf$_2$N) PIL (10^{-4} s) $\gg$ neat Hex-en (10^{-6} s) > neat En (10^{-7} s).[36,37] These high Cu^{2+} capturing abilities of both HHex-en(Tf$_2$N) and HHex-dien(Tf$_2$N) PILs as well as their hydrophobicity are applicable to liquid–liquid extraction of metal ions from aqueous solutions.[36,37]

15.5 Deep Eutectic Solvents (DESs) and Betainium-based ILs

DESs are nowadays widely acknowledged as a new class of ILs.[38–40] These are basically composed of a eutectic mixture of a quaternary ammonium halide salt and a hydrogen bond donor and have physical properties similar to those of conventional ILs, whereas they are much more practical because their components are inexpensive, sustainable, renewable, and easily prepared from readily accessible chemical reagents. The first study on this concept was reported in 2001 by Abbott *et al.*[41] Initially, DESs were composed of choline chloride (quaternary ammonium halide salt) (Scheme 15.4a)/metal chloride, where the metal ions were present as chloro complexes. The category of DESs was then extended to a range of liquids formed from eutectic mixtures of hydrogen bond acceptors (HBAs, such as quaternary ammonium salts) and hydrogen bond donors (HBDs).

DESs generally dissolve metal oxides and hydroxides well,[42] and the solubility can be strongly influenced by the HBD. This selective solubility of metal oxides depending on the HBD is useful for the metal extraction from mixed metal sources. Furthermore, DESs are convenient for metallurgy and electrodeposition applications, which are actually the processes of recovery, purification, and plating to produce useful metals from raw and recycled materials. Complex formation of metal ions in DESs is therefore a significant subject to study.

In a study conducted by Frisch *et al.*, EXAFS was used to compare the structure of a series of transition-metal complexes in diol- or urea-based DESs with classical IL systems.[43] The employed DESs were mixtures of choline chloride (HBA) and diols (1,2-ethanediol, 1,2-propanediol, and 1,3-propanediol) or urea as the HBD. Metal salts (0.1 mol dm^{-3}, mainly chlorides) were dissolved in these solvents. For diol-based DESs, most of the M(I) and M(II) metal ions were present as di- or trichloro and tetrachloro complexes, respectively, whereas Ni(II) complexes were present as unusual diol-chelating (for 1,2-diols) or monool (for 1,3-diols) complexes. When urea was used as the HBD, water (from the metal salt) and/or urea were found in the first co-ordination shell in the case of transition metals such as Mn(II), Fe(II), and Co(II), while Cu(II) formed CuCl$_4^{2-}$. On the other hand, when Ni(II) and Cu(II)

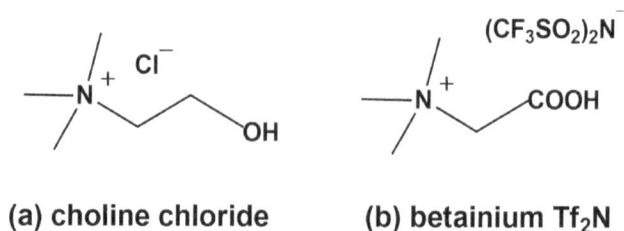

(a) choline chloride   **(b) betainium Tf$_2$N**

Scheme 15.4   Structures of (a) choline chloride and (b) betainium Tf$_2$N.

salts were dissolved in the conventional IL 1-hexyl-3-methylimidazolium chloride C_6mimCl, trichloro complexes were formed. The lower coordination numbers of the chloro-M(II) complex ions in the conventional IL (trichloro) than those (tetrachloro) in the DES are attributed to the hexyl group in the IL component.

A derivative of choline, N-trimethylglycine (betaine) (Scheme 15.4b),[44] is also a sustainable and easily obtainable component of traditional ILs. Although the melting point of the IL comprising the betainium ion (protonated betaine) with the Tf_2N^- anion is 57 °C, the liquid compound can be supercooled to room temperature. This IL dissolved metal oxides and metal salts well, similar to DESs. Since this IL is hydrophobic at room temperature, it is useful for metal ion extraction from aqueous solutions.

Problem 1

Explain why the entropy ΔS_N of complexation equilibrium can be estimated from the stability constant β_N and the enthalpy ΔH_N which were determined by calorimetry.

Problem 2

DESs are basically composed of a eutectic mixture of hydrogen bond acceptors (HBAs, such as quaternary ammonium salts) and hydrogen bond donors (HBDs). Describe a general solid–liquid phase diagram (temperature *vs.* mole fraction) of a binary mixture, where the two solids are almost immiscible and they are completely miscible in liquids. In the figure, present the melting temperatures (T_m) of components A and B and a eutectic point (E). Furthermore, show four regions in the figure for one-phase liquid (solution), two-phase solid, solution + A, and solution + B, respectively.

Problem 3

What are the molecular structural conditions required to form conventional ILs (melting points below 100 °C and composed of discrete anions and cations)? Consider that the cationic side is typically an amphiphilic ion and that there are enthalpic and entropic contributions that can lower the melting points or reduce the intermolecular interactions.

Problem 4

How do you evaluate the strength of the hydrogen bonds of the imidazolium ring with molecular liquids, such as dimethyl sulfoxide, methanol, and acetonitrile, in ionic liquid–molecular liquid mixed solvents?

Problem 5

On the basis of the following description, answer questions (a) and (b):

Conventional ILs and DESs have many common properties, *i.e.*, wide liquid range, nonflammability, low volatility, low vapor pressures, tuneable polarity, wide electrochemical windows, and hydrophilic–hydrophobic microstructures. Consequently, they are expected to be useful in similar fields of application. In recent years, however, studies on DESs have been developing more rapidly than studies on conventional ILs.

(a) What are the general advantages of DESs over ILs in applications?
(b) What are the advantageous common and different properties between ILs and DESs over molecular solvents in the following applications?
   (1) Separation of metal ions
   (2) Creation of metal nanomaterials
   (3) Electrodeposition of metals

Answer to Problem 1

The thermodynamics parameters are related by $\Delta G_N = -RT\ln\beta_N = \Delta H_N - T\Delta S_N$. Thus, when the stability constant β_N and the enthalpy ΔH_N are determined, the entropy ΔS_N can be derived from the second and third terms.

Answer to Problem 2

Figure 15.3 is an example of the answer.

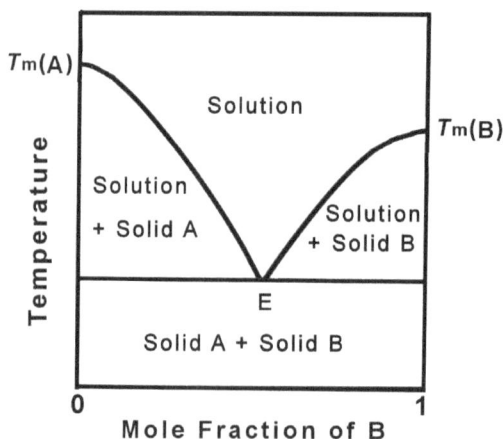

Figure 15.3   The typical temperature–composition phase diagram for the two (A and B) almost immiscible solids and their completely miscible liquids (solutions). The E point corresponds to the eutectic point. Tm(A) and Tm(B) are the melting points of the pure A and B solids, respectively.

Answer to Problem 3

(1) The high melting points of ionic compounds are attributed to the strong Coulombic forces between the counter ions. Consequently, most ILs are composed of monopositive ions, and their molecular size is large (at least for the cation). (2) If an alkyl chain is present in the hydrophobic part of the component cation, van der Waals interactions between the alkyl chains attract the ions as well; therefore, the lowest melting point (or glass transition point) is for an alkyl-chain length around C_6–C_8. (3) Since the cation typically has a hydrocarbon chain or ring, the anion containing fluorine atoms is advantageous for reducing the van der Waals attractions. (4) Concerning the entropic effect, the bulky and flexible (random conformation) structure of the component ion favors the reduction of intermolecular interactions. For example, several zinc(II) complexes with bulky ligands form RTILs despite their dipositive charge (ref. 25, 29 and 33). Unlike other metal analogues, zinc(II) complexes with alkyl-en ligands having a tetrahedral structure sometimes form RTILs (see ref. 33). (5) The ethyl-branch in the alkyl chain of the cation side has both entropic and enthalpic effects. For monopositive ILs such as Ag-IL and PILs, the 2-ethyl branch in the alkyl-en part lowers the melting points due to the entropic or bulky effect, whereas for dipositive ILs the branch often elevates the melting points due to an increase in the van der Waals attractions between the cations, depending on the counter anions (see ref. 33).

Answer to Problem 4

The simplest evaluation is the observation of molecular vibration using Raman and infrared spectroscopy. The C–H stretching vibrations of the imidazolium ring may red-shift more with strengthening of the hydrogen bonding. For example, the red-shift of the C–H stretching vibrations is more significant in the order of dimethyl sulfoxide > methanol > acetonitrile. This order agrees with the electron donicity of the molecular liquids.

Answer to Problem 5

(a) DESs are employed in various applications since they are inexpensive, biodegradable, non-toxic, and easy to prepare. Furthermore, they have widely tuneable properties by choosing the appropriate HBD reagents and offer high solubility to metal oxides and hydroxides.

(b)

(1) Hydrophobic ILs with hydrophilic cores can be used to extract metal ions from aqueous solutions and the efficiency of this extraction increases in the presence of a chelating group in the hydrophilic part. The solubility of metal ions in DESs is generally higher than that in ILs. It is remarkable that DESs have a high dissolution ability of metal oxides and the solubility can be widely

tuned by choosing the appropriate type and amount of HBD reagent. Therefore, DESs are useful to selectively extract metals from a mixed matrix and allowed them to be recovered with high efficiencies and in high purities.

(2) Since ILs have the characteristic structures of hydrophobic and hydrophilic microheterogeneity like microemulsions, they have been applied to creating metal nanoparticles (NPs) by using the microstructures as a template. They can control the morphology of metal NPs with proper additives and then provide NPs with high density. To date, more studies have been conducted on nanomaterials in ILs than in DESs. However, DESs typically have a higher ability to incorporate metal ions than ILs and can be widely designed for microstructures by choosing the appropriate HBD–HBA combinations.

(3) Electrodeposition is an electrochemical process in which an electropositive material, such as a metal or an alloy, is deposited on a cathode. ILs and DESs are considered good targets for electrodeposition due to their wide electrochemical windows, good ionic conductivity, anhydrous nature, and thermal stability. In fact, they are especially useful for the deposition of highly electropositive metals, such as Al, Li, and Mg, which normally cannot be deposited from aqueous solutions due to their reactivity. Furthermore, the advantages of ILs and DESs for electrodeposition are not only applicable to metal plating, but also to metal polishing technologies. In particular, the high solubility of metal oxides and hydroxides in DESs renders them effective in preventing passivation, which inhibits the deposition of the target metal. The issues of electrochemistry in ILs and DESs are the complex reactions at the electrical double-layer electrode and the inevitable high viscosities of the solutions. The former reactions are tuneable by the addition of appropriate complexing reagents (ILs) or HBDs (DESs).

Acknowledgements

The study on Ni^{2+} complexation with three ML molecules was supported partly by JSPS KAKENHI (Grant No. 26410018 and 18K05038) (T. T.). The density and NMR measurements for the IL systems and the single-crystal X-ray crystallographic measurements on crystals grown from the sample solutions were conducted at the Analytical Research Center for Experimental Sciences of Saga University. T. T. conducted the Raman spectroscopy experiments in the University of Lille, France as an invited professor. The SANS experiments for the MeOH–IL solutions and the DMSO–IL and AN–IL solutions were performed in joint research with the Institute for Solid State Physics, the University of Tokyo (Proposal No. 8851) and with the approval of the Neutron Program Review Committee of J-PARC MLF (Proposal No. 2017A0002), respectively. The study on the ILs of alkyl multidentate amines was financially supported by JSPS KAKENHI (Grant No. 25410191) (M. I.).

Part of the study on the silver(ɪ) ILs was supported by the "Nanotechnology Support Project of the Ministry of Education, Culture, Sports, Science and Technology (MEXT), Japan" at the Research Center for Ultra-high Voltage Electron Microscopy (Osaka University, Japan).

References

1. J. N. C. Lopes and A. A. Padua, *J. Phys. Chem. B*, 2006, **110**, 3330.
2. J. N. C. Lopes, M. F. C. Gomes and A. A. Padua, *J. Phys. Chem. B*, 2006, **110**, 16816.
3. A. A. Padua, M. F. C. Gomes and J. N. C. Lopes, *Acc. Chem. Res.*, 2007, **40**, 1087.
4. W. Jiang, Y. Wang and G. A. Voth, *J. Phys. Chem. B*, 2007, **111**, 4812.
5. M. Yamagata, Y. Katayama and T. Miura, *J. Electrochem. Soc.*, 2006, **153**, E5.
6. Y.-L. Zhu, Y. Katayama and T. Miura, *Electrochim. Acta*, 2010, **55**, 9019.
7. V. Gutmann, *The Donor–Acceptor Approach to Molecular Interactions*, Plenum Press, New York, 1978.
8. K. Fujii, T. Nonaka, Y. Akimoto, Y. Umebayashi and S. Ishiguro, *Anal. Sci.*, 2008, **24**, 1377.
9. F. Sessa, V. Migliorati, A. Serva, A. Lapi, G. Aquilanti, G. Mancini and P. D'Angelo, *Phys. Chem. Chem. Phys.*, 2018, **20**, 2662.
10. O. Borodin, G. A. Giffin, A. Moretti, J. B. Haskins, J. W. Lawson, W. A. Henderson and S. Passerini, *J. Phys. Chem. C*, 2018, **122**, 20108.
11. M. Busato, P. D'Angelo, A. Lapi, M. Tolazzi and A. Melchior, *J. Mol. Liq.*, 2020, **299**, 112120.
12. M. Busato, A. Lapi, P. D'Angelo and A. Melchior, *J. Phys. Chem. B*, 2021, **125**, 6639.
13. P. Nockemann, M. Pellens, K. Van Hecke, L. Van Meervelt, J. Wouters, B. Thijs, E. Vanecht, T. N. Parac-Vogt, H. Mehdi, S. Schaltin, J. Fransaer, S. Zahn, B. Kirchner and K. Binnemans, *Chemistry*, 2010, **16**, 1849.
14. S. Georg, I. Billard, A. Ouadi, C. Gaillard, L. Petitjean, M. Picquet and V. Solov'ev, *J. Phys. Chem. B*, 2010, **114**, 4276.
15. A. Melchior, C. Gaillard, S. Gracia Lanas, M. Tolazzi, I. Billard, S. Georg, L. Sarrasin and M. Boltoeva, *Inorg. Chem.*, 2016, **55**, 3498.
16. R. Kanzaki, S. Uchida, H. Kodamatani and T. Tomiyasu, *J. Phys. Chem. B*, 2017, **121**, 9659.
17. Y. Kawazu, H. Hoke, Y. Yamada, T. Umecky, K. Ozutsumi and T. Takamuku, *Phys. Chem. Chem. Phys.*, 2017, **19**, 31335.
18. H. Suzuki and S. Ishiguro, *Inorg. Chem.*, 1992, **31**, 4178.
19. U. Mayer, *NMR-Spectroscopic Studies on Solute-Solvent and Solute-Solute Interactions*, Elsevier Science Publishers, Amsterdam, 1983.
20. T. Takamuku, H. Sakurai, A. Ogawa, A. Tashiro, M. Kawano, Y. Kawazu, K. Sadakane, H. Iwase and K. Ozutsumi, *Phys. Chem. Chem. Phys.*, 2019, **21**, 3154.
21. T. Takamuku, H. Hoke, A. Idrissi, B. A. Marekha, M. Moreau, Y. Honda, T. Umecky and T. Shimomura, *Phys. Chem. Chem. Phys.*, 2014, **16**, 23627.

22. T. Takamuku, T. Tokuda, T. Uchida, K. Sonoda, B. A. Marekha, A. Idrissi, O. Takahashi, Y. Horikawa, J. Matsumura, T. Tokushima, H. Sakurai, M. Kawano, K. Sadakane and H. Iwase, *Phys. Chem. Chem. Phys.*, 2018, **20**, 12858.

23. T. Shimomura, K. Fujii and T. Takamuku, *Phys. Chem. Chem. Phys.*, 2010, **12**, 12316.

24. T. Takamuku, Y. Honda, K. Fujii and S. Kittaka, *Anal. Sci.*, 2008, **24**, 1285.

25. J. B. Lin and C. S. Vasam, *J. Organomet. Chem.*, 2005, **690**, 3498.

26. J. R. Harjani, T. Friščić, L. R. MacGillivray and R. D. Singer, *Inorg. Chem.*, 2006, **45**, 10025.

27. J. Sniekers, N. R. Brooks, S. Schaltin, L. V. Meervelt, J. Fransaer and K. Binnemans, *Dalton Trans.*, 2014, **43**, 1589.

28. A. C. Albéniz, J. Barberá, P. Espinet, M. C. Lequerisa, A. M. Levelut, F. J. López-Marcos and J. L. Serrano, *Eur. J. Inorg. Chem.*, 2000, 133.

29. J.-F. Huang, H. Luo and S. Dai, *J. Electrochem. Soc.*, 2006, **153**, J9.

30. M. Iida, C. Baba, M. Inoue, H. Yoshida, E. Taguchi and H. Furusho, *Chem. – Eur. J.*, 2008, **14**, 5047.

31. M. Iida, S. Kawakami, E. Syouno, H. Er and E. Taguchi, *J. Colloid Interface Sci.*, 2011, **356**, 630.

32. M. Iida, M. Inoue, T. Tanase, T. Takeuchi, M. Sugibayashi and K. Ohta, *Eur. J. Inorg. Chem.*, 2004, 3920.

33. H. Yasuda, C. Nakayama and M. Iida, *J. Mol. Liq.*, 2018, **269**, 169.

34. C. A. Angell, Y. Ansari and Z. Zhao, *Faraday Discuss.*, 2012, **154**, 9.

35. M. Watanabe, S. Takemura, S. Kawakami, E. Syouno, H. Kurosu, M. Harada and M. Iida, *J. Mol. Liq.*, 2013, **183**, 50.

36. S. Takemura, S. Kawakami, M. Harada and M. Iida, *Inorg. Chem.*, 2014, **53**, 9667.

37. C. Nakayama, M. Harada and M. Iida, *Eur. J. Inorg. Chem.*, 2017, 3744.

38. Q. Zhang, K. De, O. Vigier, S. Royer and F. Jérôme, *Chem. Soc. Rev.*, 2012, **41**, 7108.

39. E. L. Smith, A. P. Abbott and K. S. Ryder, *Chem. Rev.*, 2014, **114**, 11060.

40. B. B. Hansen, S. Spittle, B. Chen, D. Poe, Y. M. Klein and J. R. Sangoro, *Chem. Rev.*, 2021, **121**, 1232.

41. A. P. Abbott, G. Capper, D. L. Davies, H. L. Munro, R. K. Rasheed and V. Tambyrajah, *Chem. Commun.*, 2001, 2010.

42. A. P. Abbott, G. Frisch and K. S. Ryder, *Annu. Rep. Prog. Chem., Sect. A: Inorg. Chem.*, 2008, **104**, 21.

43. J. M. Hartley, C.-M. Ip, G. C. H. Forrest, K. Singh, S. J. Gurman, K. S. Ryder, A. P. Abbott and G. Frisch, *Inorg. Chem.*, 2014, **53**, 6288.

44. P. Nockemann, B. Thijs, S. Pittois, J. Thoen, C. Glorieux, K. V. Hecke, L. V. Meervelt, B. Kirchner and K. Binnemans, *J. Phys. Chem. B*, 2006, **110**, 20978.

Metal Complexes in Micellar and Liquid Crystalline Solutions

MASAYASU IIDA

Professor Emeritus, Nara Women's University, Nara 630-8506, Japan
Email: iida@cc.nara-wu.ac.jp

16.1 Metal Complexes in Micellar and Liquid Crystalline Solutions

Amphiphilic molecules or surfactants have a dual-character, that is, typically comprising a hydrophilic polar group and a hydrophobic nonpolar chain. In aqueous solutions, the water-insoluble hydrophobic moiety is transferred out of the bulk water phase and self-assembles to form association colloids. In the current chapter, the author focuses on the ionic interactions in ionic amphiphilic systems. An ionic and long alkyl-chained (at least eight alkyl chains) surfactant exhibits two ambivalent characters in water, *i.e.*, hydrophobic assembly of the nonpolar tail and electrostatic repulsion between the head groups. As a result, spherical micelles form above a critical micelle concentration (cmc).[1] It is characteristic that the hydrophilic surfaces of micelles contact with bulk water forming electrical double layers, where the counter ions concentrate. With an increase in the surfactant concentration above the cmc, the number of spherical micelles simply increases at the initial stage, after which at higher concentrations they further aggregate to form cylindrical micelles, regarded as an onset of anisotropic aggregates. With a further increase in the surfactant concentration, therefore, the

Coordination Chemistry Fundamentals Series No. 2
Metal Ions and Complexes in Solution
Edited by Toshio Yamaguchi and Ingmar Persson
© Japan Society of Coordination Chemistry 2024
Published by the Royal Society of Chemistry, www.rsc.org

micelles grow up to form lyotropic liquid crystals like hexagonally packed cylindrical micelles.

On the other hand, metal complexes have wide varieties of features at a molecular level such as hydrophilicity, hydrophobicity, electric charge number, unique molecular shape, and size. As the functionalities of metal complexes are enhanced in colloidal systems, the interactions of metal complexes with surfactants in aggregate solutions become a significant target for study.

16.1.1   Interactions of Tripositive Metal Complex Ions with Surfactants in Ionic Micellar and Liquid Crystalline Solutions

Highly-charged ions have been synthesized using cobalt(III) and chromium(III) ions. The selective and strong ionic interactions between the tripositive metal complex ions and the polar regions of ionic micelles occur by replacing the counter ions of the surfactants in the electrical double layers with the metal complex ions. In order to clarify the microscopic interactions in multicomponent solutions, multinuclear NMR spectroscopy is a particularly powerful tool because of the high receptivity and selectivity to the environment around the observed nucleus.[2] In this section, two typical studies on the interactions will be described for the trivalent complexes in the micellar and liquid crystalline solutions using multinuclear NMR spectroscopy.

16.1.1.1   *Micellar Solutions*

Selective interactions between the common surfactants and the metal complexes in solution were systematically studied using cobalt(III) and chromium(III) complexes having hydrophilic or hydrophobic ligands.[3] The metal complex ions used were $[Co(en)_3]^{3+}$ (en = ethylenediamine), $[Co(chxn)_3]^{3+}$ (chxn = *trans*-(1R,2R)-1,2-diaminocyclohexane), $[Co(phen)_3]^{3+}$ (phen = 1,10-phenanthroline), and the corresponding chromium(III) complexes. The en complex is a common hydrophilic ion having carbon atoms, the phen complex is a typical amphiphilic complex, and the chxn complex has both the hydrophilic site and hydrophobic moieties. The anionic surfactants used were head group = sulfonate, sulfate, alaninate and carboxylate, while alkyl chain = octyl, decyl, and dodecyl. The measurements were performed for the ^{59}Co NMR relaxation rates and the paramagnetic Cr(III)-induced ^{13}C NMR relaxations. The outer-sphere interactions between the metal complex ions and the micelles are significantly weaker than inner-sphere complexations of metal ions with surfactants. Accordingly, some NMR parameters can be more simply treated in the present systems than for the inner-sphere complexations.[4]

The longitudinal $(R_1 = T_1^{-1})$ and transverse $(R_2 = T_2^{-1})$ relaxation rates for the nucleus (spin quantum number $= I > 1/2$) can be expressed, to a first approximation, as[2,4]

$$R_1(\omega_0) = \frac{3\pi^2(2I+3)}{10I^2(2I-1)} \chi_{\text{eff}}^2 (0.2J(\omega_0) + 0.8J(2\omega_0)) \tag{16.1}$$

$$R_2(\omega_0) = \frac{3\pi^2(2I+3)}{10I^2(2I-1)} \chi_{\text{eff}}^2 (0.3J(0) + 0.5J(\omega_0) + 0.2J(2\omega_0)) \tag{16.2}$$

where ω_0 is the Larmor frequency, χ_{eff} is the residual quadrupole coupling constant remaining after the rapid averaging process in the frequency unit $(= e^2 q_{\text{eff}} Q/h$, where eq_{eff} is the principal value of the effective electric field gradient at the nucleus and eQ is the nuclear electric quadrupole moment) and $J(\omega_0)$ is the reduced spectral density function which describes the frequency dependence of the molecular motions modulating the quadrupolar interaction. When isotropic rotational motions dominate the relaxation, the reduced spectral densities have the following Lorentzian form:

$$J(\omega_0) = \frac{\tau_{\text{eff}}}{1 + (\omega_0 \tau_{\text{eff}})^2} \tag{16.3}$$

where τ_{eff} is the correlation time arising from the molecules of the probe nucleus in the rotationally restrictive environments of the micellar systems and follows the Arrhenius relationship.

The T_1 and T_2 values for the ^{59}Co nucleus $(I = 7/2)$ were determined over the temperature range from 20 to 45 °C, after which the χ_{eff} and temperature-dependent τ_{eff} values were calculated using eqn (16.1)–(16.3). The obtained temperature-dependent τ_{eff} (1.0–2.1 ns at 27 °C, depending on the cobalt(III) complexes and the surfactants) furthermore gave apparent activation energies (E_a). The τ_{eff} and E_a values are good indicators of the extent of the restrictive motions of the cobalt(III) complexes in the micelles. The τ_{eff}, E_a (21–35 kJ mol^{-1}), and χ_{eff} (1.5–4.4 MHz) values estimated for the ^{59}Co nucleus gave the following relationships concerning the interactions. (1) In the dodecyl chain micelles, the motional restrictions $(\tau_{\text{eff}}$ and $E_a)$ of the cobalt(III) complexes are larger than those in the decyl and octyl micelles. (2) The motional restrictions are not regularly governed by both the head groups and metal complexes. (3) The χ_{eff} values in the micelles have a trend in the order, $[\text{Co(phen)}_3]^{3+}$ (4.0–4.4 MHz) $> [\text{Co(chxn)}_3]^{3+}$ (2.1–3.3 MHz) $\gtrsim [\text{Co(en)}_3]^{3+}$ (1.5–2.5 MHz). The χ_{eff} values are slightly dependent on the head groups of the surfactants, whereas in a comparison between the complexes for the same surfactants the above order holds. This order accords with the hydrophobicity of the complexes.

The selective interactions of the complexes with the micelles were otherwise discussed from the results of the paramagnetically induced

^{13}C relaxation rates (T_{1M}^{-1}) for the C1–C8 carbons of the surfactants in the presence of chromium(III) complexes.

The obtained profiles are depicted in Figure 16.1(a); they can be classified into two categories regarding the chromium(III) complexes. One (group-1) is for $[Cr(en)_3]^{3+}$, $[Cr(chxn)_3]^{3+}$ and $[Cr(H_2O)_6]^{3+}$, and the other (group-2) is for $[Cr(phen)_3]^{3+}$ and $[Cr(acac)_3]$ (acac = acetylacetonate). The geometrical relationship between the T_{1M}^{-1} and the time-averaged distance $(\langle d \rangle)$ of ^{13}C nuclei from the center of the paramagnetic ions can be related with the relative values of the ith carbon to those of the jth carbon as follows:[2,4]

$$\frac{(T_{1M})_i^{-1}}{(T_{1M})_j^{-1}} = \frac{\langle d_i^{-6} \rangle f_i(\tau_C)}{\langle d_j^{-6} \rangle f_j(\tau_C)} \tag{16.4}$$

where $f_i(\tau_C)$ is the spectral density for the i-carbon as a function of the correlation time (τ_C) for the motion that modulates the interaction. When comparing between the two carbons in the surfactant, the $(T_{1M})_i^{-1}/(T_{1M})_j^{-1}$ ratio is governed mainly by the $\langle d_i^{-6} \rangle/\langle d_j^{-6} \rangle$ ratio. This relationship indicates that group-1 is located close to the polar head in the micelle (Figure 16.1(b), type-1), while group-2 is in the oil core region (Figure 16.1(b), type-2). The incorporation of $[Cr(phen)_3]^{3+}$ into the hydrophobic interior of micelles can be related with the significantly larger χ_{eff} values for $[Co(phen)_3]^{3+}$ than for the other two complexes. On the other hand, the motional restrictions

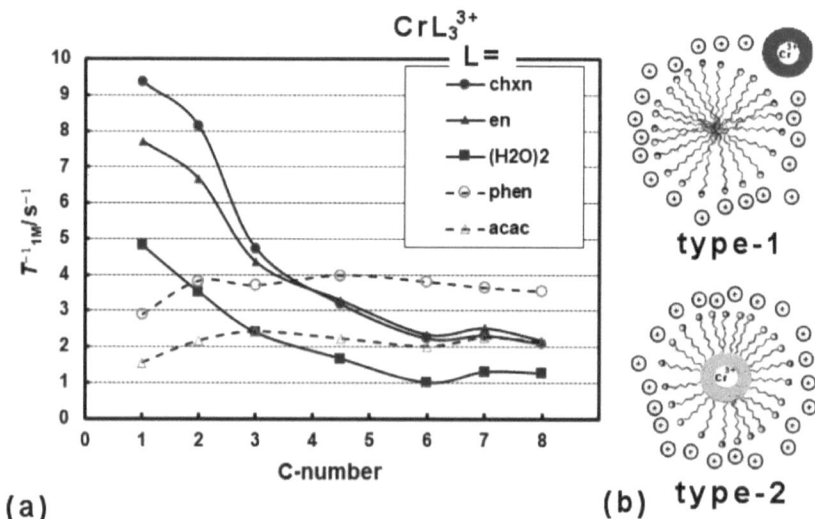

(a)

(b) type-2

Figure 16.1 (a) ^{13}C relaxation rates for each atom of the sodium 1-octanesulfonate micellar solution (0.3 mol dm^{-3}) depending on the kinds of chromium(III) complexes (5 mmol kg^{-1}) at 27 °C. (b) Two typical types of the location of tripositive metal complex ions in the spherical micelles.

(E_a and τ_{eff}) of the metal complexes did not significantly depend on the location of the metal complexes in the micelles.

16.1.1.2 Liquid Crystalline Solutions

At higher concentrations of surfactants, the structure of micelles generally changes to form lyotropic liquid crystals (LLCs), where the interactions between the dissolved components become complex. Among LLCs, a lyotropic nematic liquid crystal (LNLC), basically composed of surfactant/decanol/ aqueous alkali chloride solution, has a characteristic feature of low viscosity for concentrated surfactant solutions.[5] Thus, the LNLC spontaneously aligns in the magnetic field of the NMR spectrometer, where the sample is essentially a single crystal in terms of molecular orientation accompanied by spectrum sharpening.

When the probe ions having the nucleus with $I > 1/2$ interact with the surfactants in the LC phase, the NMR spectrum for the nucleus is split into $2I$ (Figure 16.2(a)), and the quadrupole splitting (QS), $\Delta\nu_Q$, for the nucleus can be expressed as[5-7]

$$\Delta\nu_Q = \frac{3}{2I(2I-1)} S_0 P_b S_b \chi_b \qquad (16.5)$$

where S_0 is a macroscopically oriented order parameter reflecting an angle between the LC axis (director) and the magnetic field, subscript b indicates the corresponding value concerning the binding ion, P is the fraction, and χ is the quadrupole coupling constant.

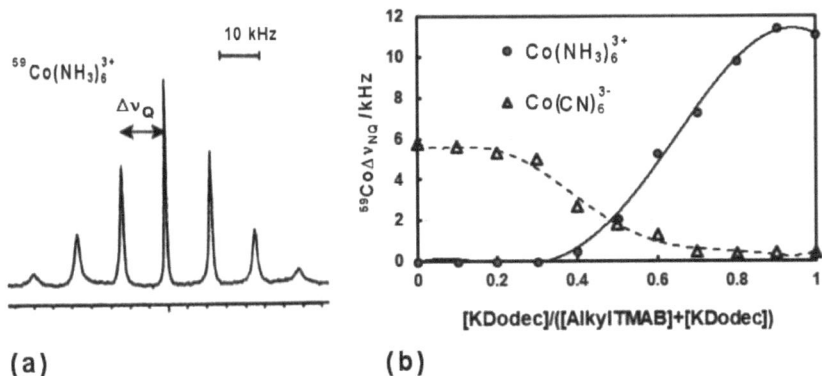

(a)   (b)

Figure 16.2 (a) ^{59}Co ($I = 7/2$) NMR spectrum of $[Co(NH_3)_6]^{3+}$ in an LNLC composed of KDodec. (b) Normalized ^{59}Co QSs ($\Delta\nu_{NQ}(^{59}$Co)/kHz) plotted as a function of the molar ratio of KDodec in the mixed KDodec/alkyl-TMAB LC system. The averaged $\Delta\nu_{1/2}(^1$H) values for the normalization were 6.8 kHz and 6.1 kHz for $[Co(NH_3)_6]^{3+}$ and $[Co(CN)_6]^{3-}$, respectively. Reproduced from ref. 6 with permission from American Chemical Society, Copyright 1991.

The variation in the molecular alignment (director) is mainly attributed to the change in the size of the mesophase and is monitored by the dipole coupling (D_{ij}) between the protons of the hydrocarbon chains. Experimentally, D_{ij} can be observed as a half-width $(\Delta\nu_{1/2}(^1\text{H}))$ of the proton dipole coupled spectral envelope and will reflect the macroscopically oriented order parameter (S_0). That is, the relationship, $\Delta\nu_{1/2}(^1\text{H}) \propto D_{ij} \propto S_0$, is useful to monitor the change in S_0.

On the basis of these principles, the interactions of the tripositive $([\text{Co(NH}_3)_6]^{3+})$ and trinegative $([\text{Co(CN)}_6]^{3-})$ ions with the surfactants in LNLCs were studied by means of multinuclear NMR spectroscopy,[6] where the head groups of the surfactants were appropriately changed.

The $^{59}\text{Co(III)}$ nucleus gave well-resolved $\Delta\nu_Q$ in the LNLC as exemplified in Figure 16.2(a). The compositions of the surfactants were changed from alkyl-trimethylammonium bromide (alkyl-TMAB, where alkyl was varying between C_{12} and C_{16} hydrocarbons to stabilize the phase) to potassium dodecanoate (KDodec) under the condition of the other components (decanol/aqueous aqueous alkali chloride solution) being kept constant. Actually, normalized $^{59}\text{Co QS}$ $(\Delta\nu_{NQ}(^{59}\text{Co})/\text{kHz})$ was obtained for each $\Delta\nu_Q(^{59}\text{Co})$ by using a slightly different $\Delta\nu_{1/2}(^1\text{H})$ at each mole fraction of KDodec and an averaged $\Delta\nu_{1/2}(^1\text{H})$ value over the mole fractions of 0–1.0. Thus, there is a relationship: $\Delta\nu_{NQ}(^{59}\text{Co}) \propto \Delta\nu_Q(^{59}\text{Co})/\Delta\nu_{1/2}(^1\text{H}) \propto P_b S_b \chi_b$. That is, $\Delta\nu_{NQ}(^{59}\text{Co})$ reflects the product of magnitude $(P_b\chi_b)$ and order parameter (S_b) for the cobalt(III) complex ion-binding to the LNLC.

Figure 16.2(b) shows the change in $\Delta\nu_{NQ}$ as a function of the mole fraction of KDodec in the mixed LNLC system. The contrast profiles of titration-like curves for the $[\text{Co(NH}_3)_6]^{3+}$ cation and $[\text{Co(CN)}_6]^{3-}$ anion indicate their strong interactions with the surfactants having the opposite signs of charges. This figure further shows that $[\text{Co(NH}_3)_6]^{3+}$ is approximately one-half bound when 60% of the alkyl-TMAB surface is replaced by KDodec. Exactly the opposite situation is obtained for $[\text{Co(CN)}_6]^{3-}$ which is one-half bound when 60% of KDodec is replaced by alkyl-TMAB. The present binding profiles for the multivalent ions differ from those of other ions such as the alkali-metal and halide ions, where the changing modes in QS were interpreted in terms of specific bindings.[7]

The $^{59}\text{Co QSs}$ were furthermore applied to discriminate the $[\text{Co(en)}_3]^{3+}$ enantiomers in cholesteric LC systems composed of N-dodecanoyl-L-alaninate (L-CsDDA).[8] The $\Delta\nu_{NQ}$ (^{59}Co) values of the Δ-$[\text{Co(en)}_3]^{3+}$ ions were larger than those of the Λ-isomer in both the mixed alkyl-TMAB/L-CsDDA and mixed alkyl-TMAB/L-KDDA LC systems, whereas their ^{59}Co chemical shifts did not exhibit a difference between the two isomers. According to eqn (16.5), this result indicates that the observed QS is sensitive to the geometry of the interactions (directly related to the S_b value) rather than to the parameters in terms of the extent of ion-binding $(P_b\chi_b)$.

16.1.2 Amphiphiles Comprising Ionic Metal Complexes

When colloidal solutions are composed of metallosurfactants, or surfactants containing transition-metals (d-block elements) in the molecular structures, the metal ions highly condense in the colloidal systems. Such kinds of colloidal solutions often exhibit unique properties together with high functionalities,[9–12] which are unattainable in conventional organic surfactant solutions. Another noticeable feature of metallosurfactants is a close relationship between the molecular structure and unique aggregation behavior in solution due to the characteristic structure of metal complex moieties. In this section, relatively simple structural metallosurfactants are targeted and their physical properties in solution are reviewed. In comparison with functionality studies, there have been much fewer studies on the aggregation of metallosurfactants in solution in the light of ion–ion and ion–solvent interactions. There are several typical examples of metallosurfactants from this viewpoint as depicted in Scheme 16.1. They can be classified into the following three categories: (i) single-chain metallosurfactants with tripositive head groups (Scheme 16.1(a) and (b)), (ii) double-chain cationic metallosurfactants (Scheme 16.1(c)–(e)) and (iii) amphiphilic dichromium(III) complexes (Scheme 16.1(f)).

The packing parameter (PP) concept roughly deduces the shape of amphiphilic aggregates in solution from the geometry of the surfactant molecule.[13] On the basis of this rule, there is the following trend. When the head group has larger effective size with higher charge density and the single chain is longer, the surfactant forms a spherical normal micelle in water, while double-chain surfactants having a smaller head group tend to form highly ordered aggregates such as liquid crystals in water and microemulsions in oil/water mixed solvents. Although the PP concept is basically useful, we observed many cases of unusual aggregations of amphiphilic metal complexes arising from their characteristic structures; they were investigated by focusing on the ion–ion and/or ion–solvent interactions for each type of the above groups (i–iii).

16.1.3 Formation of Micelles Comprising Amphiphilic Metal Complexes

16.1.3.1 Tripositive Cobalt(III) Surfactants (Type I)

Several tripositive surfactants having single chain cobalt(III) were synthesized to observe the effects of high positive charge on the aggregation process in water. This type tends to form normal micelles in water as deduced from the PP concept. The alkyl substituted sarcophagine complex was synthesized and showed antibacterial activities in *in vivo* studies.[14,15] Besides, the micellar formations of cobalt(III) complexes having *N*-alkylethylenediamine (alkyl-en, alkyl = C_8 and C_{12}) with 3,7-diaza-1,9-diaminonane (2,3,2-tet) (Scheme 16.1(a))[16] and the alkyl substituted cystamine type surfactant (Scheme 16.1(b)) in water were studied in detail using multinuclear NMR.[17,18]

The ^{59}Co and ^{35}Cl NMR relaxations were measured for the head groups and counter ions, respectively. Under the experimental conditions, both the line widths ($\Delta\nu_{1/2}(^{59}\text{Co}) = 1/(\pi T_2)$) for the surfactants and the $R_1(^{35}\text{Cl})$ ($= 1/T_1$) values for the counter ions were almost proportional to the motional

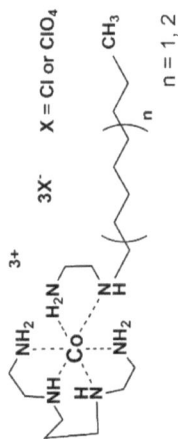

(a) Co(III) 2,3,2-tet complex

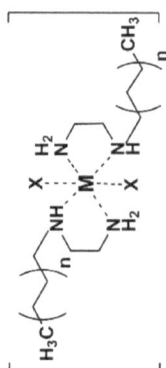

(b) Co(III) cystam complex

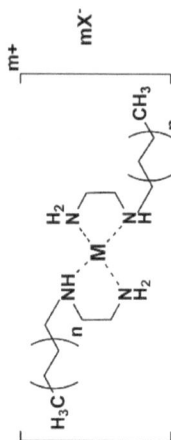

(c) M(alkyl-en)₂ complex

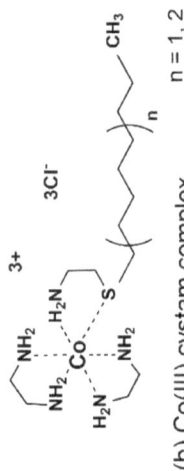

M= Zn; n= 2, 3; X=Cl⁻ or NO₃⁻
M= Cd; n= 3; X=Cl⁻

(d) M(alkyl-en)₂ complex

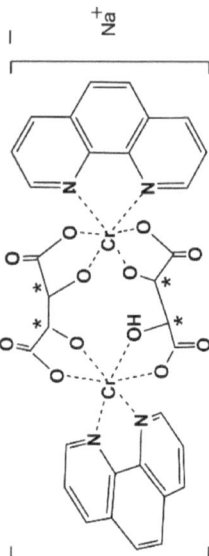

M= Pd; n= 2, 3; m= 2; X=Cl
M= Ag; n= 2, 3, 5, 6; m= 1; X=NO₃

(e) Pd(en)(dialkyl-en) complex

(f) Cr₂(Htart₂)(phen)₂ complex

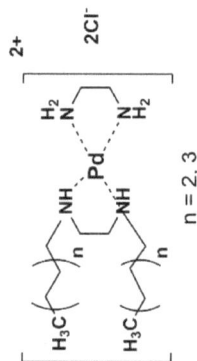

Scheme 16.1 Structures of amphiphilic metal complexes. (a) Co(III) 2,3,2-tet complex, (b) Co(III) cystam complex, (c) Zn(II) or Cd(II) (alkyl-en)₂ complex, (d) Pd(II) (alkyl-en)₂ or Ag(I) (alkyl-en)₂ complex, (e) Pd(II)(en)(dialkyl-en) complex, and (f) Cr(III)₂(HTart₂)(phen)₂ complex.

restrictions (τ_{eff}) at lower concentrations of the surfactants according to eqn (16.1)–(16.3) (here, $(\omega_0\tau_{eff})^2 \ll 1$). As a result, both the $\Delta\nu_{1/2}(^{59}Co)$ and $R_1(^{35}Cl)$ values steadily increased with an increase in the surfactant concentrations, after which at concentrations above the same value, they slightly deviated upward from the linear relationship. This breakpoint of concentration corresponds to the cmc, whose values thus obtained for the tripositive cobalt(III) surfactants are listed in Table 16.1 together with those for the common monovalent surfactants (alkyl = C_8 and C_{12}) for comparison. The cmc values of the tripositive cobalt(III) surfactants are comparable to those of the monovalent surfactants and are smaller than expected from their stronger electrostatic repulsions between the headgroups. Regarding the counter chloride ions, even below the cmc, the $R_1(^{35}Cl)$ values were significantly larger than those for the simple electrolyte solutions. This result indicates the strong counter-ion bindings of Cl$^-$ to the head groups. For the tripositive surfactants, therefore, the strong ion-bindings of the counter ions reduce the net charge of the head group and consequently gave the unexpected smaller cmc values. In addition, the larger-size head group of the metallosurfactant generates a larger van der Waals attraction between the head groups; this attraction further reduces the repulsion between the head groups and then decreases the cmc.

A closer look at Table 16.1 indicates that the cmc values of the cystam complexes are significantly larger than those of the 2,3,2-tet complexes and of the monovalent surfactants in both the C_8 and C_{12} systems. The smaller cmc values of the 2,3,2-tet complexes than those of the cystam complexes can also be attributed to the larger van der Waals attraction of the 2,3,2-tet complex owing to the additional propyl group (Scheme 16.1(a) and (b)). Therefore, for the 2,3,2-tet complexes the specific interactions are effective

Table 16.1 Critical micellar concentrations (cmc) for the tripositive cobalt(III) complexes in aqueous solutions in comparison with the common surfactants having the same alkyl chains. Adapted from ref. 17 and 18 with permission from The Chemical Society of Japan, Copyright 1996 and 2000.

Surfactant	cmc/mol kg^{-1}	T/°C
(1) Alkyl = C$_8$		
2,3,2-Tet Co(III) complex perchlorate (Scheme 16.1(a))	0.12	27
Cystam Co(III) complex chloride (Scheme 16.1(b))	0.30	27
Sodium octyl sulfate	0.14[a]	25
Sodium 1-octane sulfonate	0.16[a]	40
Octyl trimethyl ammonium chloride (OTMAC)	0.18[a]	25
(2) Alkyl = C$_{12}$		
2,3,2-Tet Co(III) complex chloride (Scheme 16.1(a))	0.010	27
2,3,2-Tet Co(III) complex perchlorate (Scheme 16.1(a))	0.008	27
Cystam Co(III) complex chloride (Scheme 16.1(b))	0.020	27
Sodium dodecyl sulfate (SDS)	0.008[a]	25
Sodium 1-dodecane sulfonate	0.012[a]	25
Dodecyl trimethyl ammonium chloride (DTMAC)	0.015[a]	25

[a] mol dm^{-3}, the difference in the units has no effect on the listed values.

on the cmc. Regarding the counter anion effect on the cmc of the 2,3,2-tet complex, a slightly smaller cmc is observed for the perchlorate salt than for the chloride salt. This is attributable to the stronger counter ion binding of the perchlorate ion despite the less electrostatic interaction by the larger size, since the hydrophobic perchlorate ion more strongly interacts with the hydrophobic 2,3,2-tet complex than the hydrophilic chloride ion.

16.1.3.2   Other Multipositive Single-chain Metallosurfactants

A lot of metallosurfactants with aza-macrocycle ligands were synthesized and their physical properties were widely studied.[10,11] The containing metals are lanthanides or first-row transition metals, and thus the surfactants have multipositive charges. They exhibited micellizations like conventional organic surfactants depending on the alkyl chain length, whereas their aggregation properties were sensitive to the head group structure as well. It is notable that their cmc values are significantly smaller than those of the common monocharged surfactants despite the multipositive head groups. This unexpected trend is due to the specific attraction between the hydroxy groups in the head group as well as to the counterion condensation.

16.1.4   Microemulsions and Lyotropic Liquid Crystals Formed from Alkyl-en Metal Complexes (Type II)

The double-chain surfactants containing metal (d-block element) ions are expected to provide more various aggregation structures than the single-chain surfactants because of the higher-ordered aggregation expected from the PP concept (Section 16.1.2) and from the moderate hydrophile–lipophile balance (HLB). Both of the features are realized in common organic surfactants such as sodium bis(2-ethylhexyl)sulfosuccinate (Aerosol-OT) and bis(2-ethylhexyl)-phosphate, which are good emulsifiers in water/oil mixed solvents. Concerning the metallosurfactants, the relationship between the molecular structure and the aggregation behavior was widely and systematically studied by using the alkyl-en complexes (Scheme 16.1(c)–(e)). A series of bis(alkyl-en) Zn(II), Cd(II), Pd(II)[19-24] and Ag(I)[25,26] complexes were synthesized from this viewpoint and their aggregations in water and water-containing organic solvents were studied by focusing on the ion–ion and ion–solvent interactions. As all of these metal complexes are diamagnetic, they are convenient to study a variety of aggregations in solution using NMR. The alkyl-en is regarded as the simplest chelate-ligand having hydrophobic moieties.

16.1.4.1   Double-alkyl Chained Zn(II), Cd(II) and Pd(II) en Complexes

Among the divalent complexes (Scheme 16.1(c)–(e)), the molecular structures of dichloro- and dinitrato-bis(oct-en)zinc(II) and of bis(oct-en)palladium(II) chloride complexes were determined by X-ray crystallography.[19,20] In the zinc(II) complexes, the anion groups and octyl chains take a *trans* configuration and a transoid conformation, respectively (Scheme 16.1(c)) in the octahedral molecular structure, while in the palladium(II) complex the two chloride ions do not coordinate to the metal. The dichloro- and dinitrato-bis(oct-en) zinc(II) complexes are sparingly soluble or insoluble in neat benzene (Bz) and water; however, their solubilities are appreciably enhanced in H_2O/Bz mixed solvents, where microemulsions (ME) are extensively formed. Figure 16.3(a) shows the one-phase diagram of $[ZnCl_2(oct-en)_2]/H_2O$/Bz. The structures of the microemulsions were investigated using freeze-fracture transmission electron microscopy (TEM), atomic force microscopy (AFM), and ^1H NMR pulse gradient spin-echo (PGSE) for self-diffusion. The self-diffusions of the metallosurfactant indicated the appreciable aggregations of the complex in the solutions. In addition, the TEM photographs (Figure 16.3(b)) directly revealed the microemulsion structures depending on the water contents. The w/o microemulsion sponge and bicontinuous structures were observed for the one phase in the ternary system of $[ZnCl_2(oct-en)_2]$ as shown in Figure 16.3(a) and (b). At the interface of the Bz and H_2O domains in the bicontinuous structure, the hydrocarbon chains are likely to be arranged as cisoid geometry (Figure 16.3(b) and (c)) despite the transoid (Scheme 16.1(c)) in the crystal state.

The solubilities of $[Zn(NO_3)_2(oct-en)_2]$, $[CdCl_2(oct-en)_2]$, and $[Pd(oct-en)_2]Cl_2$ in water/chloroform (ClF) and in water/Bz mixed solvents were also studied to relate the one-phase regions with the ionicities of the metal–anion bonds and with the hydrophilicities of the complexes. The phase diagrams for the one-phase ME region in the ternary solubility systems are shown in Figure 16.4(a)–(c). The most hydrophilic $[Pd(oct-en)_2]Cl_2$ is insoluble in Bz even in the presence of a small amount of water. The closest patterns between the $ZnCl_2$ (Figure 16.3(a)) and $CdCl_2$ complex (Figure 16.4(b)) systems indicate the similar bindings between the Zn–Cl and Cd–Cl bonds. The electrical conductivities of the ME were in the order $[Pd(oct-en)_2]Cl_2 \gg [Zn(NO_3)_2(oct-en)_2] > [ZnCl_2(oct-en)_2] > [CdCl_2(oct-en)_2]$. This order reflects the ionicity of the metal–anion bond, *i.e.*, Pd–Cl $\gg$ Zn–NO_3 > Zn–Cl > Cd–Cl. The less water incorporation for $[Zn(NO_3)_2(oct-en)_2]$ than for $[ZnCl_2(oct-en)_2]$ (comparing Figure 16.4(a) with Figure 16.3(a)) is due to the smaller hydration of the nitrate ion than that of the chloride ion, while the less solubility of $[Zn(NO_3)_2(oct-en)_2]$ in benzene than that of $[ZnCl_2(oct-en)_2]$ is attributable to the higher ionicity of the Zn–NO_3 bond than the Zn–Cl bond.

Unlike the oct-en complexes, $[ZnCl_2(hex(C_6)-en)_2]$ and $[Pd(hex-en)_2]Cl_2$ (Scheme 16.1(c) and (d), respectively; $n = 2$) are hydrophilic and well dissolved in water. The aggregations of $[ZnCl_2(hex-en)_2]$ in water and H_2O/ClF mixed solvent and of $[Pd(hex-en)_2]Cl_2$ in water were studied using vapor pressure osmometry (VPO) and ^1H NMR PGSE for the metal complexes, and ^{35}Cl NMR relaxation rates and the chloride ion-selective electrode for the counter ions.[23,24]

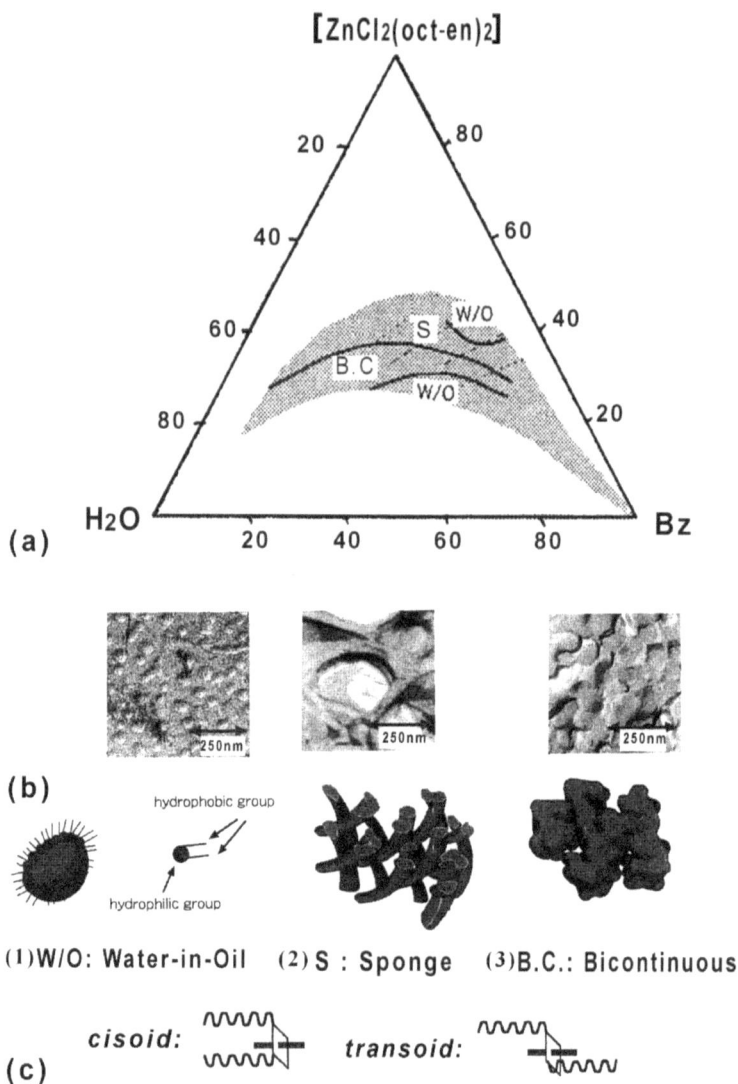

[ZnCl2(oct-en)2]

(a)

(b)

(1)W/O: Water-in-Oil   (2)S : Sponge   (3)B.C.: Bicontinuous

cisoid: *transoid:*

(c)

Figure 16.3   (a) Ternary solubility diagram for the [ZnCl$_2$(oct-en)$_2$]/water/benzene
(Bz) system. (b) Freeze-fracture TEM photographs from [ZnCl$_2$(oct-en)$_2$]/Bz = 50/50 solution with (1) 15 wt% water content, (2) 20 wt% water content and (3) 50 wt% water content. (c) Two conformations. Reproduced from ref. 21 with permission from American Chemical Society, Copyright 2000.

For [ZnCl$_2$(hex-en)$_2$] in water at concentrations above 0.25 mol kg^{-1}, about 15% of the chloride ions dissociated from the Zn(II) ion and the self-aggregation number of the complex was estimated to be 8.5 at 25 °C. Although the cmc in water was not able to be directly obtained owing to the hydrolysis below 0.2 mol kg^{-1}, an extrapolation of the vapor pressure depression

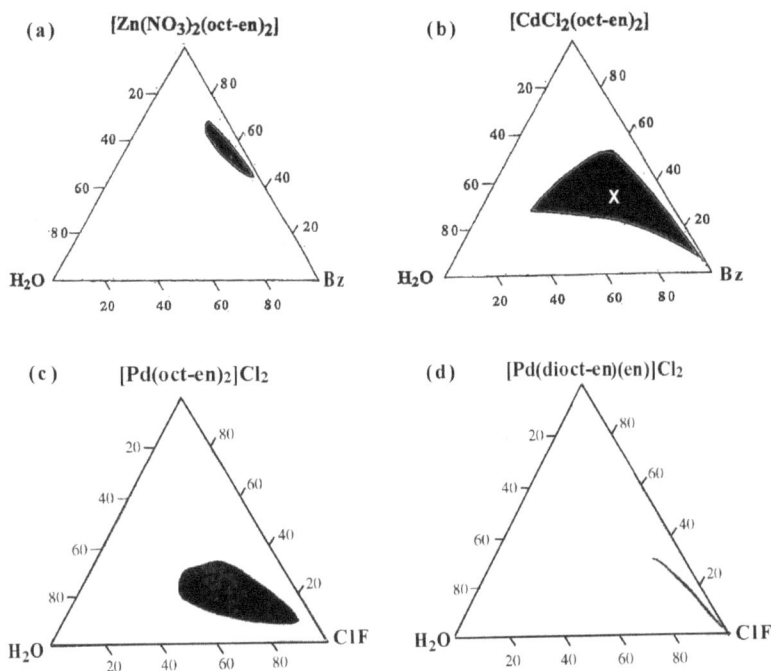

Figure 16.4 Ternary solubility diagrams for (a) [Zn(NO₃)₂(oct-en)₂]/H₂O/Bz, (b) [CdCl₂(oct-en)₂]/H₂O/Bz (for point X, see Problem 4), (c) [Pd(oct-en)₂]Cl₂/H₂O/ClF, and (d) [Pd(dioct-en)(en)]Cl₂/H₂O/ClF. Reproduced from ref. 22 and 23 with permission from Elsevier, Copyright 2003 and 2007.

vs. concentration (above 0.25 $mol\,kg^{-1}$) plot to the lower concentrations found the cmc to be approximately 0.10 $mol\,kg^{-1}$. In H₂O/ClF mixed solvent $(w_0 = [H_2O]/([H_2O] + [ClF]) = 1.0{-}1.4)$, the hydrolysis did not occur and the aggregation number was 6.8 with a cmc of 0.56 $mol\,kg^{-1}$ at 25 °C.

On the other hand, [Pd(hex-en)₂]Cl₂ is a strong electrolyte and is stable in water. The aggregation number was 4.6 with a cmc of 0.4 $mol\,kg^{-1}$. The differences in the aggregation behavior between [ZnCl₂(hex-en)₂] and [Pd(hex-en)₂]Cl₂ in water and in H₂O/ClF mixed solvent can be explained by the higher hydrophobicity of the Zn(ɪɪ) complex because of the low dissociation of the chloride ions.

16.1.4.2 The Aggregations of [Pd(N,N′-dialkyl-en)(en)]Cl₂ Complexes in Water/Organic Mixed Solvents

For comparison with the bis(alkyl-en)palladium(ɪɪ) complexes, (N,N′-dialkyl-(alkyl = hex and oct)-en)(en))palladium(ɪɪ) complexes (Scheme 16.1(e)) were synthesized, after which the solubilities in various neat solvents and water/organic mixed solvents were investigated.[23] As shown in Scheme 16.1(d) and (e), the [Pd(dialkyl-en)(en)]Cl$_2$ complex has the same molecular formula as [Pd(alkyl-en)$_2$]Cl$_2$, while the geometrical structure is different. It was clarified that the [Pd(dialkyl-en)(en)]Cl$_2$ complexes (alkyl = hex and oct) were significantly more hydrophobic than the [Pd(alkyl-en)$_2$]Cl$_2$ complexes (alkyl = hex and oct), as illustrated in the ternary solubility diagrams of [Pd(dioct-en)(en)]/ H$_2$O/ClF (Figure 16.4(d)). The solubilities of [Pd(dialkyl-en)(en)]Cl$_2$ in H$_2$O/ClF mixed solvents were very similar between the alkyl = hex (not described here) and oct complexes, differently from those of the bis(alkyl-en) complexes. In the narrow one-phase regions of both the [Pd(dioct-en)(en)]/H$_2$O/ClF and [Pd(dihex-en)(en)]/H$_2$O/ClF systems, the reverse micelles form and the obtained aggregation numbers at $w_0 = 12$ were 6.8 and 7.2, respectively, which are larger than that (4.6) for [Pd(hex-en)$_2$]Cl$_2$ in water. This difference also reflects the higher hydrophobicity of [Pd(dialkyl-en)(en)]Cl$_2$ compared to [Pd(alkyl-en)$_2$]Cl$_2$. The present results are notable since the geometrical effects on the solubilities and aggregation properties are significantly larger than the effects of alkyl chains.

16.1.4.3   The Aggregations of Other Divalent Bis(alkyl-en) Complexes in Solution

Other divalent transition metal complexes such as Cu(II)[27] and Pt(II)[28] of alkyl-en were prepared, after which the aggregation behavior[27] and the biological activities[28] were studied, respectively. Out of them, it is remarkable that CuX$_2$(alkyl-en)$_2$ (X = Cl$^-$ or ClO$_4^-$, alkyl = C$_8$–C$_{18}$) forms stable bilayer membranes in dilute aqueous dispersions and their aggregation structures are largely affected by the counter ions. When X = Cl$^-$, the Cu^{2+}-coordinated head groups take a compressed tetrahedral CuN$_4$ structure with a *trans* configuration. Alternatively, when X = ClO$_4^-$, Cu^{2+} gives a square planar CuN$_4$ structure with a *cis* configuration, where the two alkyl chains are located at the same side of the plane and parallel to each other, *i.e.* cisoid conformation.

16.1.4.4   The Aggregations of Bis(alkyl-en)Ag(I) Complexes in Water/Organic Mixed Solvents

The aggregation behavior of [Ag(alkyl-en)$_2$]$^+$ is significantly different from that of the divalent alkyl-en complexes in both the solutions and neat states. The molecular structure of [Ag(dod-en)$_2$]NO$_3$ was determined to be a dinuclear complex by X-ray crystallography.[26] Additionally, the hex-en, oct-en and C$_{16}$ (or hexd)-en complexes were prepared and isolated. Owing to the lower electric charge, the [Ag(alkyl-en)$_2$]NO$_3$ complexes are hydrophobic compared to the divalent complexes having the same alkyl-en ligands.

The [Ag(hexd-en)$_2$]NO$_3$H$_2$O crystal was firstly isolated and identified.[25] This derivative is the most hydrophobic and dissolves in *n*-heptane. The 1.5–9 mM heptane solutions created Ag-nanoparticles (NPs) by the reduction using NaBH$_4$ aqueous solution. This result indicates that the solution forms a reverse micelle where a small water pool is generated.

A variety of morphologies of the aggregates were observed in the [Ag(oct-en)$_2$]NO$_3$ and [Ag(dod-en)$_2$]NO$_3$ systems.[26] These complexes are soluble in *n*-heptane/water mixed solvents. The phase diagrams for the ternary solubility systems of [Ag(oct-en)$_2$]NO$_3$/*n*-heptane/water and [Ag(dod-en)$_2$]NO$_3$/*n*-heptane/water are depicted in Figure 16.5(a) and (b), respectively. The aggregations of these complexes in the one-phase solutions were studied mainly using ^{1}H NMR PGSE self-diffusions of the Ag-complex, which revealed high aggregation of the Ag-complex, suggesting the formation of W/O microemulsions. It is remarkable that in the lower water content region (smaller w_0) for [Ag(dod-en)$_2$]NO$_3$ formation of lyotropic LCs was clarified as given in the phase diagram (Figure 16.5(b)). Otherwise, in the neat states the melting points of [Ag(oct-en)$_2$]NO$_3$ and [Ag(dod-en)$_2$]NO$_3$ are below 100 °C; at around this temperature the formation of thermotropic LCs was observed, *i.e.*, they are also categorized as ionic liquids (Chapter 15).

16.1.5 Lyotropic Liquid Crystals Composed of Dinuclear Chiral Chromium(III) Complexes (Type III)

The other type of aggregation of metal complexes in solution arises from π–π stacking interactions in the planar-shaped complexes with aromatic group molecules, *e.g.* porphyrin derivatives and dyes. The higher-ordered aggregation by the stacking reaches a discotic liquid crystal, which was identified later than the alkyl-chained LCs. Metal complexes offer many kinds of the stacking aggregates.[9–11] Above all, an uncommon molecular structure of the chiral dinuclear chromium(III) complex gave a unique type lyotropic liquid

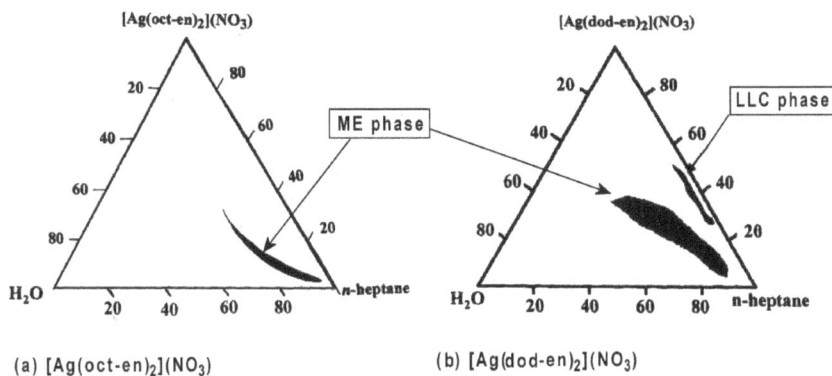

Figure 16.5 The one phase formation in the (a) [Ag(oct-en)$_2$]NO$_3$/*n*-heptane/water and (b) [Ag(dod-en)$_2$]NO$_3$/*n*-heptane/water systems, where ME is a microemulsion and LLC is a lyotropic LC. Reproduced from ref. 26 with permission from John Wiley & Sons, Copyright 2004.

crystal (LLC) in water.[29,30] The molecular structure is shown in Scheme 16.1(f), where the complex has two twisted phenanthrolines and bridging L-tart$_2$H (tart = tartrate^{4-}). It has neither a conventional amphiphilic structure like a polar head group and hydrophobic tail, nor a planar aromatic core with substituents. The formation of the LLC in water was studied in detail and it was revealed to be a particularly novel type aggregate from a viewpoint of the intermolecular interactions. The phen ligands play the role of the hydrophobic moiety, while the tart bridge acts as the hydrophilic one. At room temperature, the dinuclear chromium(III) complex highly aggregates at low concentrations in water (6×10^{-3} mol dm^{-3}) to form LLCs. An anisotropic structure was evidenced by the quadrupole splittings of NMR spectra for the ^{23}Na$^+$ counter ion and ^{2}H$_2$O solvent as well as with a polarizing microscope. The LLC phase is formed in the 6×10^{-3}–0.2 mol dm^{-3} region and remains stable up to 100 °C. The CD spectra of the LLC phase indicate the presence of large chiral domains having a macroscopic helical pitch. Furthermore, the detailed structure was later clarified using cryo-TEM, atomic force microscopy (AFM) and two-dimensional small-angle neutron scattering.[31] Figure 16.6 illustrates the proposed helix and nematic LC

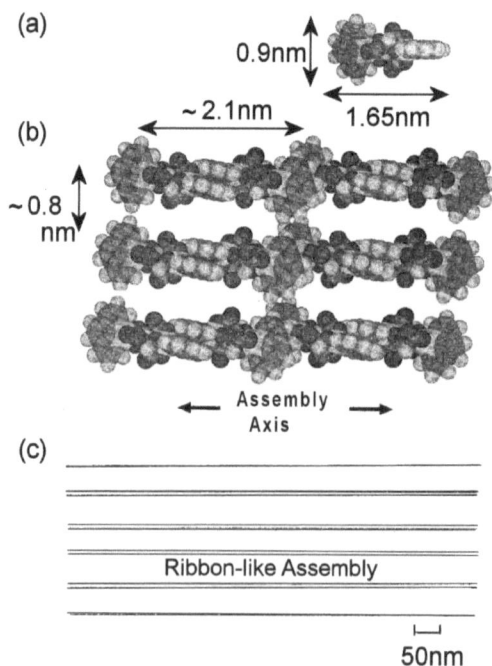

Figure 16.6 The schematic representation of (a) the Na[Cr$_2$(L-tart$_2$H)(phen)$_2$] molecule, (b) the molecular arrangement in a ribbon-like assembly and (c) the assembly arrangement in a liquid-crystalline phase. Reproduced from ref. 31 with permission from American Chemical Society, Copyright 1998.

structure along the molecular axis. Very long ribbon-like assembly consists of a lamellar structure with chromium(III) monolayers, which are separated by water layers. As the tartrate part has chirality (Scheme 16.1(f)), the complex is not planar and the molecules aggregate with each other by the π–π stacking between the phen rings. The structure was largely changed by the tripositive counter ion of $[Co(en)_3]^{3+}$,[32] where the Cr(III) complex did not exhibit the LC structure since the ribbon-like structures become significantly thinner and gather like strings or roll spirally.

Problem 1

The motional restriction of the cobalt(III) complex in solution is reflected in τ_{eff} in eqn (16.3). When the molecular restriction is small enough ($\tau_{eff} \ll \tau_0$), the T_1 and T_2 values are almost the same based on eqn (16.1)–(16.3). In the present system (Section 16.1.1.1), the ^{59}Co NMR spectra were measured at 64.6 MHz. Estimate the τ_0 value.

Problem 2

Cobalt(III) and chromium(III) complexes have several common and different properties besides the tripositive charge. Thus, we can make a comparison between the ^{59}Co NMR parameters of cobalt(III) complexes in micelles and the ^{13}C NMR spectra of the surfactants of micelles in the presence of chromium(III) complexes. List the characteristic common and different points between the cobalt(III) and chromium(III) complexes with a short explanation based on the d-electron configuration for each metal(III) ion.

Problem 3

When a nucleus with $I > 1/2$ is under an anisotropic field, the NMR spectrum is split into $2I$. In the case of $I = 1$ like ^2H, sketch the energy level diagrams and the corresponding NMR spectra with a short explanation for the following cases: (1) only Zeeman interaction and (2) the combined Zeeman and quadrupolar interactions, where the ^2H is contained in a liquid crystalline system or in an isotropic solution.

Problem 4

Figures 16.3–16.5 show the three-component phase diagrams. Using Figure 16.4(b), for example, describe the way how to estimate the composition (at X) for each component. For brevity, use the following abbreviations for each component and composition (%): $[CdCl_2(oct-en)_2] \rightarrow$ A and a%, $H_2O \rightarrow$ B and b%, and Bz $\rightarrow$ C and c%, respectively.

Answer to Problem 1

According to eqn (16.1)–(16.3), T_1 and T_2 are almost the same under the condition $\omega_0 \tau_{eff} \ll 1$. As $\omega_0 = 2\pi\nu_0 = 4.1 \times 10^8$ Hz $= 4.1 \times 10^8$ s^{-1}, the condition is $\tau_{eff} \ll 2.5$ ns $= \tau_0$.

Answer to Problem 2

Common features: octahedral amine complexes of both the metal ions are kinetically inert and stable in aqueous solutions. They form a low-spin complex, *i.e.*, the electronic configurations are $3d^3(t_{2g}^3 e_g^0)$ and $3d^6(t_{2g}^6 e_g^0)$ for the Cr(III) and Co(III) complexes, respectively. In both cases, the electrons occupy only the bonding orbitals (t_{2g}), so that the metal(III)–nitrogen bonds are strong.

Different features: for Cr(III) the t_{2g}^3 electrons are all unpaired electrons, while for Co(III) the t_{2g}^6 electrons are all paired electrons. Therefore, the former complexes are paramagnetic and the latter ones are diamagnetic.

Answer to Problem 3

In a liquid crystal, the spectrum is like (2) in Figure 16.7.

In solution, rapid isotropic tumbling of molecules containing the quadrupolar nucleus leads to an averaging of quadrupolar interactions, and in effect $\Delta\nu_Q$ becomes zero, *i.e.*, the spectrum is like (1) in Figure 16.7.

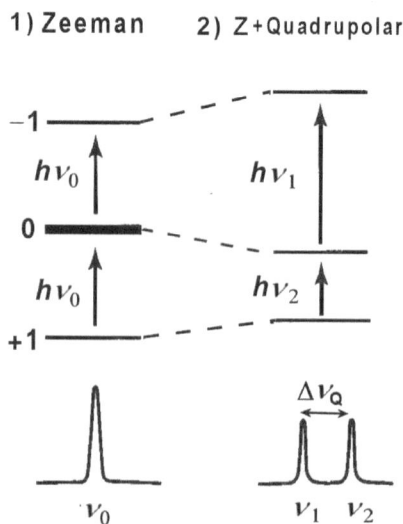

Figure 16.7   The energy level diagrams and the corresponding NMR spectra for (1) Zeeman interaction and (2) the combined Zeeman and quadrupolar interactions. Reproduced from ref. 33 with permission from Hirokawa Publishing, Copyright 1999.

Answer to Problem 4

Each of the three vertices is identified with 100% of that component, while the opposite line is with 0% of that component. Therefore, the vertical distances from X to the sides BC, AC, and AB correspond to the ratios of a, b, and c, respectively, where $a + b + c = 100\%$.

References

1. D. F. Evans and H. Wennerström, *The Colloidal Domain*, Wiley-VCH, New York, 2nd edn, 1999, ch. 1, pp. 8–10.
2. J. Mason, *Multinuclear NMR*, Plenum Press, New York, 1987, ch. 1–5, pp. 1–169.
3. Y. Mizuno and M. Iida, *J. Phys. Chem. B*, 1997, **101**, 3919.
4. C. Chachaty, *Prog. Nucl. Magn. Reson. Spectrosc.*, 1987, **19**, 183.
5. B. J. Forrest and L. W. Reeves, *Chem. Rev.*, 1981, **81**, 1.
6. M. Iida and A. S. Tracey, *J. Phys. Chem.*, 1991, **95**, 7891.
7. A. S. Tracey and T. L. Boivin, *J. Phys. Chem.*, 1984, **88**, 1017.
8. M. Iida, Y. Mizuno and N. Koine, *Bull. Chem. Soc. Jpn.*, 1995, **68**, 1337.
9. B. Donnio, *Curr. Opin. Colloid Interface Sci.*, 2002, **7**, 371.
10. P. C. Griffiths, I. A. Fallis, T. Chuenpratoom and R. Watanesk, *Adv. Colloid Interface Sci.*, 2006, **122**, 107.
11. P. C. Griffiths, I. A. Fallis, T. Tatchell, L. Bushby and A. Beeby, *Adv. Colloid Interface Sci.*, 2008, **144**, 13.
12. R. Kaur and S. K. Mehta, *Coord. Chem. Rev.*, 2014, **262**, 37.
13. J. N. Israelachvili, *Intermolecular and Surface Forces*, Elsevier Inc., Amsterdam, 3rd edn, 2011, ch. 20, pp. 538–559.
14. C. A. Behm, I. I. Creaser, B. K. Daszkiewicz, R. J. Geue, A. M. Sargeson and G. W. Walker, *J. Chem. Soc., Chem. Commun.*, 1993, 1844.
15. G. W. Walker, R. J. Geue, A. M. Sargeson and C. A. Behm, *Dalton Trans.*, 2003, 2992.
16. M. Yashiro, K. Matsumoto, N. Seki and S. Yoshikawa, *Bull. Chem. Soc. Jpn.*, 1993, **66**, 1559.
17. M. Iida, M. Yamamoto and N. Fujita, *Bull. Chem. Soc. Jpn.*, 1996, **69**, 3217.
18. M. Iida, A. Sakamoto, T. Yamashita, K. Shundoh, S. Ohkawa and K. Yamanari, *Bull. Chem. Soc. Jpn.*, 2000, **73**, 2033.
19. M. Iida, T. Tanase, N. Asaoka and A. Nakanishi, *Chem. Lett.*, 1998, 1275.
20. M. Iida, H. Er, N. Hisamatsu and T. Tanase, *Chem. Lett.*, 2000, 518.
21. Y. Ikeda, T. Imae, J. C. Hao, M. Iida, T. Kitano and N. Hisamatsu, *Langmuir*, 2000, **16**, 7618.
22. H. Er, N. Asaoka, N. Hisamatsu, M. Iida and T. Imae, *Colloids Surf., A*, 2003, **221**, 119.
23. H. Er, S. Ohkawa and M. Iida, *Colloids Surf., A*, 2007, **301**, 189.
24. M. Iida, K. Asayama and S. Ohkawa, *Bull. Chem. Soc. Jpn.*, 2002, **75**, 521.
25. A. Manna, T. Imae, M. Iida and N. Hisamatsu, *Langmuir*, 2001, **17**, 6000.

26. M. Iida, M. Inoue, T. Tanase, T. Takeuchi, M. Sugibayashi and K. Ohta, *Eur. J. Inorg. Chem.*, 2004, 3920.
27. C. Li, X. Lu and Y. Liang, *Langmuir*, 2002, **18**, 575.
28. B. Miller, S. Wild, H. Zorbas and W. Beck, *Inorg. Chim. Acta*, 1999, **290**, 237.
29. N. Koine, M. Iida, T. Sakai, N. Sakagami and S. Kaizaki, *J. Chem. Soc., Chem. Commun.*, 1992, 1714.
30. M. Iida, T. Sakai, N. Koine and S. Kaizaki, *J. Chem. Soc., Faraday Trans.*, 1993, **89**, 1773.
31. T. Imae, Y. Ikeda, M. Iida, N. Koine and S. Kaizaki, *Langmuir*, 1998, **14**, 5631.
32. Y. Ikeda, T. Imae, M. Iida, N. Koine and S. Kaizaki, *Langmuir*, 2001, **17**, 361.
33. H. Akutsu and K. Kohno, *Multinuclear NMR*, Hirokawa Bookstore, Japan, 1999, ISBN 4567184203.

CHAPTER 17

Inorganic Biochemistry

AKIRA ODANI

Emeritus Professor, Graduate School of Medical Sciences, Kanazawa
University, Kanazawa, 920-1192, Japan
Email: odani1031@gmail.com

17.1 The Interface Between Inorganic Biochemistry and Solution Coordination Chemistry

"Inorganic biochemistry" or "bioinorganic chemistry" is a new research field that seeks to elucidate why metal ions are essential for the activity of many proteins from the viewpoint of inorganic chemistry.[1] Later, it was discovered that metal ions are essential for the activity of cytochrome *c* oxidase, which converts oxygen to water, the source of biological energy, and for photosynthetic substances that produce oxygen from water, making it clear that the origin of the word "inorganic," which has nothing to do with life, was a mistake. The importance of metals in life phenomena is ubiquitous, and the field of "inorganic biochemistry" or "bioinorganic chemistry" is now expanding from metalloproteins to life science such as pharmaceuticals and genes. Here, we describe a boundary field adjacent to solution coordination chemistry, which is not covered in *"Bioinorganic Chemistry,"* a book selected by the Japan Society of Coordination Chemistry (JSCC), and may also be called "inorganic biochemistry" in this chapter.

Since the concentration of metals in human serum is correlated with the concentration of metals in the sea, it is believed that the life originated from the oceans. This is the result of the interaction of metal ions and substances such as proteins in aqueous solution during the development of life, and the fact that metal ions are soluble in water is related to the fact that the chemical reactions of life take place in aqueous solution. How metal ions

Coordination Chemistry Fundamentals Series No. 2
Metal Ions and Complexes in Solution
Edited by Toshio Yamaguchi and Ingmar Persson
© Japan Society of Coordination Chemistry 2024
Published by the Royal Society of Chemistry, www.rsc.org

bind to ligands and how they perform their functions is a problem of solution complex chemistry. The evolution of life by creating partitions called membranes has made the whole system non-equilibrium rather than equilibrium, but each part can be treated as an equilibrium system, so the research methods and results of solution complex chemistry can be applied to this system. A simple method for the study of metalloenzymes using simple low-molecular-weight metal complexes was widely used in the early inorganic biochemistry, in which the methods of solution chemistry were frequently used. In recent years, there have been increasing cases of studying metalloproteins themselves due to the fact that the coordination structure model of metalloproteins often does not reflect the actual properties and advances in technology (X-ray crystal structure analysis, *etc.*) make direct observations. However, attempts to treat proteins and nucleic acids themselves from the viewpoint of solution chemistry remain a challenge for the future.

17.2   Study of Weak Interactions is Important for Substance Recognition

Weak interactions between substances play an important role in substance recognition (enzyme–substrate, antigen–antibody, drug–receptor, *etc.*) which is essential for biological functions. However, because the interactions are weak and difficult to detect in solution, they are often discussed based on the structures in crystals obtained by X-ray crystallography, and studies from solution chemistry could be essential to see the true dynamic state.[2] The method of using mixed ligand complexes and their stability constants for systematic studies in model systems has the advantage that the weak interactions between ligands stabilize the complexes, the strength of the interactions and their structural dependence can be easily determined, and only different ligands could be added in a substitution-active metal system. The composition and stability constants of mixed ligand complexes in aqueous solution can be easily and accurately obtained using the pH titration method, providing information on the interactions. Metal ions can be used not only to bind the ligands but also as probes to reflect the interactions. This method using metal ions as a probe is useful not only for the study of intramolecular interactions, but also for the study of intermolecular associations.[3]

17.2.1   Base Selectivity of Nucleic Acid Intercalators

An intercalator that inserts its own aromatic ring into the nucleic acid bases of the DNA double helix is also an anticancer drug. Table 17.1 shows the values of the equilibrium constant (association constant) K, enthalpy change ΔH, and entropy change ΔS of the association between the platinum(II) planar complex, which is one of the nucleic acid intercalators, and the

Table 17.1 Stabilizations of adduct formations between the Pt(ɪɪ) complex and nucleic acid component involving the aromatic ring stacking interaction, $\log K$, ΔH, and ΔS (25 °C, ionic strength $I = 0.1$ (NaCl)).[1][a]

X–Y	$\log K$	$\Delta H / \text{kJ mol}^{-1}$	$\Delta S / \text{J K}^{-1} \text{mol}^{-1}$
X-adenosine			
$X = [\text{Pt(phen)(en)}]^{2+}$	2.21	−15.8	−11
$X = [\text{Pt(bpy)(en)}]^{2+}$	1.91	−14.2	−11
$X = [\text{Pt(bpy)(gly)}]^{+}$	1.72		
$X = \text{phen}$	1.33		
$[\text{Pt(phen)(en)}]^{2+}$–Y			
$Y = \text{AMP}^{2-}$	2.51	−25.6	−38
$Y = \text{GMP}^{2-}$	2.49	−26.2	−40
$Y = \text{IMP}^{2-}$	2.34	−11.9	−5

[a] 1-bpy = 2,2′-bipyridine, gly = glycinate, AMP^{2-} = adenosine-5′-monophosphate, GMP^{2-} = guanosine-5′-monophosphate, IMP^{2-} = inosine-5′-monophosphate.

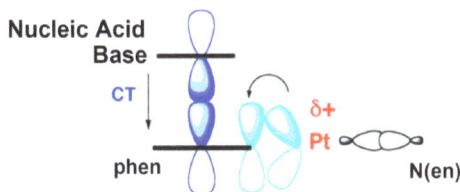

Figure 17.1 Characteristic face-to-face aromatic ring stacking interaction in Pt(ɪɪ) complexes.[11]

nucleic acid component.[2] The $\log K$ with the $[\text{Pt(phen)(en)}]^{2+}$ complex (phen = 1,10-phenanthroline, en = ethylenediamine) is larger than that with the free phen, suggesting that the coordination of Pt(ɪɪ) to phen is important for the interaction with NMP (NMP = nucleotide monophosphate). The $\log K$ values of the association constants between the $[\text{Pt(phen)(en)}]^{2+}$ complex and mononucleotides in Table 17.1 are similar to each other, but the values of ΔH are significantly different. The association with AMP^{2-} and GMP^{2-} containing NH_2 groups in the base exhibited a large negative ΔH value (-26 kJ mol^{-1}). This indicated that the electron-donating nature of the NH_2 group enhanced the association.

The mode of interaction between the $[\text{Pt(phen)(en)}]^{2+}$ complex and the mononucleotide is a face-to-face aromatic ring stacking from the X-ray crystallography, concomitant with the charge transfer (CT) absorption bands observed, and the corresponding ^{195}Pt NMR shift. These suggested that Pt(ɪɪ) is part of the source of the interaction, but is not directly involved in the interaction, and that the Pt coordinated phen-nucleobases are HOMO–LUMO interacting as shown in Figure 17.1. Since the CT interaction between the bases is strengthened by the electron-donating nature of the NH_2 group, the ΔH of the association with AMP^{2-} and GMP^{2-} is considered to reveal a large negative value. It turns out that there is a significant difference from

IMP^{2-} without the NH$_2$ group. Such base selectivity of [Pt(phen)(en)]$^{2+}$ complexes may be useful as drugs (*e.g.*, anticancer drugs).

17.2.2 Relationship Between Metalloenzyme Reaction Fields and Hydrophobic Fields

Metalloenzyme reaction fields are considered to be hydrophobic environments, but there are a few examples of evaluating their hydrophobicity. The stability constants for the binding of monocarboxylate to Zn(II) in dioxane (or ethanol)–water mixtures increased as the solvent became more hydrophobic. The hydrophobicity of some enzyme reaction fields was evaluated by using this stability enhancement as an indicator.[4] As a result, the enzymes that bind to smaller substrates were found to be more hydrophobic in the pocket, and the reaction field (metal-binding site) is located deeper from the enzyme surface. On the other hand, the correlation between the hydrophobic environment and the substrate recognition ability, which is the first step of metalloenzyme reactions, was studied by using $\Delta \log K$ as an indicator of easy formation of ternary complexes. The population of the interaction could be evaluated from the difference of $\Delta \log K$ values (a difference of 1 corresponds to 90% population). Figure 17.2 shows the dioxane–water solvent dependences of the aromatic ring stacking interaction between Cu(phen) and C$_6$H$_5$ in the Cu(phen)(C$_6$H$_5$(CH$_2$)$_n$COO$^-$) ($n = 1, 2$) complexes and the hydrophobic interaction between the isobutyl groups of the Leu side chain in the Cu(L-Leu)$_2$ complex. The interaction ratio reached a maximum at a certain value of the dioxane mole fraction. This indicated that the non-covalent

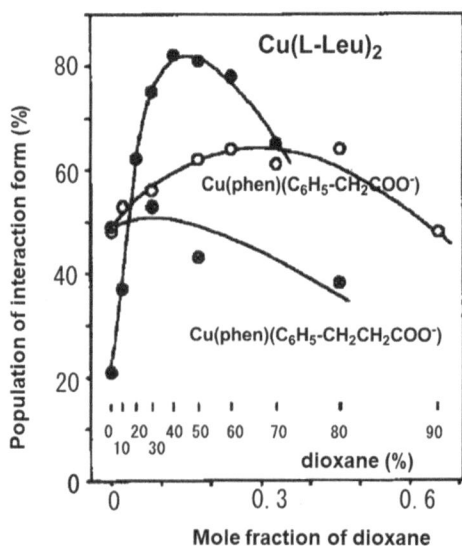

Figure 17.2   Solvent dependences of the fractional populations in ligand–ligand interactions between side chains in water–dioxane mixed solvents.[12]

interactions responsible for substrate recognition had both nonpolar and polar properties, and the interaction mode with the strongest recognition power depended on the polarity of the solvent. Thus, various types of non-covalent interaction act in the enzyme reaction field.

17.2.3 Study on the Equilibrium System in Blood

Blood consists of serum and blood clot. The serum, liquid part of blood, contains albumin and other proteins, as well as metal ions, amino acids, organic acids, and inorganic salts. For metal complexes that could be formed in serum, the populations of the complexes at pH 7.4 were evaluated based on their stability constants and the concentration of each component in serum. Table 17.2 shows the calculation results of the small molecule copper(II) complex in serum.[5] High percentages of mixed ligand Cu(II) complexes of histidine (His) and glutamine (Gln) or threonine (Thr) were observed. The reason in favour of such combinations could be the hydrogen bond between His side chain COO^- and Gln side chain $CONH_2$ groups or Thr and Ser side chain OH groups. The axial coordination of His side chain COO^- to Cu(II) and the concomitant hydrogen bond formed with the side chains of Gln or Thr or Ser could stabilize the Cu(II) mixed ligand complexes and increase their populations in serum.

Although the solution equilibrium of high molecular weight compounds was believed to be different from that of low molecular weight compounds, proteins with flexible structures such as serum albumin (molecular weight of approximately 60 000) have been found to be measured by pH titration in the same manner as low molecular weight compounds.[6] Serum albumin, which has more than 100 polar amino acids and is known as a carrier protein, could release and receive H^+ in various polar amino acids when it takes up substances. Accordingly a method to examine H^+ should be useful to study albumin–drug systems. As a result, the pH method confirmed the presence of two or more drugs binding to human serum albumin, drug–drug interactions on the albumin, and metal–drug interactions, which could not be determined by conventional fluorescence and molecular sieving methods. For example, the anti-inflammatory drug Cu(II)–salicylic acid complex is degraded to bind Cu(II) and salicylic acid to separate sites of the albumin.

Table 17.2 Fractional populations of Cu(II) complexes evaluated from the stability constants in serum.[4][a]

Cu(II) complex	%	Cu(II) complex	%
Cu(His)(Gln)	19	Cu(His)(Lys)	4
Cu(His)$_2$	16	Cu(His)(Gly)	4
Cu(His)(Thr)	15	Cu(His)(Asn)	4
Cu(His)(Ser)	8	Cu(His)(Val)	4
Cu(His)(Ala)	5	Cu(His)(Leu)	4

[a] Gln = glutamine, Lys = (protonated)lysine, Thr = threonine, Asn = asparagine, Ser = serine, Val = valine, Ala = alanine, Leu = leucine.

The serum albumin at 0.6 mM in blood plays an important role in transporting not only drugs but also nutrients such as metal ions and organic acids. The extended study of equilibrium in blood is expected to be applied to elucidate drug–drug interactions including metal–drug interactions.

17.2.4   Study of Reaction Intermediates

The study of electron transfer by Taube is well known as a typical study of reaction intermediates, and a similar study was conducted in systems involving proteins. Although the rate of electron transfer in proteins is slower than that in low-molecular-weight compounds, it is thought that there is some mechanism to increase the electron transfer rate in biological systems. Plastocyanin is an electron transfer protein that has copper as an active site. It was shown from the model experiment using a lysine peptide that, when electrostatic bonds between positive and negative charges occur on the surface, the allosteric effect distorted the coordination structure of copper, making it easier to accept electrons (Figure 17.3).[7]

In blood, Cu^{2+} is bound to the N-terminus of serum albumin, whose molecular weight is approximately 60 000, and is coordinated by N-terminal NH_2, two amide deprotonated CON^-, and His imidazole. The albumin molecule is too large to pass through the biological membrane, and donor exchange (ligand exchange) is thought to occur to transport Cu^{2+} into the cell. Cu(GlyGlyHis), which mimics the N-terminal coordination of albumin with the tripeptide GlyGlyHis, differs from Cu(GlyGlyGly) with other coordination modes in that Cu^{2+} delivery to amino acid X (X = His, Cys) is slow, taking several seconds or more (Figure 17.4).[5] This was explained by the fact that the mixed ligand complex Cu(GlyGlyHis)(X) (X: monodentate or bidentate His/Cys), which is an intermediate, becomes unstable because the fourth donor counting from the N-terminal NH_2 is His imidazole, which forms a stable coordination, and most of the intermediate returns from the

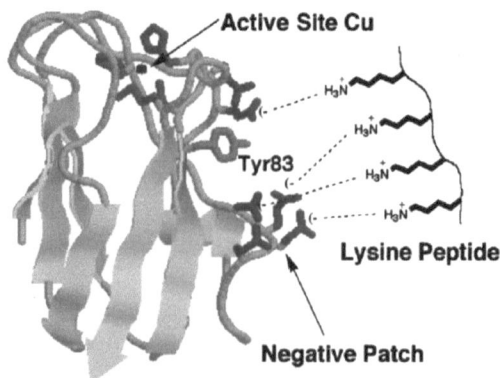

Figure 17.3   Cu enzyme plastocyanin–substrate interactions accelerated electron transfer.[13]

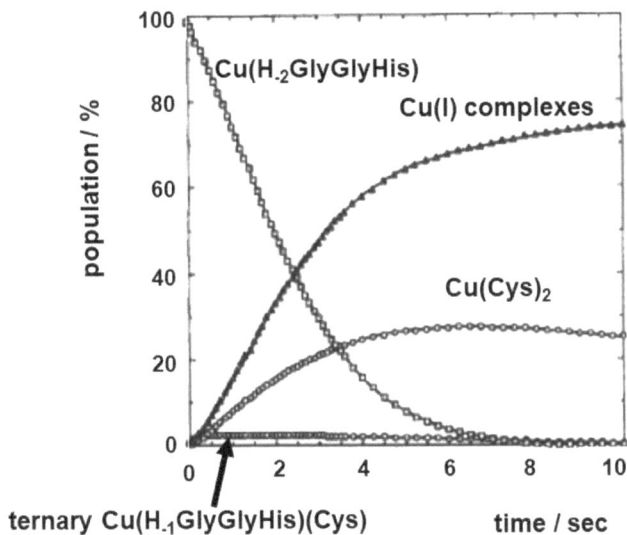

Figure 17.4 Time-dependent distribution for $Cu((H_{-2})GlyGlyHis)$–Cys systems. $[Cu((H_{-2})GlyGlyHis)] = 0.5$ mM, $[Cys] = 2.5$ mM, pH 8.7 $(I = 0.1$ $(NaClO_4)).$[14]

intermediate Cu(tridentate GlyGlyHis)(monodentate X) to the original stable Cu(GlyGlyHis), and the residue Cu(bidentate GlyGlyHis)(bidentate X) reacts with X to form the CuX_2 complex. In the case of X = Cys, the reaction rate study using the stopped flow method concluded the CuX_2 complex, followed by the reduction of Cu^{2+} to Cu^+. The intermediates governing the reaction rate were found to be slightly present when X was Cys (S^- donor), but not for His (N donor). While pH titration studies reported conflicting results – ternary Cu(GlyGlyHis)(His) complexes being the main and very small amounts being present, the ligand exchange study showed a clear conclusion that only a few ternary complexes are present. The importance of the simultaneous supply of H^+ for stable intermediate formation was inferred from the faster ligand exchange rate in acidic solutions. Therefore, it is suggested that Cu^{2+} transport *via* the cysteine donor SH is more rapid than that *via* His-imidazole because of the faster ligand exchange with the H^+ released *via* cysteine SH by the complex formation.

17.2.5 Nucleic Acid–Cation Interactions

Among the important substances in biological systems, nucleic acids are necessarily anions because of their phosphate moieties, and coexistence with metal ions such as Mg^{2+} and polycations such as polyamines is essential. In other words, nucleic acids always involve acid–base chemistry. This chemistry of charge equilibrium, which neutralizes charges, is important in transporting genes through the membrane into the cell in gene therapy, regenerative therapy, *etc*. However, the crystal structures of t-RNA and other compounds show no clear structural differences between Mg^{2+} and polyamine cations. Since Mg^{2+} is generally used in enzymatic reactions at the nucleus and polyamines are used for structural maintenance, it was believed that the strength of binding is $Mg^{2+} >$ polyamines. To study this, phytic acid $(C_6H_6-(O-PO_3H)_6)$ as a nucleic acid model was used to form an acid–base aggregate formed by the $PO_4^{2-}\cdots H^+N$ interaction between the phosphate group and the polyamine (4N, spermine; 3N, spermidine; 2N, putrescine).[8] The equilibrium constant of the interaction revealed the following order: polyamine $(4N) > Mg^{2+} >$ polyamine $(3N) >$ polyamine $(2N)$, and the interaction with polyamine (4N) was found to be stronger than with Mg^{2+}.

17.2.6 Applications for Metal Drugs

Since it became clear that metals, especially transition metals, play an important role in biological activities in minute quantities, attempts to use metal complexes as drugs to control biological activities have been the subject of ongoing research as a possible path since the famous platinum complex cisplatin $(H_3N)_2PtCl_2$ was successfully developed as an anticancer drug. Simple metal salts were the first research target; however, the metal salts were highly toxic due to their direct interaction with biological targets, and hence metal drugs are generally provided as metal complexes. The behavior of metal ions in the body can be regarded as complexation in solution, and the stability constants provide useful information on the metal complex formation at a certain concentration. For example, a $1:1$ Cu(II)–imidazole complex has $K = 10^{4.2}$ $(= [ML]/([M][L]),$[9] so under Cu(II) $=$ imidazole $= 1$ µM $(= 10^{-6}$ M $(M = mol\,dm^{-3}))$, the following equation approximately holds: $[ML]/([M][L]) = (10^{-8})/\{(10^{-6})(10^{-6})\}$. This indicated that 1% $(= (10^{-8})/(10^{-6})$ was the Cu–imidazole complex, while almost 99% of the complexes dissociated into Cu(II) and imidazole. In reality, many ligands in the biological fluid will coordinate to the metal ion, so the population present as the parent complex will be even smaller than that predicted from dissociation alone. Since the concentration of the drug in blood is generally below 10 µM, it is clear that Cu(II)–imidazole does not have appropriate properties as a drug. The order of the stability constants of divalent transition metals showed Irving–Williams series Ca $<$ Mg $<$ Mn $<$ Fe $<$ Co $<$ Ni $<$ Cu $>$ Zn. This indicated that the other divalent transition metal–imidazole complexes have lower stability constants than Cu(II). Accordingly, we can predict that transition metal(II)–imidazole complexes, except inert metal complexes, will not be suitable as drugs. In many cases, the bonds between the metal and the ligand are dissociated when the concentration of the complex, which is usually studied above mM concentrations in chemical experiments, is decreased to µM. Since highly active pharmacological compounds are active even at µM concentrations, the possibility that the metal complexes dissociate and become less active should be considered.

The part of the above (Sections 17.2.1–17.2.3, and 17.2.5) are just a few examples of studies, in which the author has been involved, in the field of both inorganic biochemistry and solution coordination chemistry. It is clear that the approach in solution chemistry is used for a more detailed understanding of the behavior of biomolecules in biological solutions. We look forward to the future development of this boundary field.

17.3 Contribution of Solution Coordination Chemistry to Life Science

In view of the recent trend of emphasizing chemical viewpoints and experiments in the life sciences, solution coordination chemistry can make a significant contribution to the life sciences when dealing with proteins and nucleic acids, which are the basis of the life sciences. For example, the equilibria of serum albumin, which is a carrier protein of metals and drugs in blood with a molecular weight of 60 000, were found to be analysed in the same way as small molecules, and their stability constants obtained could show species distribution in serum, *etc.* The electron transfer of metalloproteins is relatively well understood, but what about the proton transfer in proteins? Life science researchers have tended to avoid studying acid–base chemistry, leaving plenty of room for solution coordination chemistry. Rather, there are many issues to be solved: what concentration of DNA to be used for DNA–drug association analysis? Single-stranded? Double-stranded? What criteria should be used on the mass action law? *etc.* If you think about it, all life science reactions take place in solution, and many of them form aggregates, adducts, and complexes. This is precisely the problem of solution coordination chemistry, which differs only in that the molecular weight of the substance being handled is large. When discussing this with life scientists, I am often surprised at their ignorance of the basic knowledge and concepts in solution coordination chemistry. To prevent this from happening, it seems necessary for solution coordination chemists in the field of bioinorganic chemistry to enter the life sciences with their own research style, just as chemists study the life sciences, instead of leaving the research to biochemists. On top of that, if you share the research background just to discuss with other fields, they should definitely turn their attention to solution coordination chemistry.

In recent years, life science research has also come to be expected to contribute to human society. For example, research on the equilibria of albumin could be useful for the matching of medicines and is being conducted with the aim of returning results to society. These points seem to have been lacking in solution coordination chemistry up to now. Seen in this way, solution coordination chemistry is a discipline whose foundations have only just been laid, and is a field that has the potential to expand into the life sciences and other fields. Researchers in solution coordination chemistry can conduct research that is impossible for researchers in other fields and should respond to the demands from society in this way.

When we consider the points of contact between life science and solution coordination chemistry, there are many possible points of contact but few points of connection.[10] We cannot stop hoping that as many people as possible will increase the number of connections between the two fields.

Problem 1

The characteristic feature of biological chemical reactions, as seen in the high efficiency of enzymatic reactions and the high selectivity of reactions, is that non-covalent bonds can cause substance recognition, incorporation into the enzymatic reaction field and structural changes to the transition state.

(a) What types of non-covalent bonds exist?
(b) Why do we use non-covalent bonds and why is this not possible with covalent bonds?

Problem 2

The study of non-covalent bonds in aqueous solution is often concerned with determining the degree of interaction between interacting groups. The stability constants of metal complexes with fast ligand exchange rates are often used for this purpose, and the reasons why pH titration is widely used to determine the stability constants are discussed.

Answer to Problem 1

(a) As non-covalent bonds are weaker than covalent bonds, covalent bonds are sometimes distinguished as bonds and non-covalent bonds as interactions.

   Hydrophilic interaction: electrostatic interactions (Coulomb force), hydrogen bonds, cation–p interactions, and ion–dipole interactions.

   Hydrophobic interactions: hydrophobic interactions, aromatic ring stacking interactions, CH–p interactions, van der Waals interactions, and dipole–dipole interactions.

(b) Enzyme reactions usually undergo a cycle of the reaction (enzyme → substrate uptake → transition state (substrate reaction) → reactant release → enzyme). As each cycle usually takes place within one second, the time required to take up the substrate and fix it in the transition state structure should be less than about one second. For this purpose, weak bonds called non-covalent bonds (10–30 kJ mol^{-1}) can be used, whereas covalent bonds (300–400 kJ mol^{-1}) will result in the transition structure remaining fixed for a long time and the reactant not being released from the enzyme, so the enzymatic reaction will not take place quickly. In other words, for highly efficient enzyme reactions, rapid substrate uptake and rapid reactant release are necessary in addition to the substrate reaction, and, for this reason, there should be a certain amount of force to return to the initial state, *i.e.* the binding force (interaction) should not be strong. The low substrate recognition due to weak binding is thought to be compensated for by increasing the number of non-covalent bonds that act.

Answer to Problem 2

Experiments to determine the stability constants of metal complexes with high ligand exchange rates are simple and easy: the metal and ligands A and B are mixed and subjected to pH titration. The same operation applies to a comparative system in which non-covalent bonds are not involved. The pH titration measures the proton concentration $[H^+]$ of the solution using a glass electrode and indirectly measures the complex ML formed from the released H^+ as indicated by $M + LH \leftrightarrow ML + H^+$. The use of a pH glass electrode allows $[H^+]$ to be measured accurately over a wide range, and therefore the stability constant can be calculated by using the pH titration. Thus, stability constants can be determined more accurately than with other measurement methods. The log(stability difference), which increases due to non-covalent bonding, is estimated to be around 0.1–2, so the stability constant needs to be measured accurately.

References

1. G. Bertini, H. B. Gray, H. Gray, J. S. Valentine, E. I. Stiefel and E. Stiefel, *Biological inorganic chemistry: structure and reactivity*, University Science Books, 2007, ch. 1, pp. 1–4.
2. A. Odani, H. Masuda, O. Yamauchi and S. Ishiguro, *Inorg. Chem.*, 1991, **30**, 4484.
3. O. Yamauchi, A. Odani and M. Takani, *J. Chem. Soc., Dalton Trans.*, 2002, 3411.
4. H. Sigel, R. B. Martin, R. Tribolet, U. K. Haring and R. Malini-Balakrishnan, *Eur. J. Biochem.*, 1985, **152**, 187.
5. T. Kiss and A. Odani, *Bull. Chem. Soc. Jpn.*, 2007, **80**, 1691.
6. M. Kamei, T. Katsuno, R. Takeuchi, Y. Miyai, T. Kiwada, K. Ogawa and A. Odani, *Abstract of 33rd International Conference on Solution Chemistry (Kyoto)*, 2PD005, July 7–12, 2013.
7. O. Yamauchi, A. Odani and S. Hirota, *Bull. Chem. Soc. Jpn.*, 2001, **74**, 1525.
8. A. Odani, R. Jastrzab and L. Lomozik, *Metallomics*, 2001, **3**, 735.
9. A. Odani, H. Masuda, K. Inukai and O. Yamauchi, *J. Am. Chem. Soc.*, 1992, **114**, 6294.
10. *New-Generation Bioinorganic Complexes*, ed. R. Jastrząb and B. Tylkowski, De Gruyter, 2016.
11. A. Odani, in *Solution Chemistry of Metal Complexes*, ed. H. Yokoyama and M. Tabata, Sankyo Shuppan, Tokyo, 1st edn, 2012, ch. 8-3, p. 327.

12. A. Odani, in *Solution Chemistry of Metal Complexes*, ed. H. Yokoyama and M. Tabata, Sankyo Shuppan, Tokyo, 1st edn, 2012, ch. 8-3, p. 328.
13. A. Odani, in *Solution Chemistry of Metal Complexes*, ed. H. Yokoyama and M. Tabata, Sankyo Shuppan, Tokyo, 1st edn, 2012, ch. 8-3, p. 330.
14. A. Odani, in *Solution Chemistry of Metal Complexes*, ed. H. Yokoyama and M. Tabata, Sankyo Shuppan, Tokyo, 1st edn, 2012, ch. 8-3, p. 331.

CHAPTER 18

Solvent Extraction

CHRISTIAN EKBERG

Nuclear Chemistry, Department of Chemistry and Chemical Engineering, Chalmers University of Technology, SE-412 96 Gothenburg, Sweden
Email: che@chalmers.se

18.1 Introduction

One very clear application of knowledge about chemical speciation in different solutions is the so-called solvent extraction technique. Solvent extraction or liquid–liquid extraction is by IUPAC definition *"the process of transferring a dissolved substance from one liquid phase to another (immiscible or partially miscible) liquid phase in contact with it"*. The two phases are usually an aqueous phase and an organic phase which are more or less mutually insoluble. Thus, it is rather obvious that the solutes will have to be able to exist as dissolved species in both the phases used. The basic idea is that due to, *e.g.*, lipophilicity or hydrophilicity there is a preference of the desired solute to enrich in one of the phases. Typically enrichment in the organic phase is preferred. In principle, one may state that in the case of a metal each ion has a number of coordination sites normally occupied with, *e.g.*, hydration waters. If all these hydration waters can be replaced by more lipophilic molecules, the extraction will increase. In its simplest form, one might describe the process as in Figure 18.1.

The governing equation in solvent extraction is the expression of the distribution between the phases. This can be described either by the Nernst distribution constant, k_D, or by the distribution ratio, D, as expressed in eqn (18.1) and (18.2), respectively. In addition, it may also be convenient to express the ability of a system to separate between two different solvents

Coordination Chemistry Fundamentals Series No. 2
Metal Ions and Complexes in Solution
Edited by Toshio Yamaguchi and Ingmar Persson
© Japan Society of Coordination Chemistry 2024
Published by the Royal Society of Chemistry, www.rsc.org

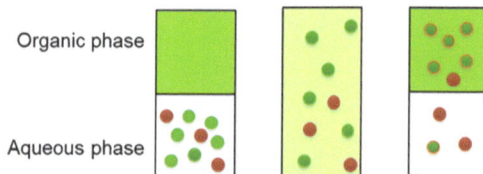

Figure 18.1   Principle of solvent extraction where the green circles are preferentially extracted to the lighter (organic) phase.

(A and B) and for this purpose the separation factor SF_{AB} is used; see eqn (18.3). For more details please see Section 18.8.

$$k_{D,A} = \frac{[A]_{org}}{[A]_{aq}} \tag{18.1}$$

$$D_A = \frac{[A]_{org,tot}}{[A]_{aq,tot}} \tag{18.2}$$

$$SF_{AB} = \frac{D_A}{D_B} \tag{18.3}$$

In the case of k_D, the ratio is valid for one species only, but in the case of D it is the total analytical concentration of the desired element regardless of the speciation that is used. SF is defined in such a way that it is always greater than unity. Using this on the system shown in Figure 18.1, we can calculate that we have $D_{green} = 4$ and $D_{red} = 0.5$ giving $SF_{green,red} = 8$.

Unfortunately, in real life things tend to be considerably more complex. In most cases, the desired element in the aqueous phase is not mainly lipophilic and a transfer by itself with a preference to the organic phase will not take place. Exceptions exist such as extraction of, *e.g.*, a carboxylic acid from an aqueous phase into toluene where no additional chemicals are needed. In most cases, however, some kind of chemical complexation is needed to facilitate the transport. The molecules used are often called extractants due to their ability to transfer the solute from one phase to the other. Depending on the chemistry, one normally differentiates between different extraction systems as outlined below.

For the simplest case suggested above, *i.e.* the extraction of a carboxylic acid (HA) from an aqueous phase into toluene eqn (18.1) becomes

$$k_{D,A} = \frac{[HA]_{org}}{[HA]_{aq}} \tag{18.4}$$

In principle this would be the solution. However, if one now adds some base, the acid will start to dissociate, and k_D will no longer describe the system and we will need to use D.

$$D_A = \frac{[HA]_{org}}{[HA]_{aq} + [A^-]_{aq}} \qquad (18.5)$$

By using the dissociation constant for the HA, usually denoted k_a, the denominator will change to $[HA](1 + k_a/[H])$ showing the dependence on pH. Clearly, also in this rather simple system knowledge of solution speciation is crucial to understand the extraction behaviour.

There are naturally a vast amount of extraction systems developed for different purposes such as separating different elements from each other or determine, *e.g.*, chemical stability constants for certain complex formations. In principle, however, one might say that they typically follow some crude denotations usually based on how the extractant behaves in the system such as acidic extractants, solvating extractants and ion exchange extractants. Clearly, this is only the mechanistic description. Chemically it is vastly more complicated including preferences in coordination, effect of chemical activity coefficient changes in the aqueous phase, effect of different properties of the selected diluents, *etc.* This is, however, not the scope of this short summary but rather the focus on the dedicated solvent extraction literature.

18.2 Extraction with Acidic Extractants

These extractants are usually similar to, *e.g.*, carboxylic acids or similar molecules with one polar and acidic side and one considerable more lipophilic side. As the charged side deprotonates it binds to the charged metal ion and the outside of the complex facing the solvent is actually rather lipophilic. The extraction system of the metal M in its ionic form $M^{\nu+}$ can then in its simplest case be described as in Figure 18.2.

As can be seen from Figure 18.2, the Nernst distribution constant is used for the extractant only, while the corresponding distribution constant for the uncharged metal–extractant complex is denoted λ_ν, where ν is the number of

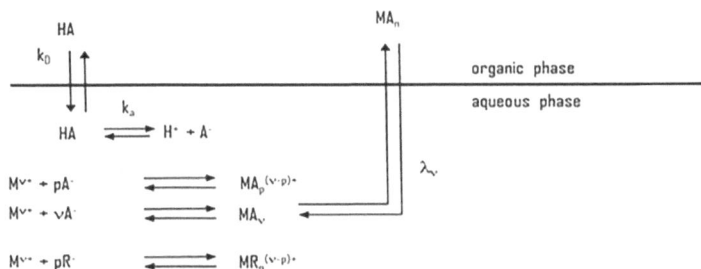

Figure 18.2 The chemical system for the extraction of a metal ion, $M^{\nu+}$, using an acidic extraction agent, HA, with a competing complexation in the aqueous phase with ligand R^-.

ligands attached to the metal to achieve a neutral complex with lipophilic surface.

The expression of the distribution ratio then becomes

$$D_M = \frac{[MA_\nu]_{org}}{\sum_{i=0}^{\nu}[MA_i]_{aq} + \sum_p [MR_p]_{aq}} \tag{18.6}$$

After using the stability constants and distribution constants for the reactions given in Figure 18.2, eqn (18.6) becomes

$$D_M = \frac{\lambda_\nu \beta_\nu [A^-]^\nu}{1 + \sum_{i=1}^{\nu} \beta_i [A^-]^{i} + \sum_{j=1} \beta_{R,j} [R^-]^j} \tag{18.7}$$

where now all concentrations are for the aqueous phase only. Please note that when doing these algebraic manipulations we have neglected any chemical activity and assumed the activity coefficient to be unity. This is a considerable simplification since the stability constants are defined in activities (unless corrected for the ionic strength used), while the distribution ratio is defined in concentrations. However, the expression given in eqn (18.7) may be used to determine conditional stability constants by fitting the obtained D as a function of the free A^- concentration.

18.3 Extraction with Solvating Extractants

One of the drawbacks with the acidic reagent is its usefulness at, *e.g.*, lower pH where the extractant will be fully protonated and thus its extraction ability will be low. One might also argue that many acidic extractants have a rather poor selectivity although this depends on too many factors. For carboxylic acids, it may certainly be true but in the case of, *e.g.*, ketones/enols there could be a very high selectivity due to filling-up of the coordination sphere. In the case of a solvating extraction mechanism, some coordination sites are occupied with some more or less lipophilic ligand, while the remaining sites (or as many as possible) are occupied by the solvating extractant, thus increasing the distribution ratio by adding one more extracted species, as shown in Figure 18.3.

The expression of the distribution ratio then becomes

$$D_M = \frac{[ML_\nu]_{org} + [ML_\nu B_p]_{org}}{\sum_{i=0}^{\nu}[ML_i]_{aq} + \sum_p [MR_p]_{aq}} \tag{18.8}$$

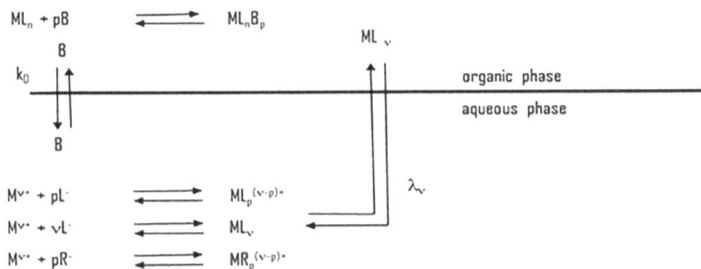

Figure 18.3 The chemical system for the extraction of a metal ion, $M^{\nu+}$, using an aqueous ligand and a solvating extractant with a competing complexation in the aqueous phase with ligand R^-.

After using the stability constants and distribution constants for the reactions given in Figure 18.3, eqn (18.8) becomes

$$D_M = \frac{\lambda_\nu \beta_\nu [L^-]^\nu (1 + \lambda_\nu \beta_\nu [L^-]^\nu k_{D,B}[B]^p)}{1 + \sum\limits_{i=1}^\nu \beta_i [L^-]^{i'} + \sum\limits_{j=1}^{} \beta_{R,j}[R^-]^j} \tag{18.9}$$

Now we have assumed that the solvating extractant B actually distributes between the phases. This is not commonly the case since it might be so highly lipophilic that there is no solubility in the aqueous phase. In such a case, eqn (18.9) has to be changed accordingly.

18.4 Ion Pair Extraction

In some cases, the organic extractant exists as a salt and can then be used to extract ions of opposite charges. Typically these extractants consist of a larger, more lipophilic ion bound electrostatically to a smaller, more hydrophilic ion. These systems exist with either being cationic or anionic. In this text we will consider anionic extraction. It is possible to consider acidic extractants as a special case of cationic extraction.

The expression of the distribution ratio then becomes

$$D_M = \frac{[ML_{\nu+p}Q]_{org}}{\sum\limits_{i=0}^{\nu+p} [ML_i]_{aq} + \sum\limits_{p} [MR_p]_{aq}} \tag{18.10}$$

Here we have assumed that the larger organic cation will not distribute to the aqueous phase but rather stay on the phase boundary and there bind to the anion for the extraction. The smaller anionic part will, however, more or less completely enter the aqueous phase. After using the stability constants and

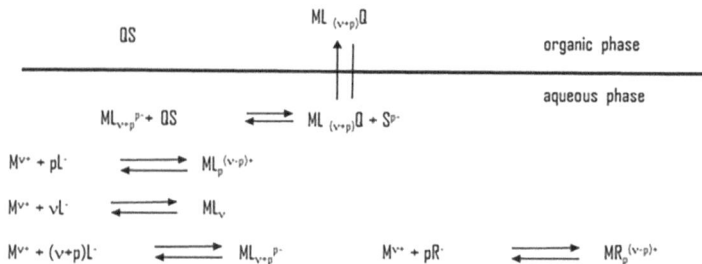

Figure 18.4 The chemical system for the extraction of a metal ion, $M^{\nu+}$, using an aqueous ligand and an anion exchange extractant with a competing complexation in the aqueous phase with ligand R^-.

distribution constants for the reactions given in Figure 18.4, eqn (18.10) becomes

$$D_M = \frac{\beta_{\nu+p}[L^-]^{\nu+p}[Q]_{\text{org}}}{1 + \sum_{i=1}^{\nu+p} \beta_i[L^-]^{i'} + \sum_{j=1}^{} \beta_{R,j}[R^-]^j} \tag{18.11}$$

18.5 Determination of Stability Constants Using Solvent Extraction

As suggested above, solvent extraction can be used for determination of conditional stability constants for a number of reactions. As an example, the determination of the hydrolysis constants of thorium(IV) will be used by using the acidic extractant acetylacetone (HA) 0.1 M dissolved in toluene. The aqueous phase is 1 M $NaClO_4$ (an inert electrolyte ensuring a constant ionic strength during the experiment).

The thorium(IV) concentration was about 10^{-5} M ensuring that the transfer during extraction will change neither the acetylacetone concentration nor the ionic strength. Referring to Figure 18.1, we get the following detailed expression for D_{Th}:

$$D_{\text{Th}} = \frac{[\text{ThA}_4]_{\text{org}}}{\sum_{i=0}^{4} [\text{ThA}_i]_{\text{aq}} + \sum_{p=1}^{4} \left[\text{Th(OH)}_p\right]_{\text{aq}}} \tag{18.12}$$

which can be written using the stability constants for the acetylacetone complexation as well as the hydrolysis constants.

$$D_{\text{Th}} = \frac{\lambda_4 \beta_{A,4}[A^-]^4}{1 + \sum_{i=1}^{4} \beta_{A,i}[A^-]^{i'} + \sum_{p=1}^{4} \beta_{\text{OH},p}[\text{OH}]^p} \tag{18.13}$$

The extraction curve as a function of "pH" is shown in Figure 18.5.

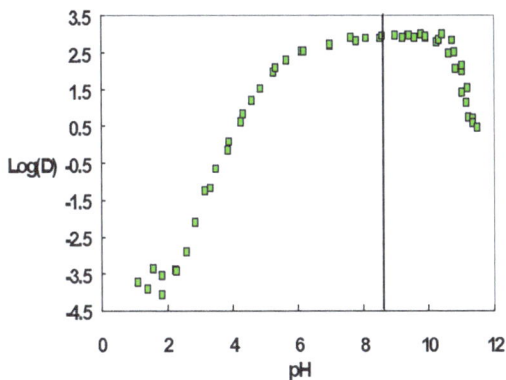

Figure 18.5 Extraction curve for the thorium–acetylacetonate system, $[HAa] = 0.1$ M, in toluene, aqueous phase: 1.0 M.[1]

Table 18.1 Determined stability constants for the thorium–acetylacetone system.[1]

Complex	$\log_{10}\beta$	Complex	$\log_{10}\beta$
ThA^{3+}	8.8 ± 0.1	$ThOH^{3+}$	10.6 ± 0.1
ThA_2^{2+}	17.1 ± 0.5	$Th(OH)_2^{2+}$	19.2 ± 0.2
ThA_3^{+}	23.5 ± 0.5	$Th(OH)_3^{+}$	28.7 ± 3.5
ThA_4	27.9 ± 0.4	$Th(OH)_4$	37.4 ± 0.4

In order to evaluate it, we need to find the relationship between free A^- and free H^+. This is done by the material balance:

$$[A]_{tot} = [HA]_{org} + [HA]_{aq} + [A^-]_{aq} \tag{18.14}$$

Using the constants given in Figure 18.2, again we get

$$[A]_{tot} = [A^-]\left(K_a k_d^{-1}[H] + K_a^{-1}[H] + 1\right) \tag{18.15}$$

Thus, the concentration of the free acetylacetonate and the hydroxide concentration can be expressed in terms of H^+ concentration, the latter through the use of K_w from the literature (which also has to be adjusted for the ionic strength used).

Given that we know the dissociation constant of the acetylacetone as well as its distribution constant (or they can be fitted but then the number of fitted parameters will be too large), we can now determine λ_4 from the plateau of the extraction curve (giving $\lambda_4 = 950$), the stability constants from the left part of the curve and the hydrolysis constants from the whole curve (using the already determined acetylacetone constants) (Table 18.1).

It is worth mentioning again here that the stability constants determined above are conditional. They are valid for 1 M ionic strength and will need to be corrected to zero ionic strength by using, *e.g.*, the specific ion interaction theory (SIT).

18.6 Separative Extraction

Industrially, the most common use for solvent extraction is to separate metals (or other solutes) from each other. Then the separation factor (SF) (see eqn (18.3)) is an important parameter. In addition, it is desired that the element that is to be extracted has $D > 1$ and the rest of the elements have $D < 1$. Since normally $\log_{10} D$ is the entity used this means that the extracted elements should have a $\log_{10} D > 0$, while the non-extracted $\log_{10} D < 0$. Generally separative extraction is used in industrial applications and the extraction is then part of an extraction system, as described in Section 18.7. Often these processes will utilise solvation extraction since it is rather robust and not as sensitive to pH as, *e.g.*, acidic extraction. Some of the important features of the extraction (once the chemical system is selected) will then be given by, *e.g.*, ligand concentration and extractant concentration.

An example of such separation as a function of some chemical parameters is given below. The aim is separation of americium from europium using a system with nitric acid as the aqueous phase and an extractant called CyMe$_4$BTBP as the organic phase. In this case, the extraction mechanisms for the two elements are the same and thus we do not expect to see any effect on the separation factor as we change the acid and extractant concentrations. Looking at eqn (18.8) and making some assumptions about the dominating species, we can elucidate some coordination chemistry from these analyses. If we plot the logarithm of the distribution ratio as a function of the logarithm of selected concentration, the slope indicates the stoichiometry of the extracted complex. In the selected case, it is possible to see from Figure 18.6 that the metal-to-extractant ratio is 1.2. Then, by using the knowledge that for best separation we want to be at a concentration giving $D < 1$ for the extracted species and preferably $D < 1$ for the rest, we can determine the best conditions for the extraction process. In this case the best extractant concentration for separation is somewhere at 0.02 M.

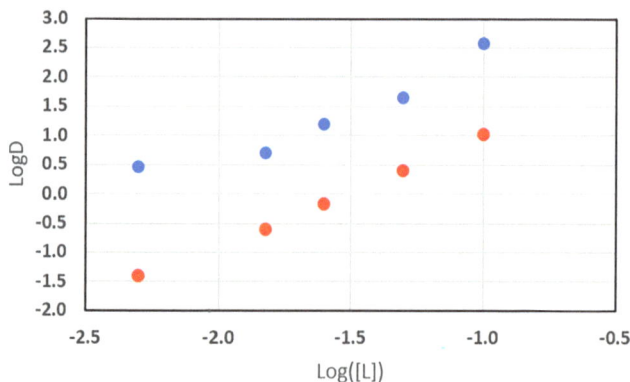

Figure 18.6   The distribution ratio as a function of solvating extractant concentration. Blue circles – Am and red circles – Eu.[2]

In the case of nitric acid dependence, the picture is somewhat more complex. From the increasing slope (Figure 18.7), we can see that the probable metal-to-acid stoichiometry is about 1.3, but now we react to the maximum extent whereafter the extraction decreases again. This behaviour is due to nitrate complexation in the aqueous phase at higher nitric acid concentrations, suggesting that the best nitric acid concentration to be selected should be about 2.5 M.

Thus, we have now shown that in the selected system the metal-to-extractant ratio is 1.2 and the metal-to-nitrate ratio is 1.3. In addition we have selected the process parameters as follows: an acid concentration of 2.5 M and an extractant concentration of 0.2 M (Figure 18.7).

18.7 Process

The simple, one-stage, extraction described above is normally not used for process implementation. In such cases, the extraction (separation) is considerably improved by connecting several extraction stages in some geometry. Typically, this geometry is so called counter current, as shown in Figure 18.8, where y_i is the concentration of the respective element after each extraction stage and x_i is the corresponding aqueous concentration.

It is also possible to construct an extraction system using, *e.g.*, co-current or cross current flow patterns depending on the goal of the process. However,

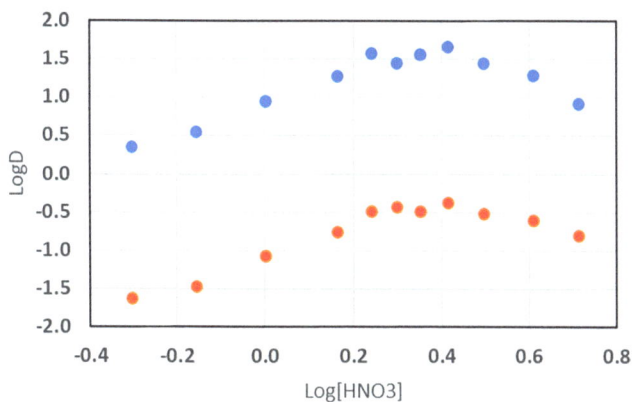

Figure 18.7 The distribution ratio as a function of nitric acid concentration. Blue circles – Am and red circles – Eu.[3]

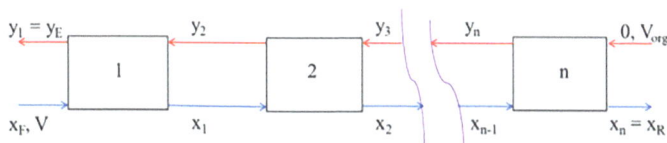

Figure 18.8 Counter current extraction.

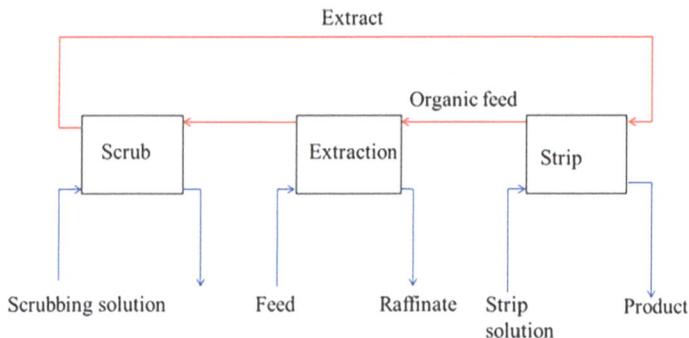

Figure 18.9 A typical set-up of an extraction system for separation of elements.

today the counter current set-up is the most common one. Then, an industrial extraction system consists typically of several parts. Firstly, we have the extraction but that may have some undesired elements in it. Therefore, it is common to have scrubbing stages where undesired elements are further washed out before the final stripping stages as shown in Figure 18.9.

The final product is then liquid which is rarely the desired form. Therefore, after the separation system there might be a precipitation stage if a salt is desired or electrodeposition if a pure metal is the final end-product. As shown in Figure 18.9, it is common to use the same organic phase in a circular manner, thus reducing the need for chemicals and decreasing the footprint of the separation system.

Optimisation of an extraction system is not straightforward since there are many parameters to take into account. One might be led to believe that a high D is the most important factor but in reality it is not. In real processing, a D value much higher than 10 is not needed since the phase separation of industrial units is typically only able to use this kind of D as a rule of thumb, for processing $0.1 < D < 10$ is a good range to be completely utilised. Then, naturally the extracted species need to be able to strip out of the organic phase as well as the extraction kinetics should match the equipment and flow rates desired and used.

Typically, solvent extraction processing of metals is currently used in, *e.g.*, the nuclear industry where the fissile materials are recovered from the rest of the waste and the copper industry where the final stage is an electrodeposition to form metallic copper.

18.8 Thermodynamics

Let us start with a tertiary system consisting of a solute and two solvents sparingly soluble in each other. It was found that the distribution of a species between them was more or less independent of the concentration of the species. Then we can define the distribution constant, k_D, as was done in eqn (18.1). Naturally the claim of independence with respect to the concentration of the species is not absolute but in a limited (but usually rather wide) concentration span. As a rule of thumb this interval is from zero up to some mole%. If the solute is soluble in both phases and for very high solute concentrations, the two-phase system collapses and we get a one-phase system, the so-called "plait point" in a ternary phase diagram. When it comes to metal extraction, we are normally more interested in the other end of the spectrum where the total concentration of the relevant species is very low.

Let us now assume that an element (A) distributes between two phases and exists in the same speciation in both phases. Then the chemical potential for the species in each phase can be denoted as $\mu_{A,aq}$ and $\mu_{A,org}$ for the aqueous and organic phases, respectively. Then at equilibrium we will have

$$\mu_{A,aq} = \mu_{A,org} \tag{18.16}$$

The relationship between the chemical activity, a_A, and the chemical potential is given by

$$\mu_A = \mu_A^0 + RT \ln a_A \tag{18.17}$$

where μ_A^0 refers to the case where the chemical activity can be set to 1 (as discussed below).

If we now substitute eqn (18.17) into eqn (18.16) and rearrange, we get

$$K_d = \frac{a_{A,org}}{a_{A,aq}} = \exp\left(\frac{\left(\mu_{A,aq} - \mu_{A,org}\right)}{RT}\right) \tag{18.18}$$

and K_d is then called the thermodynamic formulation of the distribution constant.

Before we move on we need to consider the term "ideality". Often we mean that the potential energy of a molecule is similar to the potential energy of the molecule in the pure substance at the selected pressure and temperature.

An ideal mixture is characterized by the following conditions:

(1) During mixing, the enthalpy and energy of the system do not change, *i.e.*
$\Delta H_{AB} = 0$ (constant P and T)
$\Delta U_{AB} = 0$ (constant P and T)
$\Delta V_{AB} = 0$ (constant P and T)
(2) In the mixture, there is a complete random distribution of the molecules which leads to the expression of the entropy of mixing as follows:

$$\Delta S_{AB} = -R(x_A \ln x_A + x_B \ln x_B)$$

and for the process leading to an ideal mixture we have

$$\Delta G_{AB} = \Delta H_{AB} - T\Delta S_{AB} = RT(x_A \ln x_A + x_B \ln x_B)$$

If we now look at the distribution between ideal dilute solutions, we also need to look at the concept of chemical activities. It has been shown that if the amount of a solute (A) is small enough we can make the division of the activity of an element and its concentration or molar fraction and thus define the activity coefficient, γ_A. At infinite dilution, we use the symbol γ_A^0.

If the amount of A in the diluent S is low enough that $\gamma_A = \gamma_A^0$, we say that it is an ideal dilute solution (IDS). Here we need to remember that this solution is not an ideal mixture but ideal in the sense that it follows Henry's law.

We can now keep the original standard state, *i.e.* pure A, but we see that eqn (18.16) can be written as

$$\mu_A = \mu_A^{0*} + RT \ln x_A \tag{18.19}$$

$$\mu_A^{0*} = \mu_A^0 + RT \ln \gamma_A^0 \tag{18.20}$$

The standard state corresponding to the "standard potential" μ^{0*} is hypothetical. If we use the molar fraction scale, it will even be absurd, *i.e.* a pure substance with the same activity coefficient as in an ideal dilute solution. Therefore, the term IDS is more or less exclusively used when the concentrations are expressed as molarities (c_A) or molalities (m_A).

In dilute solutions we have

$$x_A \sim \frac{c_A v_s}{1000} \tag{18.21}$$

where v_s is the molar volume of the diluent in cm^3. We can then rewrite eqn (18.19) as

$$\mu_A = \mu_A^{0**} + RT \ln c_A \quad \text{(IDS)} \tag{18.22}$$

$$\mu_A^{0**} = \mu_A^0 + RT \ln \left(\frac{\gamma_A^0 v_s}{1000}\right) \tag{18.23}$$

In this way, γ_A^0 will still be an activity coefficient expressed in the molar fraction scale. In the case of the molarity and molality scale, we usually denote the activity coefficients with γ_A and f_A, respectively. If we now create a hypothetical IDS at $c_A = 1$ M as the standard state, we can say that $\gamma_A^0 = 1$. Furthermore, if the requirements for an IDS are fulfilled within the relevant concentration interval, then also $\gamma_A \sim 1$.

We can now apply eqn (18.22) and for ideal dilute solutions reach

$$k_d = \frac{c_{A,org}}{c_{A,aq}} \sim \exp\left(\frac{\mu_{A,aq}^{0**} - \mu_{A,org}^{0**}}{RT}\right) \tag{18.24}$$

We can then express the difference between the reference potentials, Δg_A^0, as

$$\Delta g_A^0 = -\left(\mu_{A,aq}^{0**} - \mu_{A,org}^{0**}\right) \tag{18.25}$$

Thus, we can interpret Δg_A^0 as the change in Gibbs free energy upon transfer of 1 mol of A from an infinitely large pure aqueous phase to an infinite pure organic phase. With the word "pure", we normally refer to solutions that are completely pure. In practical cases, there is always some mutual solubility. It is thus convenient to redefine "pure" as the state when the mutual solubility is reached, *i.e.* then the activity factor of the two solutions is unity. This will make the expression system specific though.

We can now connect our definitions of the distribution coefficient to account for changes in non-ideality.

$$k_d = \frac{\gamma_{A,org}}{\gamma_{A,aq}} K_d \tag{18.26}$$

In practice, as we use solvent extraction for practical purposes, these detailed definitions may seem superfluous, but they do have a significant bearing when stability constants are determined using this technique. We end up in a mixture of definitions using activities and concentrations since the distribution ratio essentially is defined in concentrations, while the stability constants and distribution constants are thermodynamic entities and thus defined in chemical activities.

Problem 1

Considering the separation of americium from europium given in Figure 18.6, determine at which ligand concentration is the optimal (largest) separation achieved and calculate the separation factor at that ligand concentration.

Problem 2

If we look at the cumulative stability constants for thorium acetylacetonate given in Table 18.1, which is the strongest complex looking at the stepwise stability constants? Why is that so?

Answer to Problem 1

At $\log(L) = -2.3$, which is about 0.1 M. The SF at that ligand concentration is: $D_{Am}/D_{Eu} = 1.65/0.25 = 6.7$.

Answer to Problem 2

The first complex is the strongest. This is the normal case since for added ligands the electron density available will be lower as well as there could be steric hindrance. Exceptions exist if there is stabilization due to advantages, *e.g.* ring formation or resonances.

References

1. C. Ekberg, A. Albinsson, M. J. Comarmond and P. Brown, *J. Solution Chem.*, 2000, **29**, 63.
2. J. Halleröd, C. Ekberg, T. Authen, L. Bertolo, M. Lin, B. Gruner, J. Svelah, C. Wagner, A. Geist, P. Panak and E. Aneheim, *Solvent Extr. Ion Exch.*, 2018, **36**(4), 360–372.
3. T. L. Authen, Doctorial thesis, Chalmers, Göteborg Sweden, 2022, available at https://research.chalmers.se/publication/530583/file/530583_Fulltext.pdf.

CHAPTER 19

Complex Formation in Hyperalkaline Solutions

BENCE KUTUS* AND PÁL SIPOS*

Department of Inorganic, Organic and Analytical Chemistry, University of Szeged, 7-8 Dóm tér, H-6720 Szeged, Hungary
*Emails: kutusb@chem.u-szeged.hu; sipos@chem.u-szeged.hu

19.1 Introduction

This chapter discusses the equilibria and structures of metal complexes forming in highly alkaline aqueous solution, where the metal ion (or ions) is bound to different kinds of O-donor ligands. The most trivial of the latter is the hydroxide ion, OH^-. The stepwise hydrolysis reaction between a metal and one or several OH^- ions is one of the classical topics of inorganic solution chemistry. Here, the main aim is the elucidation of the structure, composition and thermodynamics of the hydroxido complexes formed. Formation constants for a large variety of hydroxido complexes as well as the solubility products of solid metal hydroxides are well known from various textbooks and tables.[1,2] Traditionally, the species formed in solutions with $2 < pH < 13$ have commonly been described. The structure and dynamics of hydroxido complexes present in extremely alkaline solutions ($pH > 13$) are usually unknown or poorly characterized, due to theoretical problems as well as practical hurdles; see Section 19.2.3 for more details. Nevertheless, Section 19.3 is dedicated to examples of peculiar, sometimes unexpected, or in other cases non-existing but supposed metal-hydroxido complexes. For instance, $Pb(OH)_4^{2-}$ has for a long time been claimed even in some university textbooks, but the most authoritative reviews of the field denied its existence.[1,2] On the other hand, in these extremely caustic solutions, the

Coordination Chemistry Fundamentals Series No. 2
Metal Ions and Complexes in Solution
Edited by Toshio Yamaguchi and Ingmar Persson
© Japan Society of Coordination Chemistry 2024
Published by the Royal Society of Chemistry, www.rsc.org

existence of solution species such as $Al(OH)_6^{3-}$, $Ga(OH)_6^{3-}$ and $Fe(OH)_6^{3-}$ was also postulated in university textbooks,[3–5] as well as in some publications,[6,7] but has later been proven to be experimentally undetectable.[8–10]

Needless to say, these systems become even more complicated and hard-to-describe, when a complex-forming molecule or ligand is added to a hyperalkaline solution containing a metal ion, due to the large number of conceivable complex compositions. The current short review will also deal with such systems, but will be restricted to O-donor ligands with particular emphasis on polyhydroxy carboxylates. Strongly alkaline media are generally needed for the deprotonation of the alcohol (–OH) moiety to form alcoholate (–O⁻),[11] giving rise to the formation of complex compounds with extremely high thermodynamic stability. Polyhydroxy carboxylates containing multiple –OH moieties beside carboxylate (–COO⁻) are capable of acting as bridging ligands (L) and form polynuclear complexes with various bivalent metal ions, *e.g.* Ca^{2+}.[12] In addition, the high affinity of these molecules for complexation is further enhanced by the high charge density of tri- or tetravalent metal ions.[13–15] Involvement of bivalent ions in these complexation processes gives rise to the formation of heteropolynuclear complexes with a general stoichiometry of $M(\text{II})_pM(\text{III/IV})(OH)_rL_q$. The strong binding of these mainly "hard" type metal ions has clear practical implications, *e.g.* mobilization of radionuclides in radioactive waste repositories,[16] binding of Al(III) in Bayer's alumina-trihydrate production[17] or the role of Fe(III) complexes in the cathodic reduction of vat dyes in various dyeing processes;[18] for more details, see below. Section 19.4 is dedicated to the aqueous chemistry of these environmentally sensitive and/or practically relevant systems.

19.2 Studying Complexation Equilibria in Hyperalkaline Media

19.2.1 General Scheme of Complex Formation Processes

In general, most ligands possess at least one functional group from which proton(s) can dissociate already in the acidic-neutral pH regime, yielding a negatively charged ionic moiety. The forming 'anchor' group (*e.g.* –COO⁻) acts as the primary metal-ion binding site of the molecule.[11,19] Further, secondary donor groups (*e.g.* –OH) may also take part in complexation, giving rise to the formation of chelate structures. However, such species are relatively weak due to the low electron density around the donor atom. When the pH is increased, these groups are more readily deprotonated, yielding very basic binding moieties, such as alcoholate (–O⁻). This proton displacement is also facilitated by strong Lewis acidic metal ions, as they are strong competitors to H^+ for the same binding site. Typically, transition metal, actinide, and lanthanide ions of high charge density induce ligand deprotonation already in the acidic range, whereas such equilibria take place only in the (strongly) alkaline range in the presence of alkaline earth metal ions.[20] In addition, high pH promotes the hydrolysis of metal ions, that is, the deprotonation of coordinated water molecules. Such reactions yield strong M–OH interactions, which in turn weaken the M–L ones, giving rise to the precipitation of sparingly soluble hydroxides in the extreme case. These processes are presented in Figure 19.1.

19.2.2 Methods to Probe Complexation Reactions

Given the above nature of metal complexation, a plethora of association equilibria depend strongly on solution pH. As such, potentiometry has been widely used to determine the stoichiometry and stability constant of metal complexes formed in hyperalkaline aqueous solution. However, commercially available combined glass electrodes, which are used in most complexation studies, usually fail to provide accurate measurements above pH $\approx$ 12. Above this pH, platinized platinum electrodes perform very well,[21] albeit they are commercially not yet available. Despite these technical difficulties, potentiometry is by far the most reliable method in detecting and quantifying overlapping complex formation processes.

Further methods can be divided into ligand- or metal-specific ones. The most prominent technique in the first group is nuclear magnetic resonance (NMR) spectroscopy based on the change in the chemical shift[22] of ^{1}H or ^{13}C nuclei upon deprotonation/metal-ion binding. Complexation studies benefit from the ability of NMR to provide information on the metal-binding sites of the ligand in solution. Also, it may be applied to quantify deprotonation and metal-ion-binding equilibria, although with limited sensitivity in the case of

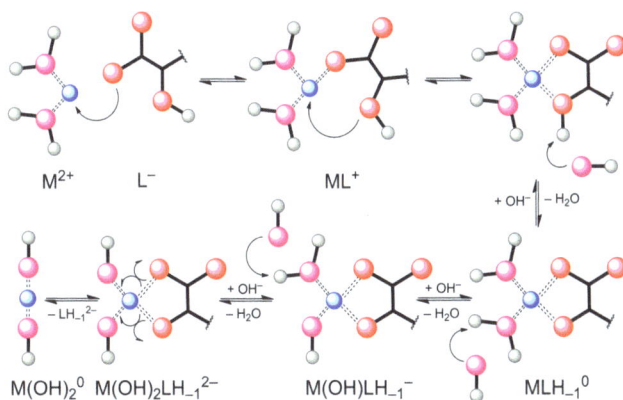

Figure 19.1 General mechanism of complex formation between a divalent metal ion (M^{2+}) and a ligand (L^-). Binding of M^{2+} by the anchor group of L^- yields the complex ML^+, whereas that of an adjacent functional group gives rise to a chelate structure. Upon increasing the pH, progressive deprotonation on the ligand and/or on coordinated water molecules yields the species MLH_{-1}^0 and mixed hydroxido-complexes, $M(OH)LH_{-1}^-$ and $M(OH)_2LH_{-1}^{2-}$. Eventually, the ligand (LH_{-1}^{2-}) might dissociate from the complex, resulting in the formation of $M(OH)_2^0$ likely to precipitate.

multiple processes. In addition to NMR, infrared and Raman[23] spectroscopies are often invoked to deduce structural information on the complexes formed in both solution and solid states. Furthermore, polarimetry is an attractive alternative for optically active ligands,[14,24] given that the abstraction of metal ions substantially alters the equilibrium conformation of the molecule.

One of the most frequently used metal-ion related methods is solubility,[25] where a sparingly soluble metal hydroxide dissolves to a higher degree in the presence of a ligand, due to complex formation. Here, the total concentration of the metal ion is measured with different techniques, *e.g.* inductively-coupled plasma optical emission spectroscopy or mass spectrometry (ICP-OES/MS), based on the system and concentration range studied. Besides solubility, spectrophotometry utilizes the fact that certain electronic transitions report sensitively on ligand binding, providing quantitative information, especially when combined with potentiometry.[26] For optically active complexes, circular dichroism (CD) detects the same transitions applying circularly polarized light.[26] Different from photometry, inner electronic or nuclear transitions are utilized for gaining structural information by extended X-ray absorption fine structure (EXAFS), X-ray absorption near-edge structure (XANES) or Mössbauer spectroscopies.[10,27] Valuable qualitative information on complexation can also be gained *via* NMR probing of the metal ion directly (*e.g.* ^{43}Ca, ^{71}Ga).[28,29]

Quantum chemistry has become a powerful tool in the past decades to estimate the structure and properties of complexes. Here, numerous density functional theory (DFT) methods have been reported offering a reasonable tradeoff between accuracy and computational cost.[27,30,31]

19.2.3 Practical and Theoretical Difficulties and Considerations

At extreme electrolyte concentrations, peculiar and sometimes unexpected pair interactions occur in solution, for example, in a 20 M solution of NaOH (which is a common laboratory stock solution, with a $H_2O:NaOH$ molar ratio of $2:1$). Therefore, the ions are only partially hydrated and the formation of ion-pairs[32] is inevitable. Such solutions do not fulfil the criterion of a strong electrolyte anymore. Moreover, even ions with identical charges are forced to exist in close proximity.

Equilibrium measurements require minimizing the variations of the mean ionic activity coefficients, $\gamma_{\pm}$, *via* keeping the ionic strength of the solution constant. For this purpose, use of a large excess of supporting electrolyte is necessary, *e.g.* up to 5 M $NaCl$[33] or up to 8 M $NaClO_4$.[21] The ions of the supporting electrolyte may also participate in ion-pairing reactions with the components of the system to be studied, thus contributing to the overall "experimental" effect observed and falsifying the results. Furthermore, ion associations and variations in activity coefficients usually occur in parallel and are often inseparable. A further key issue is the allowable extent of the substitution of the background electrolyte with the reactants studied without changing $\gamma_\pm$. As a rule of thumb, 25 mol% is considered as the maximum acceptable substitution. There is one remarkable exception though: the values of $\gamma_\pm$ of 1 M NaCl and 1 M NaOH solutions are practically identical at 25 °C;[21,34] therefore, in mixtures of these two solutions in any proportion, $\gamma_\pm$ can reasonably be assumed to be constant.

Further technical problems include the effect of impurities, which are inevitably accumulated in these concentrated solutions. This is in particular important when a given method is sensitive to a certain type of impurity: spectrophotometry for transition metal traces, NMR for paramagnetic impurities, Raman for fluorescent compounds, *etc.*[35] Thus, the use of extra pure components for the experiments is a prerequisite for avoiding artefacts. Storage of aggressive and hygroscopic (water-deficient) solutions is also problematic and these solutions might "interact" with the instruments (or with the sample holder) and damage them. Alkaline solutions also tend to absorb airborne CO_2. Hence, the first step is always the removal, or at least the minimization, of the carbonate content of the starting solution.[36]

Having said all these, the most relevant rule in the present context is that it is imperative to strictly follow all the safety regulations during the experimental work!

From these caveats, it follows that in this field it is of primary importance to check the validity of any observed experimental effect and accept the existence of a given novel solution species only when it is supported by multiple experimental observations. We think that this may possibly be the main reason why systematic investigations of such systems are relatively scarce, even though the knowledge gained can be important both scientifically and industrially.

19.3 Hydroxido Complexes of Some Selected Metal Ions Forming at pH > 13

Hydroxido complexes, which are not present or undetectable at moderate pH but have been proven to exist in extremely concentrated alkaline solutions, are presented in the following.

19.3.1 Thallium(I)

In systems containing Tl(I), besides the well-known $TlOH^0$ complex, formation of the $Tl(OH)_2^-$ species has been demonstrated at $[OH^-]_T > 0.5$ M from UV- and Tl-NMR measurements.[37] Also, experimental data suggested the formation of the $TlAl(OH)_4^0$ ion-pair in concentrated aluminate solutions.

19.3.2 Calcium(II)

Solubility experiments of portlandite, $Ca(OH)_2$, at various temperatures and background electrolyte concentrations, supported by turbidity measurements, provided sound evidence for the dominant formation of dissolved

$Ca(OH)_2^0$ above pH $\approx$ 13, whereas the well-known $CaOH^+$, which was believed to be the only hydroxido complex at this pH in earlier studies,[1,2] is also formed albeit to a much smaller extent;[28] see Figure 19.2.

19.3.3 Tin(II)

On the basis of the literature,[1,2] the last stepwise hydroxido complex of Sn(II) is the $Sn(OH)_3{}^-$ species. However, under strongly alkaline conditions, the formation of $Sn(OH)_4{}^{2-}$ or even $Sn(OH)_6{}^{4-}$ has been proposed.[3] Potentiometric titrations using a H_2/Pt electrode (up to 1 M concentration of free OH^-), as well as Raman, Mössbauer and EXAFS spectroscopies, supplemented by DFT calculations, have shown the presence of a single monomeric complex with a tin(II):hydroxide ratio of 1:3. Thus, the predominant complex is the trigonal pyramidal, *i.e.* 'piano-chair-like' $Sn(OH)_3{}^-$, whereas the formation of the other possible complex, $SnO(OH)^-$, has not been proven with experiments or simulations.[38]

19.3.4 Lead(II)

From UV-vis, [207]Pb NMR and Raman spectroscopic and polarographic studies[39] in solutions of pH > 13, it has been claimed that lead(II) ions are predominantly present as $Pb(OH)_4{}^{2-}$. The structure of this complex was

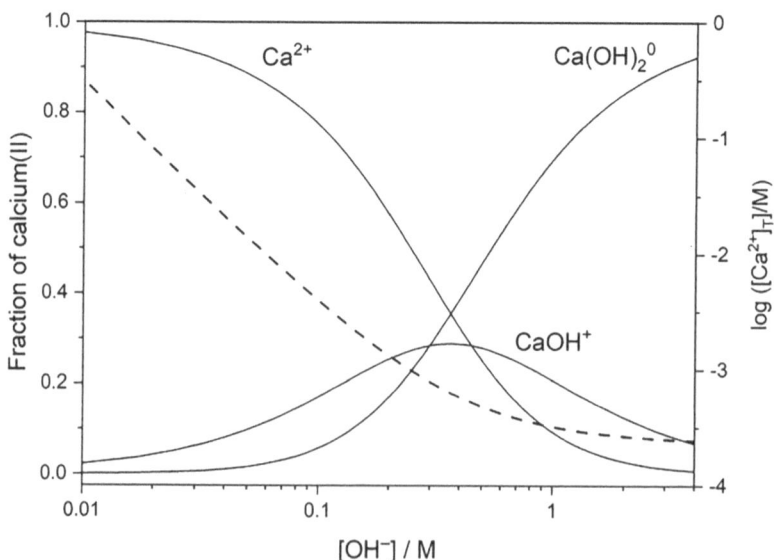

Figure 19.2 Left axis: Fraction of dissolved calcium(II) as a function of equilibrium hydroxide concentration, $[OH^-]$, for NaOH solutions saturated with $Ca(OH)_2(s)$ at 25 °C and 1 M (NaCl) ionic strength. Right axis: Dimensionless solubility of calcium(II), $\log([Ca^{2+}]_T/M)$. Reproduced from ref. 28 with permission from the Royal Society of Chemistry.

estimated by applying quantum chemical calculations.[40] Later, EXAFS as well as experimental and simulated Raman spectra prove that the last stepwise hydroxido complex of lead(II) is $Pb(OH)_3^-$, which assumes trigonal pyramidal geometry.[41]

19.3.5 Iron(III)

Preparation of a supersaturated iron(III) solution in a ~ 20 M NaOH medium gives rise to the formation of a solid Fe(III) hydroxido complex salt, which consists of highly symmetrical $Fe(OH)_6^{3-}$ building blocks, as deduced from XANES, EXAFS and Mössbauer spectroscopic data.[10] As for the solution phase, the structure of ferric ions is markedly different, as both EXAFS and XANES data suggest that only $Fe(OH)_4^-$ is formed. Consequently, $Fe(OH)_6^{3-}$ solution species are not formed in experimentally observable quantities, even at the highest possible hydroxide concentrations.

19.3.6 Aluminium(III)

Due to their clear industrial relevance, strongly alkaline aluminate solutions have been studied very intensely over the last few decades;[21,23,35,37,42–45] for a review on this topic, see ref. 46. Based on a multitude of experimental techniques, two Al-containing solution species, $Al(OH)_4^-$ and $[(OH)_3Al-O-Al(OH)_3]^{2-}$, along with their alkali-metal ion-pairs, are sufficient to account for all the experimental observations collected for concentrated alkaline aluminate solutions at temperatures up to 100 °C. The fraction of $Al(OH)_4^-$ at different total concentrations of Al(III) is shown in Figure 19.3. An extended network of these entities held together by electrostatic forces and hydrogen bonds seems also to be likely. Formation of dehydrated monomers (AlO_2^- or $AlO(OH)_2^-$) or further stepwise hydroxido complexes ($Al(OH)_5^{2-}$ and $Al(OH)_6^{3-}$) at significant concentrations can be excluded. Aluminate species formed *via* the release of hydroxide from $Al(OH)_4^-$ and subsequent oligomerization have also been conclusively eliminated. Also, hexameric aluminate complexes, believed to be the docking species during gibbsite crystallization, have been reported to be experimentally undetectable.

19.3.7 Gallium(III)

As for gallium(III), ^{71}Ga NMR and Raman spectra[29] suggest only one spectroscopically significant species to be present in strongly alkaline gallate solutions, which is the monomeric $Ga(OH)_4^-$ complex. Conclusions identical to these were drawn from EXAFS measurements in solutions of pH < 11[47] and pH < 13.[48] Further, dimerization of this complex, structurally related to the $[(OH)_3Al-O-Al(OH)_3]^{2-}$ aluminate species,[1,21,23,37–42] might not take place.[29] Solution X-ray diffraction measurements on strongly alkaline gallate solutions supplemented with DFT quantum mechanical calculations[49] confirmed this conclusion and proved the exclusive formation of $Ga(OH)_4^-$ in hyperalkaline media.

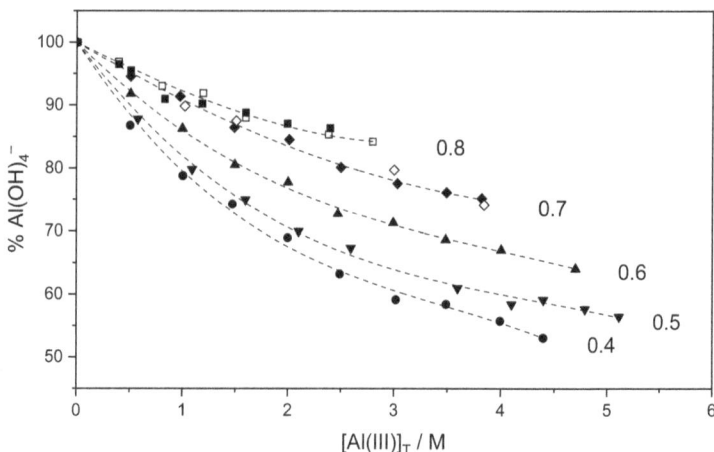

Figure 19.3   The percentage of aluminate, $Al(OH)_4^-$, of the total dissolved aluminium(III), $[Al(III)]_T$, assuming the formation of the dimeric aluminate species, $[(OH)_3Al-O-Al(OH)_3]^{2-}$. Each data series corresponds to a sample series with constant water activity, shown as numbers. Full and empty symbols represent independent reproductions, whereas dashed lines serve as a guide to the eye. Reproduced from ref. 23 with permission from the Royal Society of Chemistry.

19.3.8   Solids

Solid materials, *i.e.* metal hydroxides, oxides, oxyhydroxides or complex metal hydroxido salts, crystallizing from these solutions are also of potential interest, at least on two accounts. First, the predominant hydroxido complex, both mono- and polynuclear, may or may not be the precursor of the solid material crystallizing from the mother liquor. For example, only the tetrahedrally coordinated (T_d) M(III) species (M = Al, Fe, Ga) are seen by any experimental technique in their alkaline solutions, while the predominant geometry in the solid state is octahedral (O_h).[10,23,29,42,44–46,49] The question of whether the $T_d \rightarrow O_h$ transformation of the metal involves a (minor) solution species or it takes place on the surface of the $M(OH)_3$ during crystallization still remains unresolved.[42] The second interesting aspect is the morphology and nanostructure of the solid substances obtained from these solutions. For instance, the fabrication of titanate nanotubes *via* a hydrothermal reaction of anatase in 10 M NaOH[50–53] has been proven to proceed *via* a peculiar solution species, resulting in oriented crystal growth.[54]

19.4   Complexes with O-donor Ligands Forming in Hyperalkaline Solutions

19.4.1   Calcium(II)

The complexation of different O-donor ligands has been most intensively studied for Ca(II) due its relevance in the production of alumina, where a large amount of lime is used for the recovery of soda.[17,55] As for the ligands, the most relevant is D-gluconate, $Gluc^-$ (Figure 19.4), due to its important role in numerous industrial applications.[56] Potentiometric titrations conducted in a wide temperature range (25–75 °C) at high ionic strengths (1 or 4 M NaCl) showed the formation of $CaGlucH^0_{-1}$ and $Ca_3Gluc_2H^0_{-4}$ unequivocally,[22,57] with a third complex being either $Ca_2Gluc_2H_{-4}^{2-}$ or $Ca_2GlucH^0_{-3}$.[12,22,57] The highly stable di- and trinuclear species, which were also identified by mass spectrometry,[22] dominate the strongly alkaline region above pH $\approx$ 11.5. The stability is associated with slow ligand exchange as suggested by ^1H NMR spectra, indicating strong Ca^{2+}–alcoholate interactions with the main binding sites being the –C(2)OH and –C(3)OH groups, besides COO^-. Thus, one or two –OH groups of each anion are deprotonated, yielding the potentiometrically indistinguishable formulas of $Ca_3(OH)_2Gluc_2H^0_{-2}$ or $Ca_3Gluc_2H^0_{-4}$. A possible structure of the latter has been proposed based on quantum chemical geometry optimization; see Figure 19.5. The trinuclear complex has been reported to form with the structurally similar D-heptagluconate $(Hpgl^-)$ and L-gulonate (Gul^-) too (Figure 19.4).[12] For each complex, the ligand acts as a bridging anion;[12] see Figure 19.5. Furthermore, amorphous $Ca_3(OH)_2Gluc_2H_{-2}$ has been isolated in the solid state as well.[57]

Besides $Gluc^-$, the calcium(II) binding of the structurally related α-D-isosaccharinate, Isa^- (Figure 19.4), has also received scientific interest due to its potential impact on the solubility of actinides in concentrated $CaCl_2$ brines in underground waste repositories.[58,59] Similar to $Gluc^-$, $CaIsaH^0_{-1}$ has also

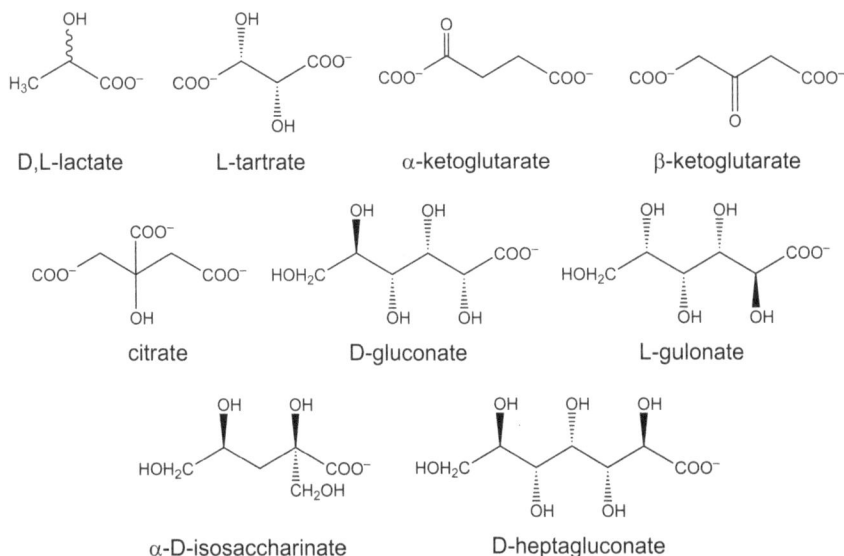

Figure 19.4 Structural formulas for O-donor ligands covered in the present contribution.

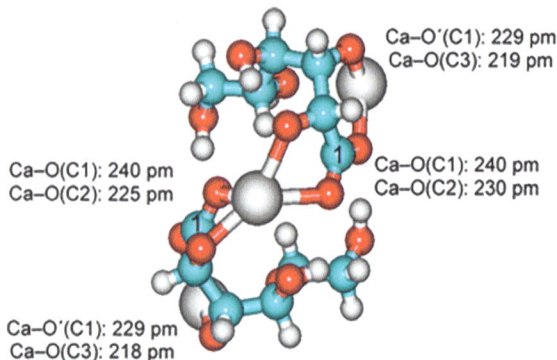

Ca–O'(C1): 229 pm
Ca–O(C3): 219 pm

Ca–O(C1): 240 pm
Ca–O(C2): 225 pm

Ca–O(C1): 240 pm
Ca–O(C2): 230 pm

Ca–O'(C1): 229 pm
Ca–O(C3): 218 pm

Figure 19.5 Proposed structure for the trinuclear gluconato complex of calcium(II), $Ca_3Gluc_2H_{-4}^0$, as obtained *via* geometry optimization at the HF/6-31G(d,p) level of theory. Also shown are some selected metal–oxygen distances; the numbering starts at the carbon atom of the COO^- group (C1). Each ligand is bound to the central Ca^{3+} ion *via* a carboxylate and a $C(2)$-O^- alcoholate oxygen, and to the other Ca^{2+} ions *via* a carboxylate and a $C(3)$-O^- alcoholate oxygen. Reproduced from ref. 22 with permission from American Chemical Society, Copyright 2014.

been detected in strongly alkaline media.[59,60] Strikingly, the multinuclear complexes found for Gluc$^-$ are not formed with Isa$^-$, suggesting the importance of an ^-OOC–$CHOH$–$CHOH$ sequence for the formation of such species; this moiety is absent in Isa$^-$.

The complexation of Ca(II) in hyperalkaline media has been extensively studied also in the presence of other small-molecular-weight ligands (Figure 19.4), applying potentiometry, spectrophotometry, NMR spectroscopy, and DFT structural optimization.[61–64] Racemic lactate[64] and L-tartrate[62] form mononuclear, deprotonated complexes, where the displacement of proton(s) takes place most likely on the water molecule(s) bound to the metal ion. Dinuclear complexes have been proposed for α- and β-ketoglutarate, where deprotonation also occurs on the $-CH_2$ moiety, yielding an enolate/carbanion.[63] Different from these, –OH deprotonation was detected for citrate upon complexing Ca^{2+} ions.[61]

19.4.2 Iron(III)

Potentiometric and amperometric titrations as well as spectrophotometric measurements showed strong binding of Fe^{3+} ions – or possibly $Fe(OH)_4{}^{2-}$ – to Gluc$^-$ above pH $\approx$ 13, yielding the $Fe(OH)GlucH_{-3}{}^{2-}$ complex.[65] The presence of a threefold deprotonated anion seems nevertheless unlikely, given the high affinity of Fe(III) toward hydrolysis. Later, the formation of 1:1 and 1:2 iron(III):gluconate and species with numerous deprotonation states and that of the $Fe_2Gluc_2H_{-7}{}^{3-}$ dimer have been reported.[66] Upon addition of Ca(II), ternary complexes with 1:1:1 and 1:1:2 calcium(II):iron(III):gluconate ratios are also formed[66] and have been proposed for use as reversible redox agents in the indirect reduction/oxidation cycles of vat dyes, such as indigo.[18] Interestingly, a basic ternary complex has been prepared also in the solid phase as early as 1936, with the composition of $CaFeGluc_2H_{-4}$.[67]

19.4.3 Aluminium(III)

Upon binding of Al^{3+} ions, $Gluc^-$ and $Hpgl^-$ exhibit ligand deprotonation already in the acidic region due to the higher Lewis acidity of Al^{3+} as compared to Ca^{2+}. In a neutral medium, hydrolysis of Al^{3+} becomes more pronounced, yielding mixed hydroxido-(hepta)gluconato complexes.[14,68] Interestingly, a strongly alkaline medium induces a condensation reaction between two –OH moieties of $Al(OH)_4^-$ and those of the ligand (catalysed probably by the OH^- ion), giving rise to the formation of $Al(OH)_4Gluc^{2-}$[24] or $Al(OH)_4Hpgl^{2-}$.[14] In addition to these, potentiometric data suggested also the formation of complexes $Al(OH)_5Hpgl^{3-}$, $Al(OH)_5Hpgl_2^{4-}$, and $Al_3(OH)_{13}Hpgl_2^{6-}$ at $pH > 13$. Further, the ^1H and ^{13}C signals of the free and complexed $Hpgl^-$ separate in the NMR spectra, indicative of very slow ligand exchange and thus strong (essentially covalent) Al–O–C bonds.[14,24] As for $Gluc^-$, significant broadening of the ^{27}Al NMR signal of $Al(OH)_4^-$ upon complexation indicates a distortion of the tetrahedral symmetry of the metal ion.[24]

In the presence of both Ca(II) and Al(III), amorphous ternary complexes precipitate from caustic solutions,[69] with compositions of $CaAlHpgl(OH)_4$, $Ca_2AlHpgl_2(OH)_5$ and $Ca_3Al_2Hpgl_3(OH)_9$, supporting an earlier finding that gluconato complexes are formed when adding NaGluc to a tricalcium aluminate suspension.[70]

19.4.4 Neodymium(III)

The complexation of trivalent actinides by various organic ligands, especially $Gluc^-$ and Isa^-, plays an important role in the mobilization of these metal ions in a reducing, hyperalkaline environment typical for cementitious-based underground repositories for low- and intermediate-level waste.[15,16] Neodymium(III), which is often used as a model for these radionuclides, forms stable mono- and dinuclear complexes with $Gluc^-$ involving ligand OH deprotonation at $pH < 8$.[71] In a strongly alkaline medium at 4 M ionic strength (NaCl), the speciation of Nd(III) is dominated by dinuclear complexes,[72] $Nd_2Gluc_4H_{-5}^{3-}$, $Nd_2Gluc_4H_{-6}^{4-}$, $Nd_2Gluc_6H_{-4}^{4-}$, and $Nd_2Gluc_6H_{-5}^{5-}$. Consistent with these findings, the monomer of $Nd_2Gluc_4H_{-6}^{4-}$, *i.e.* $NdGluc_2H_{-3}^{2-}$, has been found for praseodymium(III)[26] and numerous other lanthanides in dilute solutions.[13] However, metal–ligand interactions weaken in favour of metal hydrolysis in a strongly alkaline medium as suggested by NMR and CD spectra not only for $Gluc^-$, but also for Gul^-.[73] Thus, the above species are better described as Nd(III) hydroxido complexes bound weakly to the organic anion, *i.e.* $Nd_2(OH)_6Gluc_4^{4-}$ *vs.* $Nd_2Gluc_4H_{-6}^{4-}$.

As for systems containing also $CaCl_2$, a substantial enhancement of solubility has been detected at pH $\approx$ 12 and attributed to the formation of a ternary complex containing both Ca^{2+} and Nd^{3+} ions.[16] Given $Nd(OH)_3^0$ and $Ca_3(OH)_4Gluc_2^0$ are the dominant species in the respective binary samples,[12,22] the formation of $Ca_3Nd(OH)_7Gluc_2^0$ has been proposed,[16] resembling the stoichiometry of the mixed-metal hydroxide species $Ca_3Nd(OH)_6^+$ forming already in the absence of $Gluc^-$.[74] Moreover, DFT calculations suggested the sixfold deprotonated $Ca_3Nd(OH)Gluc_2H_{-6}^0$ species to be the most stable, with coordination sites being the COO^-, $C(3)-O^-$ and $C(4)-O^-$ groups for each ligand;[16] see Figure 19.6.

Nd^{3+} ions exhibit strong affinity for citrate (Cit^{3-}) too, forming a trinuclear $Nd_3(OH)_4Cit_4^{7-}$ complex which dominates Nd(III) speciation above pH $\approx$ 8.[75] The strong metal-ion binding has been confirmed by solubility measurements as well in the pH range of 7–14 in concentrated NaCl media.[76] However, addition of $CaCl_2$ decreases this effect significantly due to the competitive association between Ca^{2+} and Cit^{3-} ions.

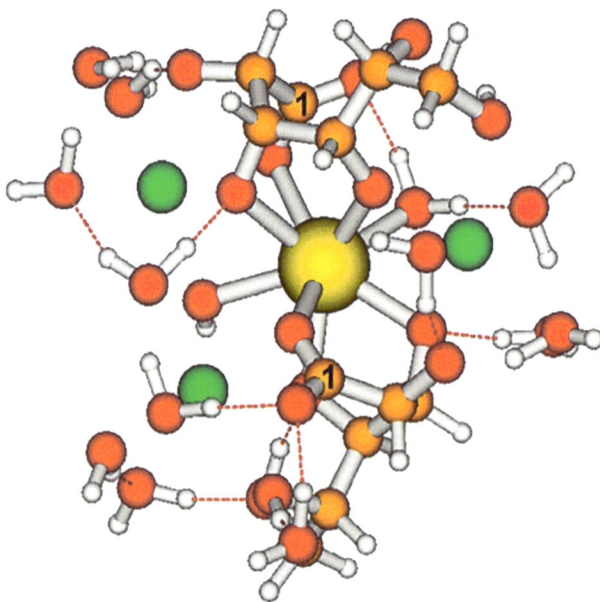

Figure 19.6 Proposed structure for the tetranuclear gluconato complex of calcium(II) and neodymium(III), $Ca_3Nd(OH)Gluc_2H_{-4}^0$, as obtained *via* geometry optimization at the BP86/def2-SVP level of theory; implicit solvation has been taken into account applying the COSMO model. The numbering starts at the carbon atom of the COO^- group (C1). Each ligand is bound to the central Nd^{3+} ion *via* a carboxylate, a $C(3)-O^-$ and a $C(4)-O^-$ alcoholate oxygen. Reproduced from ref. 16 with permission from Elsevier, Copyright 2021.

Problem 1

The molar concentration of an almost saturated aqueous CsOH solution is $c = 14.839$ mol L^{-1} and its density is $\rho = 2.8014$ g cm^{-3} (25 °C). (a) Calculate the molar ratio of H_2O : CsOH in this system. $M(CsOH) = 149.91$ g mol^{-1}. (b) Calculate the weight of CsOH·H_2O ($M = 167.93$ g mol^{-1}) and that of H_2O ($M = 18.02$ g mol^{-1}) that has to be mixed to prepare 1 L of such solution.

Problem 2

We measure the solubility of calcium hydroxide in a hyperalkaline solution (e.g. $c_{NaOH} = 1$ mol L^{-1}). Under these conditions, the hydroxido complexes $CaOH^+$ and $Ca(OH)_2^0$ are formed with equilibrium constants β_1 and β_2, respectively. Derive an equation for the dependence of the solubility of $Ca(OH)_2(s)$, $[Ca^{2+}]_T$, on c_{NaOH} using the solubility product of $Ca(OH)_2$, K_{sp}. ($[Ca^{2+}]_T$ equals the total concentration of dissolved Ca^{2+} ions.) What is the limiting value of $[Ca^{2+}]_T$ at very high base concentrations?

Problem 3

We conduct a study with a hyperalkaline aqueous solution consisting of Ca(II), Al(III) and D-gluconate ions. Which experimental means can be used for studying such a system? Specify what kind of information can be obtained from those different techniques.

Problem 4

We have a solution containing aluminium(III) and sodium D-gluconate at pH = 13, which we want to titrate with HCl and follow the changes by measuring the pH of the solution. We find that applying different waiting times between two added aliquots of acid, we obtain somewhat different titration curves. What is the main reason for this phenomenon and how could this issue be solved?

Answer to Problem 1

(a) Take 1 L of this solution; its weight is 2.8014 kg ($m = \rho \cdot V$). The dissolved CsOH weighs 2.2245 kg since $m(CsOH) = c \cdot V \cdot M(CsOH)$. The rest of this solution, 0.5768 kg, is equal to 32.01 mol of the solvent water. The ratio is 2.16; thus, a bit more than 2 moles of water per one CsOH unit is able to keep the solute in solution. (b) Likewise, the weight of the CsOH·H_2O is 2.4919 kg since $m(CsOH·H_2O) = c \cdot V \cdot M(CsOH·H_2O)$. Therefore, the amount of water is $0.5768 - (2.4919 - 2.2245) = 0.3094$ kg. The weight of the solute is *ca.* 8 times as much as that of the added solvent.

Answer to Problem 2

In equilibrium, the mass balance equations for Ca^{2+} and OH^-: $[Ca^{2+}]_T = [Ca^{2+}] + [CaOH^+] + [Ca(OH)_2^0]$ and $[OH^-]_T = [OH^-] + [CaOH^+] + 2[Ca(OH)_2^0]$, where square brackets denote equilibrium concentrations. The equilibrium concentrations of the hydroxido complexes can be expressed using the formation constants: $[CaOH^+] = \beta_1[Ca^{2+}][OH^-]$ and $[Ca(OH)_2^0] = \beta_2[Ca^{2+}][OH^-]^2$, respectively. Combining these with the mass balance for Ca^{2+}, we obtain

$$[Ca^{2+}]_T = [Ca^{2+}] + \beta_1[Ca^{2+}][OH^-] + \beta_2[Ca^{2+}][OH^-]^2.$$

Using this relationship and substituting $K_{sp} = [Ca^{2+}][OH^-]^2$, we obtain

$$[Ca^{2+}]_T = K_{sp}[OH^-]^{-2} + \beta_1 K_{sp}[OH^-]^{-1} + \beta_2 K_{sp}.$$

Note that $[Ca(OH)_2^0] = \beta_2 K_{sp}$ is constant.

The only source of calcium(II) is its dissolution from the solid phase, whereas hydroxide stems from both the dissolution of $Ca(OH)_2$ and addition of NaOH to the system. Thus, the total concentration of hydroxide, $[OH^-]_T$, equals exactly $c_{NaOH} + 2[Ca^{2+}]_T$. However, given the large excess of NaOH as compared to calcium(II), $[OH^-]_T \approx [OH^-] \approx c_{NaOH}$. Thus, substituting $[OH^-]$ for c_{NaOH} in the above equation, we obtain

$$[Ca^{2+}]_T = K_{sp} c_{NaOH}^{-2} + \beta_1 K_{sp} c_{NaOH}^{-1} + \beta_2 K_{sp}.$$

At very high base concentrations, the first two terms become negligible as compared to $\beta_2 K_{sp}$, hence the limiting value of the $[Ca^{2+}]_T$ reads as

$$[Ca^{2+}]_T = \beta_2 K_{sp}.$$

Answer to Problem 3

Potentiometry (with the H_2/Pt electrode): composition and formation constant of the species in solution; solubility: composition and formation constants of the species formed; UV spectrophotometry: equilibria and changes associated with C=O double bonds (limited use); polarimetry and CD spectroscopy: structural changes in the organic ligand upon complexation; NMR spectroscopy (^1H, ^{13}C, ^{27}Al), identification of binding sites, dynamics of the chemical exchange (qualitative); EXAFS and XANES: the local structure of the metal ions; Raman spectroscopy: equilibria and structure of the complexes formed; ESI-MS: composition of the species formed (limited use); freezing point depression and conductometry: model validation.

Answer to Problem 4

Al^{3+} ions are known as inert ions, that is, their water exchange, complex formation and association reactions are slow as compared to those of labile metal ions (*e.g.* Ca^{2+}). Thus, monitoring such slow processes *via* measuring solution pH is strongly time-dependent. Consequently, applying different waiting times will result in different potentials. To tackle this problem, titrations can be performed with systematically increasing the dwelling times at each titration step. If the differences between two cell potentials (recorded with different waiting times) at the same added volume of HCl are similar to the error of the electrode (*e.g.* 1 mV), then the thermodynamic equilibrium has likely been attained. However, if the duration of the titration is too long, the pH electrode, which is continuously in contact with the solution, might start to wear out (too long response times, signal drifts, *etc.*). In this case, batch measurements can be applied: separate samples with different added volumes of HCl are prepared and the pH is monitored *via* a certain time period for each sample. Here, the electrode needs to be immersed into the solution only for a short time to read the actual pH.

Acknowledgements

The authors acknowledge the work of the members of the Materials and Solution Structure Research Group at the Department of Analytical and Inorganic Chemistry, University of Szeged, which greatly contributed to the content of the current review.

References

1. C. F. Baes and R. E. Mesmer, *The Hydrolysis of Cations*, Wiley Interscience, New York, US, 1976.
2. C. Ekberg and P. L. Brown, *Hydrolysis of Metal Ions*, Wiley-VCH, Weinheim, Germany, 2016.
3. F. A. Cotton and G. Wilkinson, *Advanced Inorganic Chemistry*, Wiley Interscience, New York, USA, 1988, p. 217.
4. *Comprehensive Coordination Chemistry*, ed. G. Wilkinson, Pergamon, Oxford, UK, 1987, vol. 2, p. 305.
5. I. A. Sheka, I. S. Chaus and T. T. Mityureva, *The Chemistry of Gallium*, Elsevier, London, UK, 1966, p. 46.
6. W. N. Perera and G. T. Hefter, *Inorg. Chem.*, 2003, **42**, 5917.
7. H. R. Watling, S. D. Fleming, W. van Bronswijk and A. Rohl, *J. Chem. Soc., Dalton Trans.*, 1998, 3911.
8. P. Sipos, G. T. Hefter and P. M. May, *Talanta*, 2006, **70**, 761.
9. P. Sipos, T. Megyes and O. Berkesi, *J. Solution Chem.*, 2008, **37**, 1411.
10. P. Sipos, D. Zeller, E. Kuzmann, A. Vértes, Z. Homonnay, M. Walczak and S. E. Canton, *Dalton Trans.*, 2008, 5603.
11. B. Gyurcsik and L. Nagy, *Coord. Chem. Rev.*, 2003, **203**, 81.
12. B. Kutus, X. Gaona, A. Pallagi, I. Pálinkó, M. Altmaier and P. Sipos, *Coord. Chem. Rev.*, 2020, **417**, 213337.
13. S. Giroux, S. Aury, B. Henry and P. Rubini, *Eur. J. Inorg. Chem.*, 2002, 1162.

14. Á. Buckó, B. Kutus, G. Peintler, Z. Kele, I. Pálinkó and P. Sipos, *J. Mol. Liq.*, 2020, **314**, 113645.

15. X. Gaona, V. Montoya, E. Colàs, M. Grivé and L. Duro, *J. Contam. Hydrol.*, 2008, **112**, 217.

16. H. Rojo, X. Gaona, T. Rabung, R. Polly, M. García-Gutiérrez, T. Missana and M. Altmaier, *Appl. Geochem.*, 2021, **126**, 104864.

17. S. Rosenberg, D. A. Wilson, and C. A. Heath, in *Essential Readings in Light Metals, Alumina and Bauxite*, ed. D. Donaldson and B. E. Raahauge, John Wiley & Sons, Hoboken, US, 2013, vol. 1, p. 210.

18. M. Abdelileh, A. P. Manian, D. Rhomberg, M. B. Ticha, N. Meksi, N. Aguiló-Aguayo and T. Bechtold, *J. Cleaner Prod.*, 2020, **266**, 121753.

19. J. A. Rendleman, *Food Chem.*, 1978, **3**, 127.

20. M. van Duin, J. A. Peters, A. P. G. Kieboom and H. van Bekkum, *Recl. Trav. Chim. Pays-Bas*, 1989, **108**, 57.

21. P. Sipos, M. Schibeci, G. Peintler, P. M. May and G. Hefter, *Dalton Trans.*, 2006, 1858.

22. A. Pallagi, É. G. Bajnóczi, S. E. Canton, T. Bolin, G. Peintler, B. Kutus, Z. Kele, I. Pálinkó and P. Sipos, *Environ. Sci. Technol.*, 2014, **48**, 6604.

23. P. Sipos, P. M. May and G. Hefter, *Dalton Trans.*, 2006, 368.

24. A. Pallagi, Á. G. Tasi, G. Peintler, P. Forgo, I. Pálinkó and P. Sipos, *Dalton Trans.*, 2013, **42**, 13470.

25. M. Altmaier, X. Gaona and T. Fanghänel, *Chem. Rev.*, 2013, **113**, 901.

26. S. Giroux, P. Rubini, B. Henry and S. Aury, *Polyhedron*, 2000, **19**, 1567.

27. A. Tasi, X. Gaona, D. Fellhauer, M. Böttle, J. Rothe, K. Dardenne, R. Polly, M. Grivé, E. Colàs, J. Bruno, K. Källström, M. Altmaier and H. Geckeis, *Appl. Geochem.*, 2018, **98**, 247.

28. B. Kutus, A. Gácsi, A. Pallagi, I. Pálinkó, G. Peintler and P. Sipos, *RSC Adv.*, 2016, **6**, 45231.

29. P. Sipos, T. Megyes and O. Berkesi, *J. Solution Chem.*, 2008, **37**, 1411.

30. A. Nandy, C. Duan, M. G. Taylor, F. Liu, A. H. Steeves and H. J. Kulik, *Chem. Rev.*, 2021, **121**, 9927.

31. Y. Kwon, H.-K. Kim and K. Jeong, *Molecules*, 2022, **27**, 1500.

32. Y. Marcus and G. Hefter, *Chem. Rev.*, 2000, **106**, 4585.

33. I. Kron, S. L. Marshall, P. M. May, G. Hefter and E. Königsberger, *Monatsh. Chem.*, 1995, **126**, 819.

34. H. S. Harned, and B. B. Owen, *The Physical Chemistry of Electrolytic Solutions*, Reinhold Publishing Corporation, New York, US, 1950.

35. P. Sipos, P. M. May, G. T. Hefter and I. Kron, *J. Chem. Soc., Chem. Commun.*, 1994, 2355.

36. P. Sipos, P. M. May and G. Hefter, *Analyst*, 2000, **125**, 955.

37. P. Sipos, S. G. Capewell, P. M. May, G. T. Hefter, G. Laurenczy, F. Lukács and R. Roulet, *J. Solution Chem.*, 1997, **26**, 419.

38. É. G. Bajnóczi, E. Czeglédi, E. Kuzmann, Z. Homonnay, S. Bálint, G. Dombi, P. Forgo, O. Berkesi, I. Pálinkó, G. Peintler, P. Sipos and I. Persson, *Dalton Trans.*, 2014, **43**, 17971.

39. W. N. Perera, G. T. Hefter and P. M. Sipos, *Inorg. Chem.*, 2001, **40**, 3974.

40. S. Soralová and M. Breza, *Collect. Czech. Chem. Commun.*, 2008, **73**, 59.
41. É. G. Bajnóczi, I. Pálinkó, T. Körtvélyesi, S. Bálint, I. Bakó, P. Sipos and I. Persson, *Dalton Trans.*, 2014, **43**, 17539.
42. K. R. Bauspiess, T. Murata, G. Parkinson, P. Sipos and H. Watling, *J. Phys. IV France*, 1997, **7**, C2-485–C2-487.
43. P. Sipos, G. Hefter and P. M. May, *Aust. J. Chem.*, 1998, **51**, 445.
44. T. Radnai, P. M. May, G. T. Hefter and P. Sipos, *J. Phys. Chem. A*, 1998, **102**, 7841.
45. P. Sipos, G. T. Hefter and P. M. May, *Talanta*, 2006, **70**, 761.
46. P. Sipos, *J. Mol. Liq.*, 2009, **1**, 146, and references therein.
47. G. S. Pokrovski, J. Schott, J.-L. Hazemann, F. Farges and O. S. Pokrovski, *Geochim. Cosmochim. Acta*, 2002, **66**, 4203.
48. E. Dooryhee, G. N. Greaves, A. T. Steel, R. P. Townsend, S. W. Carr, J. M. Thomas and C. R. A. Catlow, *Faraday Discuss. Chem. Soc.*, 1990, **89**, 119.
49. T. Radnai, S. Bálint, I. Bakó, T. Megyes, T. Grósz, A. Pallagi, G. Peintler, I. Pálinkó and P. Sipos, *Phys. Chem. Chem. Phys.*, 2014, **16**, 4023.
50. E. Horváth, Á. Kukovecz, Z. Kónya and I. Kiricsi, *Chem. Mater.*, 2007, **4**, 927.
51. Á. Kukovecz, M. Hodos, Z. Kónya and I. Kiricsi, *Chem. Phys. Lett.*, 2005, **411**, 445.
52. M. T. Byrne, J. E. McCarthy, M. Bent, R. Blake, Y. K. Gun'ko, E. Horvath, Z. Konya, A. Kukovecz, I. Kiricsi and J. N. Coleman, *J. Mater. Chem.*, 2007, **17**, 2351.
53. M. Hodos, E. Horváth, H. Haspel, Á. Kukovecz, Z. Kónya and I. Kiricsi, *Chem. Phys. Lett.*, 2004, **399**, 512.
54. Á. Kukovecz, M. Hodos, E. Horváth, G. Radnóczi, Z. Kónya and I. Kiricsi, *J. Phys. Chem. B*, 2005, **129**, 17781.
55. B. I. Whittington, *Hydrometallurgy*, 1996, **43**, 13.
56. S. Ramachandran, P. Fontanille, A. Pandey and C. Larroche, *Food Technol. Biotechnol.*, 2006, **44**, 185.
57. Á. Buckó, B. Kutus, G. Peintler, I. Pálinkó and P. Sipos, *Polyhedron*, 2019, **158**, 117.
58. C. Bube, V. Metz, E. Bohnert, K. Garbev, D. Schild and B. Kienzler, *Phys. Chem. Earth*, 2013, **64**, 87.
59. K. Vercammen, M. A. Glaus and L. R. Van Loon, *Acta Chem. Scand.*, 1999, **53**, 241.
60. C. Dudás, B. Kutus, É. Böszörményi, Z. Kele, G. Peintler, I. Pálinkó and P. Sipos, *Dalton Trans.*, 2017, **46**, 13888.
61. A. Gácsi, B. Kutus, Á. Buckó, Z. Csendes, G. Peintler, P. Pálinkó and P. Sipos, *J. Mol. Struct.*, 2016, **1118**, 110.
62. A. Gácsi, B. Kutus, Z. Csendes, T. Faragó, G. Peintler, I. Pálinkó and P. Sipos, *Dalton Trans.*, 2016, **45**, 17296.
63. C. Dudás, B. Kutus, G. Peintler, I. Pálinkó and P. Sipos, *Polyhedron*, 2018, **156**, 89.

64. C. Dudás, B. Kutus, É. Böszörményi, G. Peintler, A. A. A. Attia, A. Lupan, Z. Kele, P. Sipos and I. Pálinkó, *J. Mol. Struct.*, 2019, **1180**, 491.
65. R. L. Pecsok and J. Sandera, *J. Am. Chem. Soc.*, 1955, **77**, 1489.
66. T. Bechtold, E. Burtscher and A. Turcanu, *J. Chem. Soc., Dalton Trans.*, 2002, 2683.
67. W. Traube and F. Kuhbier, *Ber. Dtsch. Chem. Ges. (A and B Ser.)*, 1936, **69**, 2655.
68. A. Lakatos, T. Kiss, R. Bertani, A. Venzo and V. B. di Marco, *Polyhedron*, 2008, **27**, 118.
69. Á. Buckó, Z. Kása, M. Szabados, B. Kutus, O. Berkesi, Z. Kónya, Á. Kukovecz, P. Sipos and I. Pálinkó, *Molecules*, 2020, **25**, 4715.
70. C.-E. Kim, S.-H. Lee and S.-K. Lee, *J. Korean Ceram. Soc.*, 1990, **27**, 883.
71. B. Kutus, N. Varga, G. Peintler, A. Lupan, A. A. A. Attia, I. Pálinkó and P. Sipos, *Dalton Trans.*, 2017, **46**, 6049.
72. É. Böszörményi, Z. Kása, G. Varga, Z. Kele, B. Kutus, G. Peintler, I. Pálinkó and P. Sipos, *J. Mol. Liq.*, 2022, **345**, 117047.
73. É. Böszörményi, O. Dömötör, B. Kutus, G. Varga, G. Peintler and P. Sipos, *J. Mol. Struct.*, 2022, **1261**, 132894.
74. V. Neck, M. Altmaier, T. Rabung, J. Lützenkirchen and T. Fanghänel, *Pure Appl. Chem.*, 2009, **81**, 1555.
75. D.-R. Svoronos, S. Boulhassa, R. Gullaumont and M. Quarton, *J. Inorg. Nucl. Chem.*, 1981, **43**, 1541.
76. N. Adam, K. Hinz, X. Gaona, P. J. Panak and M. Altmaier, *Radiochim. Acta*, 2021, **109**, 431.

CHAPTER 20

Resource Separation from Salt Lake Brine

YONGQUAN ZHOU,* ZHONG LIU AND MIN WANG

Key Laboratory of Comprehensive and Highly Efficient Utilization of Salt Lake Resources, Key Laboratory of Salt Lake Resources Chemistry of Qinghai Province, Qinghai Institute of Salt Lakes, Chinese Academy of Sciences, Xining, Qinghai 810008, China
*Email: Yongqzhou@163.com

20.1 Salt Lake Brine and Resources

Salt lakes usually refer to the most concentrated natural waters with salt contents over 50 $g L^{-1}$.[1] Salt lakes contain abundant mineral resources such as sodium (Na), magnesium (Mg), potassium (K), lithium (Li), boron (B), *etc.* which are considered as the storehouse of inorganic mineral salts. Salt lake brines are extremely complex and concentrated electrolyte solutions, which contain electrolyte ions such as Li^+, Na^+, K^+, Rb^+, Cs^+, Mg^{2+}, $Ca^{2+}//SO_4^{2-}$, Cl^- and $B_4O_7^{2-}$.[1] The pH values of the lakes tend to decrease from the carbonate type, sodium sulfate subtype, magnesium sulfate subtype to the chloride type.[2] Many methods have been employed to extensively exploit salt lakes, and there are many books[3] and review papers[2,4] on this topic. The overall status of the salt lake resources and the exploration in China are shown in Figure 20.1. As shown in Figure 20.1, technical development has made it possible to produce high purity chemical raw materials, such as NaCl, $Na_2SO_4 \cdot xH_2O$, $MgCl_2 \cdot xH_2O$, $MgSO_4 \cdot xH_2O$, $CaCl_2 \cdot xH_2O$, *etc.* through water–salt phase equilibrium and the well-constructed solar evaporation process.[1] In this chapter we will ignore these raw materials and focus on other elements in the following context. Recently, researchers found that ion

Coordination Chemistry Fundamentals Series No. 2
Metal Ions and Complexes in Solution
Edited by Toshio Yamaguchi and Ingmar Persson
© Japan Society of Coordination Chemistry 2024
Published by the Royal Society of Chemistry, www.rsc.org

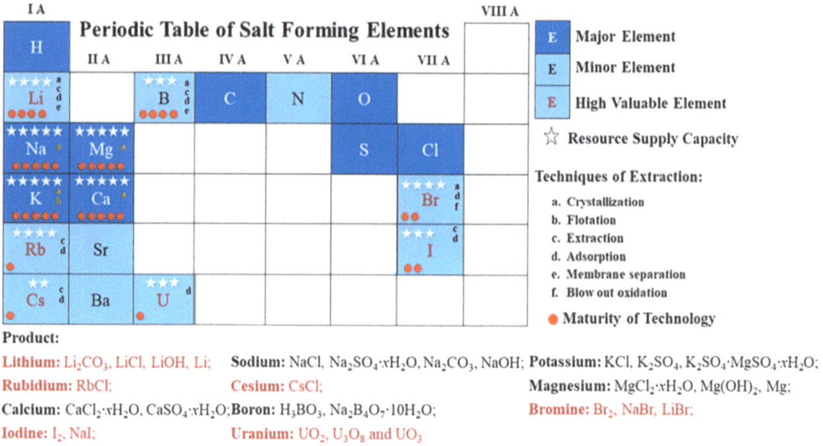

Periodic Table of Salt Forming Elements

I A								VIII A

Elements shown: H (I A); Li, B, C, N, O (with II A, III A, IV A, V A, VI A, VII A groups); Na, Mg, S, Cl; K, Ca, Br; Rb, Sr, I; Cs, Ba, U.

Legend:
- E Major Element
- E Minor Element
- E High Valuable Element
- ☆ Resource Supply Capacity

Techniques of Extraction:
- a. Crystallization
- b. Flotation
- c. Extraction
- d. Adsorption
- e. Membrane separation
- f. Blow out oxidation
- ● Maturity of Technology

Product:

Lithium: Li_2CO_3, LiCl, LiOH, Li; Sodium: NaCl, $Na_2SO_4 \cdot xH_2O$, Na_2CO_3, NaOH; Potassium: KCl, K_2SO_4, $K_2SO_4 \cdot MgSO_4 \cdot xH_2O$;

Rubidium: RbCl; Cesium: CsCl; Magnesium: $MgCl_2 \cdot xH_2O$, $Mg(OH)_2$, Mg;

Calcium: $CaCl_2 \cdot xH_2O$, $CaSO_4 \cdot xH_2O$; Boron: H_3BO_3, $Na_2B_4O_7 \cdot 10H_2O$; Bromine: Br_2, NaBr, LiBr;

Iodine: I_2, NaI; Uranium: UO_2, U_3O_8 and UO_3

Figure 20.1 The overall status of the salt lake resources and the exploration state in China.

hydration and complex formation play very important roles in salt lake resource separation, and in the present section we will focus on this topic mainly based on works from the last decade. In fact, salt lake resource separation covers a wide topic; due to the page limitation and the authors' limited vision, it is a general picture and far from being complete.

20.2 Potassium (K)

Potash as the third indispensable fertilizer for agriculture can improve grain yield to meet the growing huge food demands.[5] Therefore, the demand of potassium is also increasing dramatically, and the global potash demand did exceed 41 Mt (K_2O) in 2020 (www.fertilizer.org). Over 75% of the world's potash fertilizer is produced from solid sylvite ore.[6] However, in some countries such as China, salt lakes supply over 90% domestical potash resources. Potassium chloride accounts for most of the potash used in worldwide agriculture and represents 91% of the world potash capacity, while the remaining 4% includes potassium sulfate (K_2SO_4), potassium nitrate (KNO_3) and potassium–magnesium salts. Many potential technologies have been developed for potassium production from salt lakes, among which selective flotation is the most efficiently and widely used one. Every year, over 7.5 million tons of potassium is produced in China from salt lakes. The K^+ content in salt lake brine with the level of content of 6.0 $g\,L^{-1}$ (average) is relatively low compared with sodium, magnesium and calcium. Briefly, the salt lake brine is concentrated and the solid potash ore (usually carnallite) is obtained through solar evaporation followed by flotation. Salt–water phase equilibrium is the key fundament for solar evaporation, and

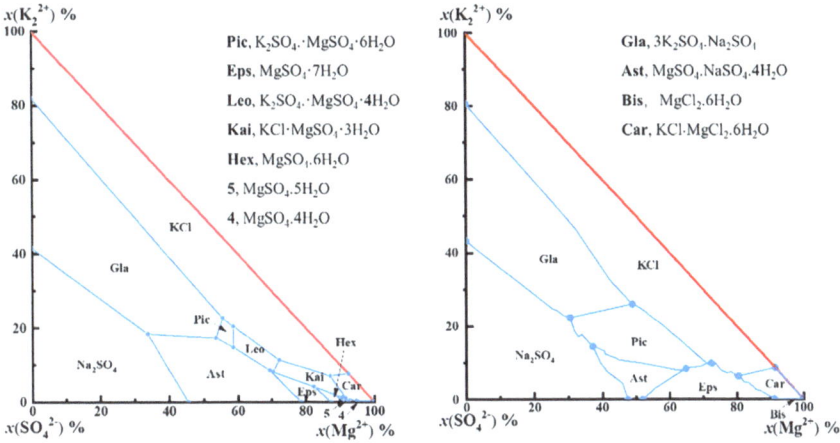

Figure 20.2 The most widely used water–salt phase equilibrium for potash production from salt lakes. The van't Hoff stable equilibrium (left) and the metastable equilibrium (right)[1] pentamery system Na^+, K^+, $Mg^{2+}//Cl^-$, $SO_4^{2-}-H_2O$ at 25 °C.

Figure 20.2 shows the most widely used water–salt phase equilibrium phase diagram used for potash production from salt lakes.[7]

Taking the potash production in China as an example, two flotation technological processes are developed the traditional one in which KCl flotation from NaCl is performed with an amine as the collector after selective decomposition of carnallite, and the second one, in which NaCl is floated from carnallite with an alkyl-morpholine as the collector prior to carnallite decomposition for KCl production.[8] Despite the successful application of flotation technology in the potash industry, the mechanism for collector adsorption has not yet been fully developed to satisfactorily explain soluble potash flotation. The general mechanism for soluble salt bubbling flotation can be found in Figure 20.3. Hancer *et al.*[9] studied the flotation behaviours of soluble salts with dodecyl amine hydrochloride (DAH) and sodium dodecyl sulfate (SDS) collectors. They demonstrated that the interfacial water structure and hydration states of soluble salt surfaces have significant impact in soluble salt flotation. For structure making salts (NaF, NaCl, KF, *etc.*), the brine completely wets the salt surface and no contact angle can be measured. For structure breaking salts (KCl, KI, CsI, *etc.*), the brine does not completely wet the salt surface and a finite contact angle can be measured. The soluble salt flotation with either the cationic DAH or anionic SDS collector is possible only if the salt is a structure breaker. The flotation efficiency also deeply depends on the method used to generate the bubbles: dispersed-air flotation (DispAF), dissolved-air flotation (DAF), and electro-flotation.[10,11] Besides the salt ion hydration, mineral particle surfaces, collector molecules or ions, and collector aggregates play important roles in water-soluble mineral flotation.[9,12]

The interactions between the collectors and the minerals include electrostatic and hydrophobic interactions, hydrogen bonding, and specific interactions, with electrostatic and hydrophobic interactions being the common mechanisms. Some studies have been performed in this field, in which the mechanism of the interaction between the octadecylamine (ODA)-monoacid and KCl/NaCl in flotation separation has been investigated by experiments and molecular dynamics (MD) simulations.[13] Another unavoidable problem is the organic collector (octadecylamine or dodecylmorpholine) dissolution in the brine, which causes environmental risk, affects the follow-up process and decreases the quality of products. Further work in this field such as more effective and more environmentally friendly collector design is desired.

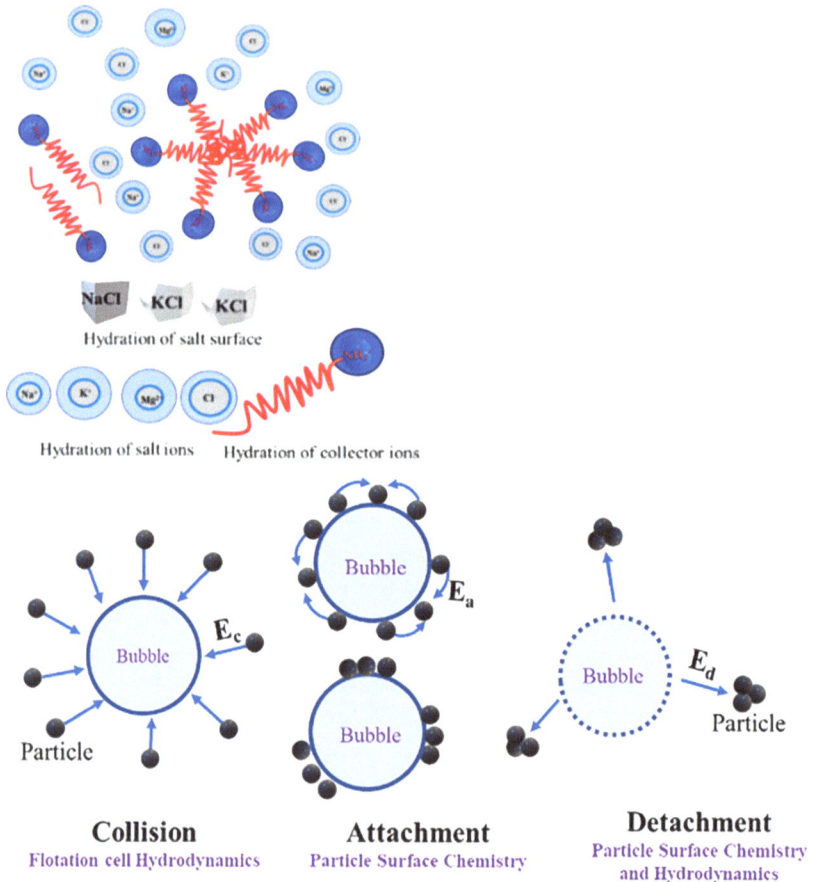

Figure 20.3   Schematic diagram of the hydration phenomenon in a direct KCl flotation system (upper);[12] collision, attachment and detachment processes controlling the collection efficiency in mineral flotation (lower).[14]

20.3   Lithium (Li)

Lithium plays an increasingly important role in scientific and industrial processes. Particularly, with the widespread application of lithium-ion battery (LIB) energy storage devices, the demand for Li resources is increasing sharply.[15] The recent exploitation boom in salt lake brine could have contributed to the "lithium fever". Over 75% of the global Li production comes from Australia (spodumene solid ore), Chile, Argentina and China (brine). Salt lake brine is one of the most substantial and potential lithium resources. Salt lakes are regarded as the primary source of lithium.[4] The salt lakes on the central Andes mountains account for more than 69% of the world's lithium reserves in Li.[4] Developing cost-effective Li extraction technology from the liquid phase is urgent with the gradual depletion of Li mineral ores.

The Li^+ content varies between different salt lakes, with an average content of ~ 60.0 mg L^{-1} and an enrichment factor larger than 358.[1] Besides the relatively low concentration, the high Mg/Li ratio sharply increases the difficulty, which can be attributed to the nearly identical properties of Mg^{2+} and Li^+ (the well-known diagonal rule). Unfortunately, the majority of known brines around the globe have high Mg/Li ratios. In China, around 75% lithium exists in brines which often have high Mg/Li ratios (the highest reaching up to 1832 : 1[16]). In the past decade, a mature general flow sheet for lithium recovery from salt lake brine was established as shown in Figure 20.4. After solar evaporation and sylvite production, the effect of Mg^{2+} has become the most significant impact factor.

Overall, solar evaporation, precipitation, calcination, solvent extraction, adsorption and membrane separation were developed.[17] Among them,

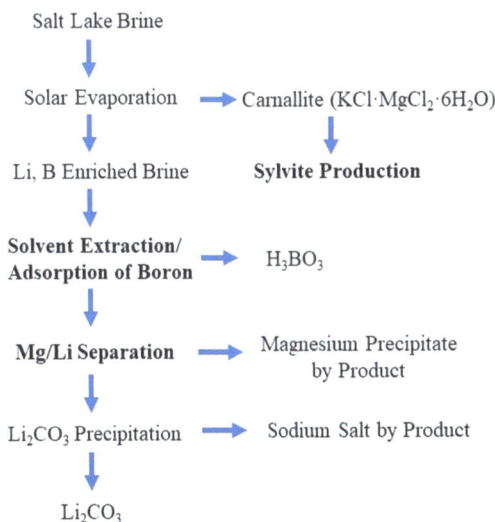

Figure 20.4   The general flow sheet for lithium recovery from salt lake brine.

solvent extraction, adsorption and membrane separation are the three main methods used in industrial applications. Both Li^+ and Mg^{2+} are classified as strongly hydrated and structure making ions,[18] as ion hydration plays a very important role in their separation.

20.3.1   Solvent Extraction

Solvent extraction (or liquid–liquid extraction) has now successfully developed into a highly-effective and commercial technology for the extraction and separation of Li^+ from salt lake brine. There are several studies on the extraction and separation of Li from brines with novel organic systems and the process mechanisms have also been reported, wherein synergistic solvent extraction systems,[19] room temperature ionic liquid systems,[20] crown ether systems,[21] *etc.* are included. Among them, tributyl phosphate (TBP)/kerosene + $FeCl_3$ has received significant attention.[21,22] In this solvent extraction system, TBP serves as the neutral organophosphorus extractant, kerosene as the diluent, and $FeCl_3$ is the coextraction reagent. The extraction mechanism of Li has been shown to be a cation exchange process with the pre-generated species $[Li(TBP)_x]\cdot[FeCl_4]$ (where x is always equal to 2[23]), which is only formed in a high chloride (Cl^-) concentration environment (eqn (20.1)–(20.3)).

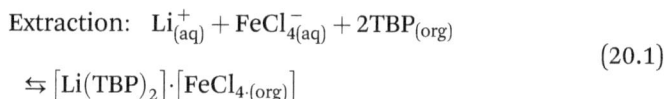

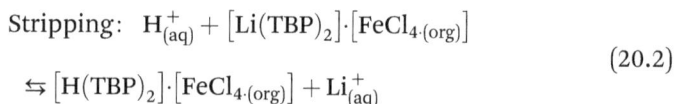

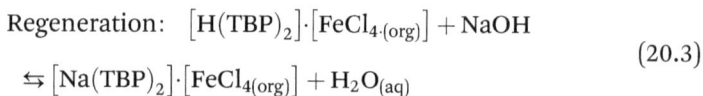

$$\text{Extraction:}\quad Li^+_{(aq)} + FeCl^-_{4(aq)} + 2TBP_{(org)}$$
$$\leftrightarrows \left[Li(TBP)_2\right]\cdot\left[FeCl_{4\cdot(org)}\right] \tag{20.1}$$

$$\text{Stripping:}\quad H^+_{(aq)} + \left[Li(TBP)_2\right]\cdot\left[FeCl_{4\cdot(org)}\right]$$
$$\leftrightarrows \left[H(TBP)_2\right]\cdot\left[FeCl_{4\cdot(org)}\right] + Li^+_{(aq)} \tag{20.2}$$

$$\text{Regeneration:}\quad \left[H(TBP)_2\right]\cdot\left[FeCl_{4\cdot(org)}\right] + NaOH$$
$$\leftrightarrows \left[Na(TBP)_2\right]\cdot\left[FeCl_{4(org)}\right] + H_2O_{(aq)} \tag{20.3}$$

In the extractant, Li^+ coordinates to the oxygen atom of P=O in TBP molecules. The chloride anion is an important factor for Li extraction. For instance, it has been reported that $MgCl_2$ as a chloride source increases the recovery of Li, which can be attributed to the influence of the significant salting-out effect of $MgCl_2$.[24] Thus, the organic system of TBP/kerosene + Cl^- is especially suitable for Li extraction from brines with high Mg/Li ratios because the $MgCl_2$ in the brine provides a sufficient chloride ion environment where a stronger complex ($[Li(TBP)_2]\cdot[FeCl_4]$) can be formed. Extraction, washing, stripping and regeneration processes are included. $6\ mol\,L^{-1}$ hydrochloric acid solution is often applied in the washing and stripping processes, which causes a severe corrosion of materials in the oxidative hydrochloric acid environment. In addition, sometimes a third phase and emulsification also reduce the overall process efficiency. Tremendous efforts have been made to solve these existing problems. Systems with ketones as polar solvents and TBP as the neutral extractant can increase Li^+ partition coefficients and prevent the formation of the third phase. For instance, in the system of TBP/methyl isobutyl ketone (MIBK) + $FeCl_3$, the solvent MIBK plays a significant role in prohibiting the formation of the third phase by promoting the solubility of the extracted complex.[25] Some nonpolar solvents exhibit more extraordinary efficiencies than polar solvents, where solvents with hydrogen bonding functional groups can decrease the extraction coefficient. To avoid the use of severe corrosion causing solutions (always 6 mol L^{-1} hydrochloric acid), a mixed ternary solvent extraction system consisting of TBP, $FeCl_3$ and P506 (2-ethylhexyl phosphonic acid mono-2-ethylhexyl ester) was developed by Su *et al.*[19,23] This newly developed system demonstrated good selectivity of Li over Mg and efficient lithium stripping simply using water. Particularly, during stripping, the Fe^{3+} remained fully in the organic phase, enabling the organic phase to be directly used in the next extraction without regeneration. The high lithium selectivity, lithium recovery and efficient lithium stripping with water promote the use of the TBP/$FeCl_3$/P507 system as a cost-effective and sustainable method for recovering lithium from brines with high Mg/Li ratios.

Another important system that should be noted is the crown ether system.[26] Under the guidance of Pearson's hard–soft acid–base (HSAB) principle, "hard acids prefer to coordinate hard bases and soft acids to soft bases", and based on the size concept, crown ethers such as 12-crown-4, benzo-12-crown-4, *etc.*, and their derivatives with ether oxygens as hard bases have shown very special reactivity with lithium.[21,26] It is worth mentioning that crown ethers are also very promising candidates for lithium isotope separation.[27]

Although various innovative systems have been proposed and huge progresses have been made, an important downside of solvent extraction is that the extractant will dissolve in the brine partially, resulting in the organic extractant loss and environmental risks. Salt lakes are usually located in the eco-fragile regions; although solvent extraction has shown extraordinary performance, the above mentioned drawbacks constrain its large-scale application.

20.3.2 Ion-sieve Adsorption

The adsorption approach has now developed into an excellent candidate method for lithium extraction from brines. Lithium can be easily selected over alkaline and alkaline-earth metal ions from brines by using lithium-ion sieves (LIS). Recently, the synthesis processes and lithium separation performances of ion-sieves have been extensively studied. LIS can be roughly divided into three types: the spinel lithium manganese oxide (LMO) type with a high adsorption rate and superior lithium selectivity, the lithium titanium oxide (LTO) type with low dissolution loss and long recyclability, and the aluminium hydroxide (AH) type with a moderate adsorption rate and capacity, and they have a relatively good selectivity and recycling performance with a low preparation cost.[28]

Due to the specific properties of the spinel lattice structure and framework of LMO and LTO, LIS have high selectivity and capacity, and good special memory. The memory effect of the exchange sites can be attributed to two aspects: the similar ionic radius and the approximate energy of dehydration. As shown in Figure 20.5, ions larger than Li^+ (0.74 Å), such as Na^+ (1.02 Å), K^+ (1.38 Å), and Ca^{2+} (1.00 Å), cannot occupy these sites; while as for Mg^{2+} (0.72 Å, it has a very similar bare ion radius to Li^+, and it is the most competitive with lithium ions in selective adsorption), it requires much higher energy for dehydration: $\Delta G^0_{H,Mg^{2+}} = -1980$ kJ mol^{-1}, $\Delta G^0_{H,Li^+} = -475$ kJ mol^{-1} to enter these sites[29,30] (see Figure 20.5). In Zhao *et al.*'s work,[31] flower-shaped and octahedral spinel lattice LMO ($Li_4Ti_5O_{12}$) were prepared with exposed (110) and (111) faces and investigated by experimental and computational methods to clarify how the exposed surface on the adsorbent affect the adsorption behaviour. They found that $[Li(H_2O)_4^+]$ forms a stable complex and its adsorption energy on the (111) surface is higher than that on the (110) one, which leads to a larger adsorption capacity and better reusability. Aluminium hydroxide (AH)-based adsorbents are the only adsorbents that are used for the production of lithium salts from brines at an industrial scale. Lithium cations are probably adsorbed by entering the octahedral cavities within the aluminium hydroxide layers. In order to balance the charge, chloride anions enter the interlayers of

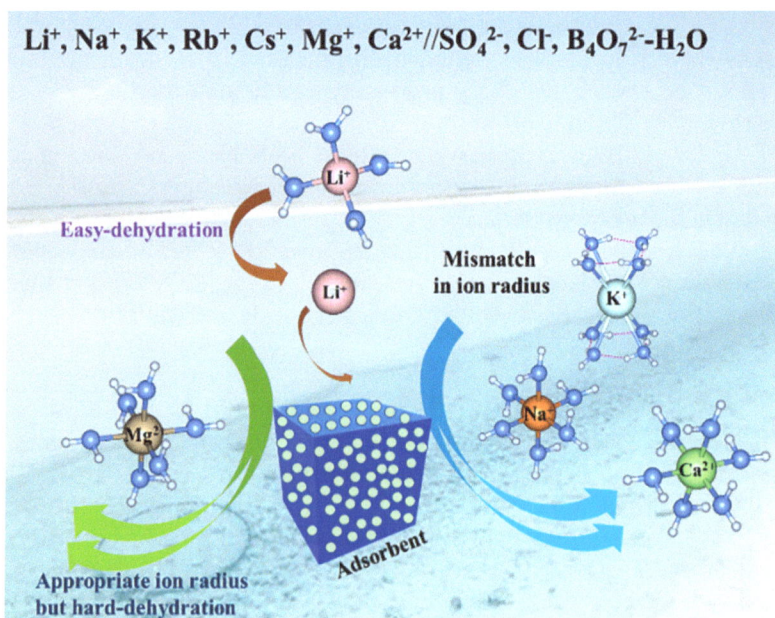

Figure 20.5   Schematic diagram of selective adsorption of various cations in brine using an adsorbent.

aluminium hydroxide. The sieve effect and hydration differences between Li^+ and the other cations are the key factors for the good selectivity of the adsorbent.

Other innovative absorbent systems, such as layered double hydroxides (LDHs),[32] electrochemical processes,[33-35] carbon based materials,[36,37] and ion-imprinted polymers,[38] and some new organic sorbents also show good performances for lithium separation from salt lake brine. Taking novel carbon materials as an example, Jiang *et al.*[36] performed the selective adsorption of Li^+ in thermally reduced (at 139 °C) graphene oxide (rGO) membranes over other metal ions such as Mg^{2+}, Co^{2+}, Mn^{2+}, Ni^{2+}, and Fe^{2+}. The adsorption strength of Li^+ reaches up to 5 times the adsorption strength of Mg^{2+}, and the mass ratio of a mixed Mg^{2+}/Li^+ solution at a very high value of 499 : 1 can be effectively reduced to 0.7 : 1 within only six treatment cycles, demonstrating the excellent applicability of rGO membranes in the Mg^{2+}/Li^+ separation[36] from brine. This unexpected selectivity is attributed to the competition between cation–π interactions[39] and steric exclusion when hydrated cations enter the confined space of the rGO membranes.

20.3.3 Membrane Separation Technology

Membrane technology is a rapidly growing and cost-effective green research area for ionic separation, with significant progress and wide applications.[40] Hydrometallurgical membrane processes can be divided into three categories based on the driving force of the processes: nanofiltration (NF; pressure driven), liquid membrane (chemical potential driven), and electrodialysis and membrane electrolysis (electrical potential driven).[17,41] Among them, NF is the most effective in industrial applications. Both the sieving and the Donnan effect play important roles in the separation performance. Physical properties such as the pore radius, membrane thickness, and surface charge density also affect the separation process. For example, the slightly charged surface of a NF membrane makes it possible to separate ions with different valences. The rejection behaviours of anions, such as sulfate and borate, were also investigated. Inchoate, a Desal-5 DL membrane element, was adopted to treat diluted salt lake brine containing lithium chloride and high concentrations of magnesium, borate and sulfate by Wen *et al.*,[42] which showed a Li/Mg separation coefficient of 3.6. Sun *et al.*[43] investigated the separation of Li/Mg from brines using a Desal nanofiltration (DL-2539 NF) membrane. It has been concluded that the competitive coefficient of Mg^{2+} remains nearly constant even with the Mg/Li ratio greater than 20, and the process can be favourable for the treatment of the West Taijinaier brine in China. The commercial DK polyamide composite membrane was employed by Li *et al.*[44] to recover lithium from salt lake brine with a high Mg/Li mass ratio and evaluated under different operating pressures, temperatures, multi-ion presence, and concentration ratio conditions. It was found that the DK membrane shows considerable promise as a method to separate magnesium and lithium and to recover lithium from salt lake brine with high Mg/Li ratios. Besides the commercially available membranes, Li *et al.*[45] synthesized a composite NF membrane with a positively charged surface layer and then modified it with EDTA. This membrane showed excellent separation behaviour and stability. Zhang *et al.*[46] synthesized an NF membrane modified with hydroxyl-containing multi-walled carbon nanotubes. This membrane, which was prepared with PEI/M-CNTs (MPMC), exhibited excellent rejection for divalent ions (over 96%) and high permeability. Besides the single nanofiltration process, the integrated membrane process is also a notable technology. In an instructive work by Zhao *et al.*,[47] bipolar membrane electrodialysis (BMED), NF, reverse osmosis (RO), and electrodialysis (ED) were combined to prepare LiOH from a salt lake brine with a Mg/Li ratio higher than 30.0.

In summary, there are many published reports on membrane-based separation processes for lithium extraction, both at laboratory and at a large industrial scale.[41] Selective separation of Li^+ and other ions requires molecular scale control of nanomaterials of chemical/physical homogeneity. A full understanding of the dehydration of Li^+ ions and the interaction of the pore geometry/chemistry with Li^+ ions is hampered by the lack of experimental analytical methods within the nanoscale spatial resolution. Membranes of high ion permselectivity are significant for the separation of ion species at the sub-nanometer scale. For example, Xu *et al.*[48] reported porous organic cage (*i.e.*, CC3) membranes with hierarchical channels including discrete internal cavities and cage-aligned external cavities connected by subnanometer-sized windows. The windows of CC3 sieve monovalent ions from divalent ones and dual nanometer-sized cavities provide pathways for fast ion transport with a flux of $1.0 \ mol \ m^{-2} \ h^{-1}$ and a mono-/divalent ion selectivity (*e.g.*, K^+/Mg^{2+}) up to 10^3, several orders of magnitude higher than those of previously reported membranes. This work shows that ion-hydration and dehydration play important roles in ion permselectivity, and offers guidelines for developing membranes with hierarchical channels for efficient ion separation.

Noteworthily, plenty of industrial experience shows that adsorption combined with membrane technology will give full potential to the advantages of the two technologies and it is the most promising way for Li separation from salt lake brine.

20.4   Boron (B)

Borate is widely used in many fields such as glass, ceramics, agriculture, wood preservation, flame retardance, nonlinear optical materials, nuclear electric power, *etc.*[49] The Qinghai–Tibet plateau of China contains hundreds of salt lakes that are abundant boron resources.[3] Aqueous borate solutions are surprisingly intricate compared to many other aqueous inorganic salt solutions. One of the well-known questions is the polyborate distribution in aqueous borate solutions. Various techniques including conductometric and potentiometric titrations, and ^{11}B NMR, Raman and mass spectroscopy were used to unravel the polyborate distribution in solutions. To date, the polyborate distribution[50] and solid–liquid equilibrium prediction[51] in aqueous borate solutions are satisfactory. There are at least four types of polyborates: $B_3O_3(OH)_4^-$, $B_3O_3(OH)_5^{2-}$, $B_4O_5(OH)_4^{2-}$ and $B_5O_6(OH)_4^-$, and five equilibria (eqn (20.4)–(20.8)) can be deduced in aqueous borate solutions:[52]

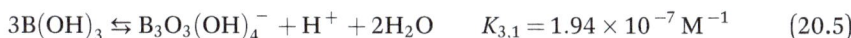

$$B(OH)_3 + H_2O \leftrightarrows B(OH)_4^- + H^+ \qquad K_a = 5.06 \times 10^{-10}\ M \qquad (20.4)$$

$$3B(OH)_3 \leftrightarrows B_3O_3(OH)_4^- + H^+ + 2H_2O \qquad K_{3,1} = 1.94 \times 10^{-7}\ M^{-1} \qquad (20.5)$$

$$3B(OH)_3 \leftrightarrows B_3O_3(OH)_5^{2-} + 2H^+ + H_2O \quad K_{3,2} = 3.83 \times 10^{-18} \qquad (20.6)$$

$$4B(OH)_3 \leftrightarrows B_4O_5(OH)_4^{2-} + 2H^+ + 3H_2O \quad K_{4,2} = 1.82 \times 10^{-15}\ M^{-1} \qquad (20.7)$$

$$5B(OH)_3 \leftrightarrows B_5O_6(OH)_4^- + H^+ + 5H_2O \qquad K_{5,1} = 2.17 \times 10^{-6}\ M^{-3} \qquad (20.8)$$

Boric acid $(B(OH)_3)$ is a weak acid with $pK_a = 9.29$ which transform into the borate ion $(B(OH)_4^-)$ at high pH. In seawater of pH ≈ 8.0, boron is mainly present in the molecular form of $B(OH)_3$ with an average concentration of 4.5 mg L^{-1}.[53] In a highly concentrated or old brine, boron can be enriched to over 10 g L^{-1}, where the polyborate distribution is a non-negligible problem (see Figure 20.6). According the view of Gao *et al.*,[3] the boron present in salt lake brine can be expressed as an aggregate statistical form of "$B_4O_5(OH)_4^{2-}$". Several methods can be used to extract boron from brine, including acid precipitation, electrodialysis, solvent extraction,[54] adsorption and crystallization.[53] Among them, solvent extraction and adsorption are the most widely studied.

The general routine for the solvent extraction separation of boron from boron-containing brine comprises adjusting the pH to the range of 1 to 4 (in this condition, boron mainly exists as $B(OH)_3$). This is followed by adding

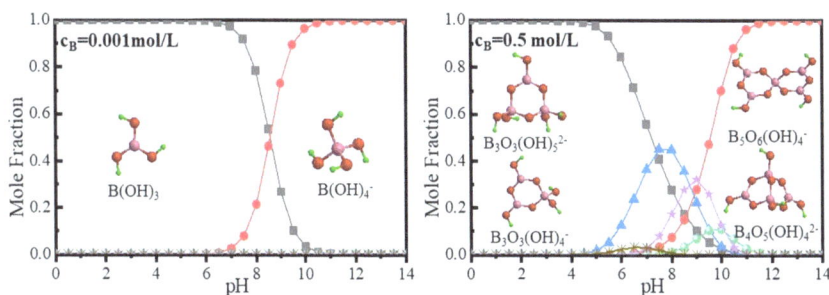

Figure 20.6 Polyborate distribution in aqueous borate solutions with various pH values and a total boron concentration (c_B) of 0.001 mol L^{-1} (left) and 0.5 mol L^{-1} (right) (■, H$_3$BO$_3$; ●, B(OH)$_4^-$; ▲, [B$_3$O$_3$(OH)$_4$]$^-$; ●, B$_3$O$_3$(OH)$_5^{2-}$; ★, B$_4$O$_5$(OH)$_4^{2-}$; ※, B$_5$O$_6$(OH)$_4^-$).

an extracting agent, *i.e.* monohydric alcohol and/or dihydric alcohol, and a diluting agent such as alkane to obtain the extracting organic phase, adjusting the pH, extracting to obtain the boron-containing organic phase, stripping the boron-containing organic phase to obtain the organic phase and water phase, where the stripping agent is deionized water or an alkaline aqueous solution, evaporating the water phase, concentrating, and crystallizing to obtain boric acid or borate such as $Na_2B_4O_7$. The use of boron selective chelating ion exchange resins seems to have still the greatest importance. Some studies showed that boron selective resins based on macroporous polystyrene matrices with the *N*-methyl glucamine ligand are the best sorbents for boron separation and removal.[53] The binding mechanism of boron by functionalization with an *N*-methyl-D-glucamine (NMDG) type chelating resin is shown in Figure 20.7. The adsorption separation process includes frequent resin elution and regeneration, resulting in costly wastewater disposal, and the boron concentration in the eluent is always too low to get the borate products. Even trace amounts of boron seriously affect the performance of lithium batteries, and thus it is important to remove the boron down to a very low concentration (usually less than $1–3$ $mg\,L^{-1}$) by adsorption before Li separation. Untill now, the use of macroporous crosslinked polystyrenic resins is the most reliable way for boron removal in industry. Recently, efficient separation of boron from salt lake brine using a novel flotation agent synthesized from NMDG and 1-bromooctadecane[55] was developed.

The World Health Organization (WHO) currently recommends limiting the concentration of boron in drinking water to 2.4 $mg\,L^{-1}$.[56] A lot of new absorbents such as rGO membranes,[57] covalent organic framework (COF) multilayers,[58] *etc.* have also been reported. Notably, the presence of hydroxide moieties (boron hydroxyl) in $B(OH)_3$ implies strong H-bonds with

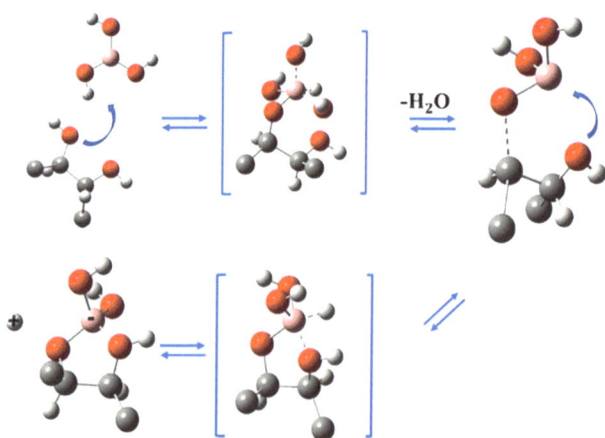

Figure 20.7   The binding mechanism of boron by functionalization with an *N*-methyl-D-glucamine type chelating resin.

water.[59] The hydration environment of $B(OH)_3$ plays a very important role in saline water separation also.[57-59]

In China, salt lakes in the Qinghai–Tibetan plateau are often located at a high altitude (more than 3000 m). The ecosystems in this region are fragile and there is no railway transportation to date. Thus, a more economical, environmentally friendly (chemical reagent free) technology is urgently needed. Due to the high concentration of salt in salt lake brine, a very distinct and interesting phenomenon called "crystallization by dilution" occurs that is completely different from "crystallization by evaporation". Some borate in concentrated salt lake brine does not precipitate even if it is supersaturated several times, but it can be crystallized by heat. In the past years, a green dilution method[60] was developed and promoted industrially. Recently, we tried to understand the "crystallization by dilution" phenomenon through X-ray diffraction,[61] neutron diffraction,[59] Raman spectroscopy, computer simulations, *etc.* The polyborate distribution calculation shows that the dilution changes the water activity and the chemical equilibria among the polyborates, which makes the aqueous borate crystallize. For the brine in which Mg^{2+} and Li^+ are highly concentrated, both Li^+ and Mg^{2+} have a stable first hydration layer, and it is difficult to form contact ion pairs (crystal nuclei) between the cations and polyborates by removing water molecules in the first hydration shell of Mg^{2+} and Li^+. Thus, more severe conditions must be adopted to promote the substantial contact between the cations and borate ions (to form the crystal nucleus). According to Yamaguchi *et al.*'s[62] work, which aimed to explore the structure of Li salt solution at high temperature and pressure by neutron diffraction, the cations have a relatively loose hydration structure at high temperature, which is conducive to lithium borate crystallization. Based on the characteristics of Li^+ at high temperature and pressure, Ge *et al.* have developed a more effective way for $Li_2B_4O_7$ recrystallization and purification through the hydrothermal method.[63]

20.5 Rubidium (Rb), Cesium (Cs) and Other Rare Dispersion Elements

Besides the resources which have already been relatively well exploited, Rb, Cs, bromine (Br), iodine (I), strontium (Sr), uranium (U), *etc.* are also widely enriched in salt lakes. Salt lake brine is one of the important Rb and Cs resources, which has not been industrially exploited. Cs is also of great environmental concern as ^{136}Cs and ^{134}Cs are the fission products that are typically present in nuclear waste. Separation of Cs and Rb from brine systems has been one of the most difficult problems due to their close similarity in physical and chemical properties for other macro-elements such as Na^+ and K^+. Methods such as precipitation,[64] electrochemical methods,[65] membrane processes[66] and solvent extraction are potential technologies. Solvent extraction and adsorption are the two most promising ways for Rb^+ and Cs^+ separation from salt lake brine.

Substituted phenols such as 4-*sec*-butyl-2-(α-methylbenzyl)phenol (*s*-BAMBP) and 4-*tert*-butyl-2-(α-methylbenzyl)phenol (*t*-BAMBP) have been used in trace amounts in Cs^+ and Rb^+ extraction. It is considered as the most promising solvent extraction system in industrial applications due to its good Rb^+ and Cs^+ recognition, and Rb^+ and Cs^+ can be eluted from the organic phase easily. We found that the extraction complex is MOR·3ROH ($M = Cs^+$, Rb^+, K^+), and the electrostatic attraction between M^+ and the oxygen of phenolic hydroxyl is significant, and the cation–π interaction has a true effect.[67] Furthermore, the use of pre-saponified *t*-BAMBP can sharply reduce the consumption of alkali solution compared with the traditional technological process.[67] The synergistic effect of ether oxygen and phenolic hydroxyl groups has been successfully used in the new separation system design. Xu *et al.*[68] designed and fabricated a hypercrosslinked polymer (HCP) with the synergistic effect of ether oxygen and phenolic hydroxyl groups for ultra-high-selective and fast Cs^+ isolation from salt lake brine. PO-HCPs-6 possesses an ultra-high separation factor of 57 and an extremely fast kinetic adsorption rate (reaches 90% of the equilibrium adsorption capacity quickly in just 1 min). After 5 times adsorption–desorption experiments, it is proved that PO-HCPs-6 has inspiring application potential for Cs^+ isolation in the salt lake brine, which shows a remarkable elution rate (>91%) and reusable adsorption capacity (>90%).

In practice, adsorption is the most effective method to extract or remove rare dispersion alkali metal ions from saline solutions due to its advantages of low cost, easy operation and high efficiency and has gained considerable attention in radioactive Cs^+ removal from aqueous solutions.[69,70] Compared with radioactive Cs^+ removal, where only the ion storage is needed, the Rb and Cs separation absorbents used for the brine system find the balance of the high selectivity and the possibility of elution (regeneration) paradox as the main challenge. Currently, various adsorbents including zeolites,[71] clays,[72] ammonium molybdophosphates (AMP),[73] metal sulfides,[74] Prussian blue analogs (PBAs),[75] and titanosilicate (HK_3)[76] have been developed. Among them, PBAs are the most promising materials because of their good adsorption selectivity and affinity toward Cs^+. Most of them are radioactive Cs^+ removal oriented, and some good adsorption separation works have also been reported. In one of our recent works,[77] carboxyl Fe_3O_4 nanoparticles (Fe_3O_4@R–COOH) modified with 18-crown-6 ether functional groups have been prepared *via* an amidation reaction and used as a bifunctional adsorbent for Cs^+. The adsorbent exhibited a high selectivity for Cs^+ in the solution with co-existing cations (NH_4^+, Rb^+, K^+, Na^+ and Li^+). The adsorbent underwent ten consecutive adsorption/desorption cycles and could retain more than 89% of its original Cs^+ adsorption capacity, with 0.1 mol L^{-1} NH_4Cl as the desorbing agent. It is a promising adsorbent for Cs^+ extraction from wastewater, and the methodology may be applied to extracting other useful ion resources from salt lakes and seawater.

In another interesting work by Cheng $et\ al.$,[78] a novel Prussian blue analog-based layered double hydroxide (PBA@ZnTi-LDH) was $in\ situ$ synthesized and used for radioactive Cs^+ removal from wastewater. It was found that the PBA@ZnTi-LDH exhibited an outstanding adsorption capacity $(242.9\ mg\,g^{-1})$ and low dissolution loss, which had promising industrial applications, and showed high selectivity and reusability. The detailed mechanism has not been discussed. However, we can deduce that the confinement effect of the LDH may play an important role.

Tremendous work has been done on the separation of other high value rare dispersion elements, such as Br, I, Sr, and U.[79] However, due to high salt concentrations, effective rare dispersion element separation from salt lakes is not an easy task, and there is a huge gap from the fundamental to the industrial application.

20.6   Prospects and Outlook

In a muti-component aqueous system, the target resources are usually relatively small compared with the micro-components such as Na^+, Mg^{2+}, Cl^-, SO_4^{2-}, $etc.$, and the high salt concentration and high viscosity of the brine make the separation of salt lake resources a challenging task. The following points in fundamental research require special attention as shown in Figure 20.8.

20.6.1   Advancing the Understanding of Brines

Fundamental knowledge of brines (concentrated salt solutions) in realistic conditions is essential for the design and manufacture of highly selective

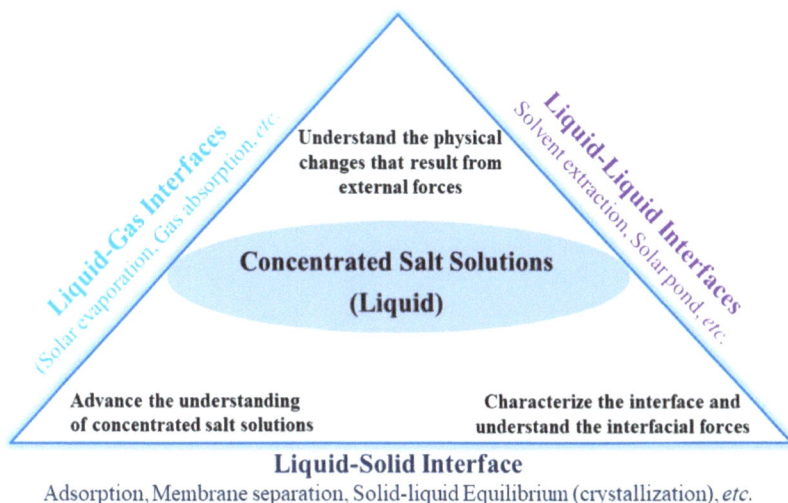

Figure 20.8   The key fundamental sciences and the ubiquitous interfaces in salt lake brine separation.

separation systems that have high capacity and throughput. A deep understanding of the ion hydration structure and water dynamics in the hydration sphere is extremely import for the novel separation system development. Furthermore, the use of molecular receptors as second-sphere ligands enables significant modulation of the chemical and physical properties of coordination complexes.[80]

20.6.2   Characterizing the Interface of Concentrated Salt Solutions and Understanding the Interfacial Forces

When discussing novel materials used for salt lake resources with specific separation properties, we mean their interactions with concentrated salt solutions *via* the interface.[81,82] However, the available methods are too limited to improve our understanding of these complex physical and chemical phenomena. Numerous mechanisms should be explored to achieve chemical separation in the interface zone with specially designed materials, such as electrostatic interactions, hydrophobic interactions, molecular sieving and size exclusion, and reversible chemical reactions. "How to measure interface phenomena at the microscopic level" is a difficult task, which was selected as the updated top 125 questions we do not know the answers to by the "SHJT-SCIENCE".[83]

20.6.3   Understanding the Physical and Chemical Changes That Result from External Forces

External forces, such as temperature, pressure, microwaves, optical fields, magnetic fields, *etc.*, cause physical changes in a separation material, and these changes affect the affinity of species for that material. An adequate understanding of these phenomena presents the possibility of controlling the structure of separation materials on length scales that allow better control over molecular separation. Besides the solutions under ambient conditions, a better understanding of the solutions in nano-confinement[84,85] at high temperature or high pressure may give some important clues for salt lake resource separation.

20.6.4   Green Chemistry Should be a Consistent Strategy During Salt Lake Resource Exploration

Salt lakes are usually located in regions with high altitude, arid and semi-arid climates, and far from civilization, where the ecosystems are fragile and transportation and energy supplements are imperfect or even unserviceable. Green, chemical free and water saving technologies are urgently needed in the salt lake chemical industry.

Problem 1

Are there salt lakes in your country? On a world map, mark the locations of salt lakes, and you may find why those salt lakes are located there.

Problem 2

Li separation from salt lakes is a very promising task. Besides the technologies discussed in the present chapter, please find some innovative ones in the literature. Also, find how ion hydration affects Li^+ separation from salt lake brine.

Problem 3

Why does boron have such a complex and diverse polyborate form in brine? You can find some clues in your inorganic book.

References

1. Qinghai Institute of Salt Lakes, Chinese Academy of Sciences, *Sixty years of research on Salt lakes by Chinese Academy of Sciences*, Science Press, Beijing, 2015, pp. 28–100.
2. M. P. Zheng and X. Liu, *Aquat. Geochem.*, 2009, **15**, 293.
3. S. Y. Gao, S. P. Xia, P. S. Song and M. P. Zhen, *Salt Lake Chemistry-new type borate salt lake*, Science Press, Beijing, 2007, pp. 21–82.
4. J. K. Warre, *Earth-Sci. Rev.*, 2010, **98**, 217.
5. S. Prakash and J. P. Verma, Global Perspective of Potash for Fertilizer Production, in *Potassium Solubilizing Microorganisms for Sustainable Agriculture*, ed. V. S. Meena, B. R. Maurya and J. P. Verma, *et al.*, Springer India, New Delhi, 2016, pp. 327–331.
6. G. J. Orris, M. Cocker, P. Dunlap, J. Wynn, G. T. Spanski and D. A. Briggs, Global Mineral Resource Assessment: Potash-A Global Overview of Evaporite-Related Potash Resources, Including Spatial Databases of Deposits, Occurrences, and Permissive TractsU, Scientific Investigations Report, 2014, pp. 4–65.
7. Z. M. Jing, X. Z. Xiao and S. M. Liang, *Acta Chim. Sin.*, 1980, **38**, 313.
8. X. M. Wang, J. D. Miller, F. Q. Cheng and H. G. Cheng, *Miner. Eng.*, 2014, **66**, 33.
9. M. Hancer, M. S. Celik and J. D. Miller, *J. Colloid Interface Sci.*, 2001, **235**, 150.
10. C. Li and H. Zhang, *J. Ind. Eng. Chem.*, 2022, **106**, 37.
11. G. Z. Kyzas, A. C. Mitropoulos and K. A. Matis, *Processes*, 2021, **9**, 1287.
12. Z. Wu, X. Wang, H Liu, H. Zhang and J. D. Miller, *J. Colloid Interface Sci.*, 2016, **235**, 190.
13. L. Wang, Y. Xu, H. Chen and X. Song, *Appl. Surf. Sci.*, 2021, **554**, 149613.
14. B. Wang and Y. Peng, *Miner. Eng.*, 2014, **66**, 13.

15. B. Diouf and R. Pode, *Renewable Energy*, 2015, **76**, 375.

16. J. F. Song, L. D. Nghiem, X. M. Li and T. He, *Environ. Sci.: Water Res. Technol.*, 2017, **3**, 593.

17. G. Liu, Z. Zhao and A. Ghahreman, *Hydrometallurgy*, 2019, **187**, 81.

18. Y. Marcus, *Chem. Rev.*, 2009, **109**, 1346.

19. H. Su, Z. Li, J. Zhang, Z. Zhu, L. Wang and T. Qi, *Hydrometallurgy*, 2020, **197**, 105487.

20. C. Shi, Y. Jing, J. Xiao, X. Wang and Y. Jia, *Hydrometallurgy*, 2017, **169**, 314.

21. B. Swain, *J. Chem. Technol. Biotechnol.*, 2016, **91**, 2549.

22. H. Li, L. Li, W. Li and Y. Zhou, *Chem. Phys. Lett.*, 2020, **754**, 7.

23. H. Su, Z. Li, Z. Zhu, L. Wang and T. Qi, *Sep. Sci. Technol.*, 2020, **55**, 1677.

24. Z. Zhou, W. Qin, Y. Liu and W. Fei, *J. Chem. Eng. Data*, 2012, **57**, 82.

25. W. Xiang, S. Liang, Z. Zhou, W. Qin and W. Fei, *Hydrometallurgy*, 2017, **171**, 27.

26. R. E. C. Torrejos, G. M. Nisola, H. S. Song, L. A. Limjuco, C. P. Lawagon, K. J. Parohinog and W. J. Chung, *Chem. Eng. J.*, 2017, **326**, 921.

27. J. Xiao, Y. Jia, C. Shi, X. Wang, S. Wang, Y. Yao and Y. Jin, *J. Mol. Liq.*, 2017, **241**, 946.

28. W. T. Stringfellow and P. F. Dobson, *Energies*, 2021, **14**, 6805.

29. G. He, L. Zhang, D. Zhou, Y. Zou and F. Wang, *Ionics*, 2015, **21**, 2219.

30. X. Pu, X. Du, P. Jing, Y. Wei, G. Wang, C. Xian, K. Wu, H. Wu, Q. Wang, X. Ji and Y. Zhang, *Chem. Eng. J.*, 2021, **425**, 130550.

31. B. Zhao, Z. Qian, M. Guo, Z. Wu and Z. Liu, *Chem. Eng. J.*, 2021, **414**, 128729.

32. Y. Sun, R. P. Yun, Y. F. Zang, M. Pu and X. Xiang, *Materials*, 2019, **12**, 1968.

33. H. Kanoh, K. Ooi, Y. Miyai and S. Katoh, *Sep. Sci. Technol.*, 1993, **28**, 643.

34. S. Perez-Rodriguez, S. D. S. Fitch, P. N. Bartlett and N. Garcia-Araez, *ChemSusChem*, 2022, **15**, e202102182.

35. X. Zhao, M. Feng, Y. Jiao, Y. Zhang, Y. Wang and Z. Sha, *Desalination*, 2020, **481**, 114360.

36. J. Jiang, L. Mu, Y. Qiang, Y. Yang, Z. Wang and R. Yi, *Chin. Phys. Lett.*, 2021, **38**, 116802.

37. L. Chen, G. Shi, J. Shen, B. Peng, B. Zhang, Y. Wang, J. Wang, D. Li, Z. Qian, G. Xu, G. Liu, J. Zeng, L. Zhang, Y. Yang, G. Zhou, M. Wu, W. Jin, J. Li and H. Fang, *Nature*, 2017, **550**, 380.

38. J. Lu, Y. Qin, Q. Zhang, Y. Wu, J. Cui, C. Li, L. Wang and Y. Yan, *Appl. Surf. Sci.*, 2018, **427**, 931.

39. L. Mu, Y. Yang, J. Liu, W. Du, J. Chen, G. Shi and H. Fang, *Phys. Chem. Chem. Phys.*, 2021, **23**, 14662.

40. Y. Zhang, L. Wang, W. Sun, Y. H. Hu and H. H. Tang, *J. Ind. Eng. Chem.*, 2020, **81**, 7.

41. S. Xu, J. Song, Q. Bi, Q. Chen, W.-M. Zhang, Z. Qian, L. Zhang, S. Xu, N. Tang and T. He, *J. Membr. Sci.*, 2021, **635**, 119441.

42. X. Wen, P. Ma, C. Zhu, Q. He and X. Deng, *Sep. Purif. Technol.*, 2006, **49**, 230.
43. S. Y. Sun, L. J. Cai, X. Y. Nie, X. Song and J. G. Yu, *J. Water Process Eng.*, 2015, **7**, 210.
44. Y. Li, Y. Zhao, H. Wang and M. Wang, *Desalination*, 2019, **468**, 114081.
45. W. Li, C. Shi, A. Zhou, X. He, Y. Sun and J. Zhang, *Sep. Purif. Technol.*, 2017, **186**, 233.
46. H. Z. Zhang, Z. L. Xu, H. Ding and Y. J. Tang, *Desalination*, 2017, **420**, 158.
47. Y. Zhao, H. Wang, Y. Li, M. Wang and X. Xiang, *Desalination*, 2020, **493**, 114620.
48. T. Xu, B. Wu, L. Hou, Y. Zhu, F. Sheng, Z. Zhao, Y. Dong, J. Liu, B. Ye, X. Li, L. Ge, H. Wang and T. Xu, *J. Am. Chem. Soc.*, 2022, **144**, 10220.
49. D. M. Schubert, Borates in industrial use, in *Group 13 Chemistry III: Industrial Applications, Structure and Bonding*, Springer-Verlag Berlin, 2003, vol. 105, pp. 1–40.
50. Y. Zhou, C. Fang, Y. Fang and F. Zhu, *Spectrochim. Acta, Part A*, 2011, **83**, 82.
51. P. Wang, J. J. Kosinski, M. M. Lencka, A. Anderko and R. D. Springer, *Pure Appl. Chem.*, 2013, **85**, 2117.
52. J. E. Spessard, *J. Inorg. Nucl. Chem.*, 1970, **32**, 2607.
53. N. Kabay, E. Güler and M. Bryjak, *Desalination*, 2010, **261**, 212.
54. R. Zhang, Y. Xie, J. Song, L. Xing, D. Kong, X. M. Li, X. M. Li and T. He, *Hydrometallurgy*, 2016, **160**, 129.
55. C. Bai, K. Li, D. Fang, X. Ye, H. Zhang, Q. Li, J. Li, H. Liu and Z. Wu, *Colloids Surf., A*, 2021, **627**, 127178.
56. World Health Organization, Guidelines for drinking-water quality, 2011, http://whqlibdoc.who.int/publications/2004/9241546387.pdf
57. F. Risplendi, F. Raffone, L.-C. Lin, J. C. Grossman and G. Cicero, *J. Phys. Chem. C*, 2020, **124**, 1438.
58. X. Zhang, M. Wei, Z. Zhang, X. Shi and Y. Wang, *Desalination*, 2022, **526**, 115548.
59. Y. Zhou, T. Yamaguchi, W. Zhang, K. Ikeda, K. Yoshida, F. Zhu and H. Liu, *Phys. Chem. Chem. Phys.*, 2020, **22**, 17160.
60. J. Y. Peng, S. J. Bian, B. Zhang, Y. P. Dong and W. Li, *Hydrometallurgy*, 2017, **174**, 4.
61. Y. Zhou, S. Higa, C. Fang, Y. Fang, W. Zhang and T. Yamaguchi, *Phys. Chem. Chem. Phys.*, 2017, **19**, 27878.
62. T. Yamaguchi, H. Ohzono, M. Yamagami, K. Yamanaka, K. Yoshida and H. Wakita, *J. Mol. Liq.*, 2010, **153**, 2.
63. H. Ge, Y. Fang, C. Fang and Y. Zhou, *Purification lithium tetraborate by hydrothermal*, China, CN201510078259, 2015.
64. J. Y. Su, G. P. Jin, T. Chen, X. D. Liu, C. N. Chen and J. J. Tian, *Electrochim. Acta*, 2017, **230**, 399.
65. Y. Zheng, J. Qiao, J. Yuan, J. Shen, A. J. Wang and L. Niu, *Electrochim. Acta*, 2017, **257**, 172.

66. S. Ding, L. Zhang, Y. Li and L. A. Hou, *J. Hazard. Mater.*, 2019, **368**, 292–299.
67. D. Pang, Z. Zhang, Y. Zhou, Z. Fu, Q. Li, Y. Zhang, G. Wang and Z. Jing, *Chin. J. Chem. Eng.*, 2022, **46**, 31.
68. Z. Xu, M. Rong, S. Ni, Q. Meng, L. Chen, H. Liu and L. Yang, *Sep. Purif. Technol.*, 2022, **284**, 120285.
69. E. C. Escobar, J. E. L. Sio, R. E. C. Torrejos, H. Kim, W. J. Chung and G. M. Nisola, *J. Ind. Eng. Chem.*, 2022, **107**, 1.
70. D. Alby, C. Charnay, M. Heran, B. Prelot and J. Zajac, *J. Hazard. Mater.*, 2018, **344**, 511.
71. E. Han, Y. G. Kim, H. M. Yang, I. H. Yoon and M. Choi, *Chem. Mater.*, 2018, **30**, 5777.
72. S. M. Park, D. S. Alessi and K. Baek, *J. Hazard. Mater.*, 2019, **369**, 569.
73. S. Chen, J. Hu, J. Shi, M. Wang, Y. Guo, M. Li, J. Duo and T. Deng, *J. Hazard. Mater.*, 2019, **371**, 694.
74. J. H. Tang, J. C. Jin, W. A. Li, X. Zeng, W. Ma, J. L. Li, T. T. Lv, Y. C. Peng, M. L. Feng and X. Y. Huang, *Nat. Commun.*, 2022, **13**, 658.
75. J. Wang, S. Zhuang and Y. Liu, *Coord. Chem. Rev.*, 2018, **374**, 430.
76. Y. Qian, D. Ding, K. Li, D. Fang, Y. Wang, Y. Hu, H. Liu, Z. Wu and J. Li, *Hydrometallurgy*, 2022, **208**, 105802.
77. Z. Liu, Y. Zhou, M. Guo, B. Lv, Z. Wu and W. Zhou, *J. Hazard. Mater.*, 2019, **371**, 712–720.
78. S. Chen, X. Yang, Z. Wang, J. Hu, S. Han, Y. Guo and T. Deng, *J. Hazard. Mater.*, 2021, **410**, 124608.
79. N. Liu, C. Li, J. Bai, H. Liang, Q. Gao, N. Wang, R. Guo, Z. Qin and Z. Mo, *J. Environ. Chem. Eng.*, 2021, **9**, 106688.
80. W. Liu, P. J. Das, H. M. Colquhoun and J. F. Stoddart, *CCS Chem.*, 2022, **4**, 755.
81. Z. We, X. Ye, H. Liu, H Zhang, L. Zhong, M. Guo, Q. Li and J. Li, *Pure Appl. Chem.*, 2020, **92**, 1655.
82. I. Persson, J. Werner, O. Björneholm, Y. S. Blanco, Ö. Topel and É. G. Bajnóczi, *Pure Appl. Chem.*, 2020, **92**, 1553.
83. S. Sanders, *125 Questions: Exploration and Discovery*, Science/AAAS Custom Publishing Office, Washington, DC, USA, 2021, p. 6.
84. W. Xin, J. Fu, Y. Qian, L. Fu, X.-Y. Kong, T. Ben, L. Jiang and L. Wen, *Nat. Commun.*, 2022, **13**, 1701.
85. G. Wang, Y. Zhou, Z. Jing, Y. Wang, K. Chai, H. Liu, F. Zhu and Z. Wu, *J. Mol. Liq.*, 2022, **361**, 119409.

Subject Index